Ori Lahav, Liat Birnhack
Aquatic Chemistry

Also of interest

Drinking Water Treatment.
An Introduction
Worch, 2019
ISBN 978-3-11-055154-9, e-ISBN 978-3-11-055155-6

Wastewater Treatment.
Application of New Functional Materials
Chen, 2018
ISBN 978-3-11-054278-3, e-ISBN 978-3-11-054438-1

Environmental Engineering.
Basic Principles
Tomašić, Zelić (Eds.), 2018
ISBN 978-3-11-046801-4, e-ISBN 978-3-11-046803-8

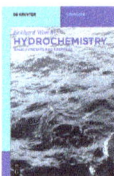

Hydrochemistry.
Basic Concepts and Exercises
Worch, 2015
ISBN 978-3-11-031553-0, e-ISBN 978-3-11-031556-1

Ori Lahav, Liat Birnhack

Aquatic Chemistry

For Water and Wastewater Treatment Applications

DE GRUYTER

Authors
Prof. Ori Lahav
Department of Environmental, Water and Agricultural
Technion – Israel Institute of Technology
32000 Haifa, Israel
agori@cv.technion.ac.il

Guangdong Technion Israel Institute of Technology (GTIIT)

Dr Liat Birnhack
Faculty of Civil and Environmental Engineering
Technion – Israel Institute of Technology
32000 Haifa, Israel
birbirit@gmail.com

ISBN 978-3-11-060392-7
e-ISBN (PDF) 978-3-11-060395-8
e-ISBN (EPUB) 978-3-11-060409-2

Library of Congress Control Number: 2018965709

Bibliographic information published by the Deutsche Nationalbibliothek
The Deutsche Nationalbibliothek lists this publication in the Deutsche Nationalbibliografie;
detailed bibliographic data are available on the Internet at http://dnb.dnb.de.

© 2019 Walter de Gruyter GmbH, Berlin/Boston
Typesetting: Integra Software Services Pvt. Ltd.
Printing and binding: CPI books GmbH, Leck
Cover image: Dimitris66v/ iStock / Getty Images Plus

www.degruyter.com

Preface

We are happy to introduce this textbook, which is the culmination of many years of teaching and research in the fields of aquatic chemistry and water treatment process development.

The textbook was written not only with both university-level undergraduate and graduate students in mind, but also to serve as a tool for water, chemical and environmental engineers in their quest for correct and optimal design and interpretation of phenomena occurring in the aqueous phase.

The book differs from most other water-chemistry-related books in the method developed and adopted for problem-solving and for the design of engineered water treatment systems. The main idea throughout the text is to employ mass balances on conservative parameters (e.g., alkalinity and acidity species) in order to compute the values of non-conservative parameters (e.g., pH, the concentration of individual weak acid/base species) thereby solving most of the issues related to weak acid water quality, whether in natural or engineered systems. The data required for solving all the problems presented in the book can be obtained by relatively simple analytical procedures, and the solutions do not require strong computing abilities.

The book starts (Chapters 1 and 2) by reciting classical aquatic chemistry material, then shifts to the definition of conservative weak acid/base parameters (alkalinity/acidity in their various manifestations) in the aqueous, gas and solid phases (Chapters 3 to 6) and then to the techniques by which they are used to obtain information on the water quality and design water treatment processes (e.g., mixing of streams, softening, remineralization) (Chapter 8 and 9). The book also introduces (Chapter 7) a free computerized tool (the software "Stasoft") as an easy tool to design and simulate processes occurring in the aqueous phase upon the dosage of chemicals and the interaction between the aqueous, solid and gas phases. The book finishes with a comprehensive questions and answers session, which encompasses the whole range of materials covered in the text.

We wish to thank De Gruyter for giving us a venue for publishing this book, which we believe is important to any water and environmental/chemical-engineering professional. We also wish to thank Guangdong Technion Israel Institute of Technology (GTIIT) and the Technion International School (TI) for their financial support and Mr Yehuda Cohen (B.Sc. in Environmental Engineering) for his invaluable work in editing and translating a part of this text, which originally appeared in Hebrew. Dr Lahav would want also to thank the late Prof. Richard Loewenthal from the University of Cape Town for his invaluable contribution in making aquatic chemistry science what it is today.

We hope you will find this text both interesting and educative.

Prof. Ori Lahav
Dr Liat Birnhack
April 2019

https://doi.org/10.1515/9783110603958-201

Contents

1 Water chemistry fundamentals

1.1 Introduction

This chapter aims to clarify/recap fundamental concepts concerning the chemistry of aqueous solutions. A large portion of the concepts and terms addressed in this chapter are taught in introductory chemistry textbooks, and the text appearing here does not intend to replace them. Rather, the review of the basic concepts in this chapter is meant to remind the reader of certain definitions and tools that are used in solving simple and complex water-chemistry related problems, which appear in the following chapters.

1.2 Solutions

The word "solution" describes a system in which one or more substances (the "solute") are dissolved and distributed uniformly in another substance (the "solvent"). Both solute and solvent can be either solid, gas or liquid. Water chemistry, as relating to water treatment processes, is concerned mainly with reactions in the aqueous phase (and interactions thereof with the gas and solid phases) pertinent to the treatment of water or wastewater. Examples of such reactions may be dissolution (or precipitation) of solids in aqueous solutions, oxidation-reduction reactions or stripping of a gas (e.g., CO_2) from a solution and its consequences on solution characterization.

1.3 The chemical structure of the H_2O molecule

Water (H_2O) is a molecule that is made up of two hydrogen atoms bonded by a covalent bond to an oxygen atom. The sharing of the two hydrogen atoms' electrons with the oxygen causes a net positive charge in their vicinity and a net negative charge in the far end of the oxygen atom (Fig. 1.1). As a result, the water molecule has a dipole moment and is "polar." Due to its strong polar nature, water molecules are bonded to each other by hydrogen bonds, which are derived from intra-molecular attractive forces that are electrostatic by nature. The hydrogen bonds give water its unique characteristics that are so vital to the existence of life on earth (being fluid at ambient temperature, having dissolution capacity for other polar substances and ions in particular, buffering capacity at the extreme end of the 0 to 14 pH scale, higher density of the liquid phase relative to the solid phase, and more).

https://doi.org/10.1515/9783110603958-001

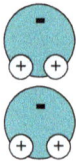

Fig. 1.1: Polar attraction between water molecules.

1.4 Dissolution

The ability of water to dissolve many chemical species is a direct result of the polar nature of the water molecules. When an ion or another polar substance is added to water, it is immediately surrounded by water molecules, with the negative pole of the water molecules facing toward the cation or toward the positive section of the dissolved species and vice versa (Fig. 1.2). This process is termed "dissolution".

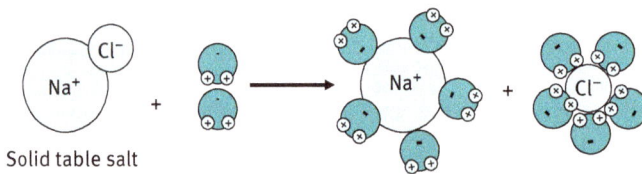

Solid table salt

Fig. 1.2: Schematic of the dissolution of table salt (NaCl) into Na^+ and Cl^-.

In most cases, for ionic or polar substances (such as salts, alcohols and organic acids), the dissolved state is energetically preferred over the initial state (simple energy considerations are elaborated upon further in this chapter). These substances are called hydrophilic (from Greek, literal translation "water loving"). Non-polar substances such as oils and fats do not dissolve and are thus called hydrophobic (literal translation "water fearing"). Certain ions such as sodium (Na^+) or chloride (Cl^-) are very stable in water-based solutions, and therefore dissolve entirely in water up to very high concentrations, and are thus involved only in a few reactions. The most common ions on Earth are also those that are most stable in aqueous solution: Na^+, K^+, Mg^{2+}, Cl^-, SO_4^{2-}, NO_3^- and HCO_3^-.

1.5 Solubility

The maximum mass of a substance that can be dissolved in the pure state of another substance is defined as the solubility of that substance. For example, 0.015 g of calcium carbonate ($CaCO_3$) will dissolve in 1000 g of (pure or distilled) water at a temperature of zero degrees Celsius. The solubility of calcium carbonate is therefore 0.015 g per 1000 g of water at 0 °C. In certain cases (as in the case of calcium carbonate itself), the solubility of a substance is also dependent upon the pH, the pressure at which the solution is maintained, the general ionic strength of the solution (a term defined later in this chapter) and/or its specific ionic composition. The topic of solubility is discussed in detail in the chapter dealing with equilibrium with the solid phase also referred to as precipitation/dissolution of solids (Chapter 6). At this stage, general guiding principles for the solubility of selected ions in water are given in Table 1.1.

Table 1.1: General guiding principles for the solubility of different ions in aqueous solution.

Ion	Solubility characteristics
Nitrates (NO_3^-)	All the compounds are soluble.
Chlorides (Cl^-)	All chloride compounds are soluble except for AgCl, PbCl and Hg_2Cl_2.
Sulfates (SO_4^{2-})	Sulfate compounds are generally soluble except for $BaSO_4$ and $PbSO_4$. Ag_2SO_4, $CaSO_4$ and Hg_2SO_4 are slightly soluble.
Carbonates (CO_3^{2-}), phosphates (PO_4^{3-}) and silicates (SiO_4^{4-})	Carbonate, phosphate and silicate compounds are not soluble, except for salts containing ammonium, sodium and potassium (see bottom row of this table).
Calcium (Ca^{2+}) and magnesium (Mg^{2+}) ions	The solubility of most compounds which comprise these ions is low (both cations are defined as "hardness" in water).
Hydroxides (OH^-)	Most of the compounds are not soluble. However, NaOH, LiOH, KOH and NH_4OH are soluble. $Ca(OH)_2$, $Mg(OH)_2$ and $Ba(OH)_2$ are characterized by low to medium solubility.
Sulfides (S^{2-})	Very insoluble apart from Na_2S, K_2S, $(NH_4)_2S$, MgS, CaS, BaS.
Sodium (Na^+), potassium (K^+) and ammonium (NH_4^+)	In general, compounds containing these ions are soluble.

1.6 Electrolytes and non-electrolytes

Electrolytes are substances that conduct electrical charge in solutions. Electrolytes may be ions or substances that release ions in an aqueous solution. In general, acids, bases and salts are electrolytes.

Examples of electrolytes: H_2SO_4 – sulfuric acid; H_3PO_4 – phosphoric acid; HCl – hydrochloric acid; NaOH – caustic soda; NH_4OH – ammonium hydroxide; NaCl – sodium chloride (table salt).

Examples of non-electrolytes: $C_6H_{12}O_6$ – glucose; C_2H_5OH – ethanol; $CO(NH_2)_2$ – urea.

Strong electrolytes break down (the chemical term is dissociation or ionization) almost entirely in water and can be written with a one-directional arrow ($\rightarrow$):

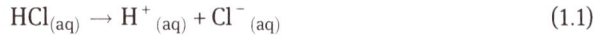

$$HCl_{(aq)} \rightarrow H^+_{(aq)} + Cl^-_{(aq)} \qquad (1.1)$$

Weak electrolytes dissociate partially and, accordingly, are represented by a bidirectional arrow ($\leftrightarrow$):

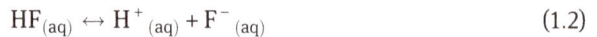

$$HF_{(aq)} \leftrightarrow H^+_{(aq)} + F^-_{(aq)} \qquad (1.2)$$

Examples of strong electrolytes: $Ca(OH)_2$, $NaOH$, KOH, HNO_3, HCl, H_2SO_4 and most salts.

Examples of weak electrolytes: NH_4OH, $HOCl$, HNO_2, HF, H_2S, H_2CO_3.

Differentiation between strong and weak electrolytes is based on the value of the equilibrium constant of the corresponding dissociation reaction; this will be elaborated upon further in this chapter. Differentiation between strong and weak acids is discussed in Chapter 2.

1.7 Expressing solute concentrations in environmental engineering

In the chemical/environmental engineering disciplines, there are several accepted forms by which a solute's concentration in aqueous solutions can be expressed:

a. Weight fraction (% weight, promil/ppt = parts per thousand, parts per million – ppm, parts per billion – ppb).
b. Mass per unit volume (e.g., grams per liter).
c. Molarity (moles per liter or molar, also denoted M) and Molality (moles per kg).
d. Normality (equivalents per liter or normal, also denoted N).
e. Expressing concentration in units of mass per volume as another substance (grams per liter as X, where the molecular/atomic weight of the substance X is used to express the concentration).
f. The p-notation, which is the negative logarithm of a certain value.

The abovementioned concentration expressions are thoroughly discussed in the following paragraphs.

1.8 Weight fraction (mass per unit mass)

This form expresses the ratio between the mass of the solute and the mass of the entire solution. For example, in a solution that contains 40 g of ethanol and 60 g of water, the concentration of the ethanol is 40% by weight.

Commercial concentrations of acids and bases are usually expressed in these units, in addition to the information of the density of the solution.

For example: hydrochloric acid (HCl) at a concentration of 16% (16 g of pure HCl per 100 g of solution) at a density of 1.0776 g/mL.

In order to convert this unit to the form of weight per volume (see following section), multiply the weight percentage by the density of the solution:

$$\frac{16 \text{ g HCl}}{100 \text{ g solution}} \cdot \frac{1.0776 \text{ g solution}}{1 \text{ mL solution}} \cdot \frac{1000 \text{ mL}}{L} = 172.41 \frac{\text{g HCl}}{L}$$

Weight percentage is sometimes denoted as w/w (weight/weight percentage).

Converting to different weight concentration units (per mill, ppm, ppb and others) requires merely changing the order of magnitude:

$15\% = 150‰ = 150{,}000 \text{ ppm} = 150 \cdot 10^6 \text{ ppb}$

Since the density of dilute, water-based solutions at room temperature is approximately 1.00 g/ml, the weight concentration is approximately equal to the weight per volume concentration (e.g., $1 \text{ ppm} \approx 1 \text{ mg/L}$). Nevertheless, for higher precision, it is always better to use weight per volume units, as this is normally the result obtained in analytical chemistry analyses.

1.9 Weight per volume

This unit form for expressing concentrations is very common in engineering fields, where it is common to dissolve salt of a known weight into a solution of a known volume. For example, dissolution of 1 g of table salt (NaCl) in 1 L (final volume) of water yields a solution whose salt concentration is 1 g/L. In order to find the concentration of the chloride or sodium ions solely, one must first find the number of moles of salt that were added. Since table salt is a strong electrolyte, it undergoes complete ionization in water:

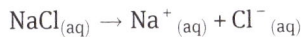

$$NaCl_{(aq)} \rightarrow Na^+{}_{(aq)} + Cl^-{}_{(aq)}$$

Since the stoichiometric ratio between the salt and each of the dissolved ions is 1:1, the number of moles of each of the ions that form in solution is identical to the number of moles of salt that were added, which is calculated by dividing the mass by the molecular weight (denoted MW). The atomic weight of sodium is 23.0 g/mol, and the atomic weight of chlorine is 35.45 g/mol, thus:

$$\frac{W_{NaCl}}{MW_{NaCl}} = \frac{1\,g\,NaCl}{23.0\frac{g}{mol} + 35.45\frac{g}{mol}} = 0.0171\,mol$$

Accordingly, there are 0.0171 moles of each of the two ions in 1 L of solution. Multiplying this number by the molecular weight of each ion yields the mass of each ion per 1 L of solution.

Sodium concentration = $(0.0171\,mol \cdot 23.0\,g/mol)\,/1\,L = 0.393\,g/L$

Chloride concentration = $(0.0171\,mol \cdot 35.45\,g/mol)\,/1\,L = 0.607\,g/L$

Note that adding the concentrations of the two ions must bring us back to the total concentration of the salt (1 g/L).

1.10 Molarity (M)

The molar concentration describes the number of moles of a certain substance per given volume of solution. In the above example, the molar concentration of the table salt was 0.0171 mol/L. Molar concentration is denoted by square brackets.

Example 1.1 Calculate the concentration, in molar and grams per liter, of a solution with a volume of 200 mL in which 2 g of NaCl were dissolved.

Solution

$$\frac{2\,g\,NaCl}{200\,mL} \cdot 1000\frac{mL}{L} = 10\,g/L$$

$$\frac{10\,\frac{g}{L}}{23\,\frac{g}{mol} + 35.45\,\frac{g}{mol}} = 0.17\,M$$

1.11 Normality (N = eq/L)

One normal is defined as 1 "equivalent weight" of a substance dissolved in 1 L of solution. An equivalent weight is defined as the molecular weight of a substance divided by its "valence". The question of what constitutes the "valence" depends on the use of the substance and the relevant chemical reactions, and sometimes more than one definition can exist. The reason for using this form of concentration units is to create a common measure for different substances that can react in similar reactions or for a similar purpose. The concentrations of the following groups of substances are commonly expressed using the "normal" (or equivalent per liter) units:

– Salts/ions (valence is defined as the ionic charge).
– Acids or bases (valence defined by number of protons, H^+, or OH^- ions released to solution).

– Oxidizing or reducing agents (valence defined by the number of electrons transferred in considered reaction).

Topics related to normal units as manifested in acids and bases are covered in detail in Sections 3.2, 3.3 and 3.4.

Mathematical definitions

Normality (N = eq/L):

$$N = \frac{\text{solute weight (g)}}{\text{equivalent weight} \left(\frac{g}{eq}\right) \times \text{solution volume (L)}} \tag{1.3}$$

Where the equivalent weight is:

$$\text{equivalent weight} \left(\frac{g}{eq}\right) = \frac{\text{molecular weight}\left(\frac{g}{mol}\right)}{C_c \left(\frac{eq}{mol}\right)} \tag{1.4}$$

Where C_c, the valence, may be (1) the number of protons available in the acid (n in the acid H_nA); (2) the number of hydroxide ions available in the base (n in the base $B(OH)_n$); (3) the number of positive charges or the number of negative charges in a molecule (mole) of a salt; or (4) the number of electrons transferred per molecule (mole) by an oxidizing or reducing agent. C_c is always a positive integer number.

The relationship between normality and molarity is:

$$N = C_c M \tag{1.5}$$

Example 1.2 What is the normal concentration of 0.5 L of water in which 0.3 g of gypsum were dissolved ($CaSO_4$)?

Solution In a water-based solution, gypsum dissociates into calcium and sulfate, both are divalent ions. Therefore, the valence of the substance (C_c) is 2.
The molecular weight of $CaSO_4$ is 136 g/mol.

Equivalent weight of $CaSO_4$: $\dfrac{136 \, g CaSO_4/mol CaSO_4}{2 \, eq/mol CaSO_4} = 68 \, \dfrac{g CaSO_4}{eq}$

Normal concentration:

$$N = \frac{\text{solute weight (g)}}{\text{equivalent weight (g/eq)} \times \text{solution volume (L)}} = \frac{0.3 \, g \, CaSO_4}{2 \, g/eq \times 0.2 \, L} = 0.009 \, eq/L$$

Example 1.3 Determination of the valence of a given substance is not always unambiguous; it may sometimes depend on the type of the reaction (process) for which the normal units are used, as demonstrated in the following example:

Determine the equivalent weight of the strong electrolyte potassium dichromate ($K_2Cr_2O_7$) in each of the following reactions (you can consider only the dichromate ion since the potassium ions are not involved in the following reactions).

a. $Cr_2O_7^{2-} + 2Pb^{2+} + H_2O \rightarrow 2PbCrO_4 + 2H^+$

b. $Cr_2O_7^{2-} + 14H^+ + 6e^- \rightarrow 2Cr^{3+} + 7H_2O$

Solution In reaction (a) there is no transfer of electrons (i.e., this is not an oxidation-reduction reaction), and therefore C_c should refer to the electrical charge of the dichromate, which is 2 (eq/mol). The equivalent weight is thus:

$$\text{equivalent weight } K_2Cr_2O_7 = \frac{294.2 (g/mol)}{2 (eq/mol)} = 147.1 (g/eq)$$

On the other hand, in reaction (b) there is a transfer of electrons (i.e., this is a half-cell reduction reaction). Since the oxidation state of chromium atoms drops from +6 to +3 (a transfer of three electrons for every mole of chromium), and since there are two atoms of chromium per mole of dichromate, it would make sense to give C_c a value of 6 (eq e^- per mol of $Cr_2O_7^{2-}$) in this reaction, and the equivalent weight in this case would be:

$$\text{equivalent weight } K_2Cr_2O_7 = \frac{294.2 g/mol}{6 eq/mol} = 49.03 g/eq$$

1.12 Weight per volume expressed as a different substance (units: g/L as X)

This form of concentration is very common in environmental and chemical engineering applications, however, it is also the source of much confusion. This method is based on expressing the concentration of a given substance using the molecular weight of a different substance. The objective of this method is to have a common measure amongst different groups of chemical species with a similar characteristic. It is common to use this method for compounds that contain the atoms S, P, N, species that represent hardness in water and species that make up the acidity and alkalinity equations (elaborated in Chapter 3).

There are various ways to express these units. Using phosphorous as an example, the units can be written as (1) mg/L as P; (2) mgP/L; (3) mg/L PO_4^{3-}–P or in words: mg/L of the species PO_4^{3-} expressed as P (phosphorous).

To explain the reasoning behind this technique let us consider, for example, the nitrogen cycle in wastewater. Wastewater contains dissolved nitrogenous compounds of varying molecular weights (NH_3, NH_4^+, NO_2^-, NO_3^- and more). The transfer between the various nitrogenous compounds in common wastewater treatment processes occurs at a molar stoichiometric ratio tending toward 1:1. Thus, expressing the concentration of each nitrogenous compound using similar units allows for simple inspection of the biological, physical and chemical processes that occur in the water, that involve these species. Example 1.4 demonstrates this occurrence.

Example 1.4 Given is water with an ammonia concentration of 1 M. What is the concentration of the ammonia using units of grams per liter ammonia as nitrogen?

Solution Instead of multiplying the molar concentration by the molecular weight of ammonia (17 g/mol), it is multiplied by the atomic weight of nitrogen (14 g/mol). Note that it must also be multiplied by the number of nitrogen atoms in one mole of ammonia. The calculation yields:

$$1\frac{mol\ NH_3}{l} \times 1\frac{mol\ N}{mol\ NH_3} \times 14\frac{gN}{mol\ N} = 14\frac{gNH_3 - N}{l}$$

Example 1.5 Given a solution with an ammonia concentration of 1 mg/L. What is the concentration in units of mg/L ammonia as nitrogen (or mgN/L)?

Solution First, the concentration units must be converted to molar units:

$$\frac{\frac{1mg\ NH_3}{l}}{\frac{17mg\ NH_3}{mmol}} = 0.06\frac{mmol\ NH_3}{l} = 0.00006\frac{mol\ NH_3}{l}$$

Then, the above expression is multiplied by the molecular weight of nitrogen:

$$0.00006\frac{mol\ NH_3}{l} \times 1\frac{mol\ N}{mol\ NH_3} \times 14\frac{g\ N}{mol} = 0.00082\frac{g\ NH_3 - N}{l} = 0.82\frac{mg}{l}NH_3\ as\ N$$

Example 1.6 Nitrogenous compounds in wastewater undergo several transformations in wastewater treatment plants that are designed to almost entirely remove them to benign nitrogen gas. Assuming that in the first stage of the treatment, 36 mg/L ammonium undergo aerobic oxidation to form the nitrite ion (NO_2^-) and thereafter nitrite is oxidized to form the nitrate ion (NO_3^-) in the nitrification process. In the second stage, the nitrate that was formed in the nitrification process is reduced under anoxic conditions (absence of oxygen) to nitrogen gas in a process known as denitrification.

Assuming all the reactions are carried out completely, write the concentrations of the different components in the process using the units M, mg/L and mg/L as nitrogen.

Solution The biochemical equations for nitrification and denitrification (neglecting biomass growth):
Nitrification: $NH_4^+ + 1.5O_2 \rightarrow NO_2^- + H_2O + 2H^+$

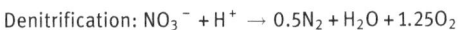

$$NO_2^- + 0.5O_2 \rightarrow NO_3^-$$

Denitrification: $NO_3^- + H^+ \rightarrow 0.5N_2 + H_2O + 1.25O_2$

Based on the stoichiometric ratios in these equations, the concentrations of the various species can be found, given in the following table (Table 1.2).

Table 1.2: The molecular weights of all nitrogen species as well as their calculated concentrations in varying units.

Species	Molecular weight g/mol	Molar concentration m mol/L	Weight per volume mg/L	Concentration as nitrogen mg/L as N
NH_4^+	$14 + 4 \cdot 1 = 18$	$36/18 = 2$	36	$2 \cdot 1 \cdot 14 = 28$
NO_2^-	$14 + 16 \cdot 2 = 46$	2	$2 \cdot 46 = 92$	$2 \cdot 1 \cdot 14 = 28$
NO_3^-	$14 + 16 \cdot 3 = 62$	2	$2 \cdot 62 = 124$	$2 \cdot 1 \cdot 14 = 28$

Note that despite the differing molecular weights of the three substances, their concentrations in mg/L as nitrogen (right column) are identical, therein lies the advantage of using this form of units.

1.13 The p-notation

The use of the p-notation is customary in expressing the concentration of protons (H^+) and hydroxide ions (OH^-) in systems involving acids or bases (see elaboration in Chapter 2), as well as equilibrium constants of chemical reactions (elaborated upon further in this chapter). Since the concentrations of protons and hydroxides are, for the most part, very small and may shift easily by several orders of magnitude, it is convenient to express them as a negative logarithm. Similarly, it is common to express equilibrium constants in the same fashion. For example, for active concentration of protons (definition of the term "activity" is given later in this chapter):

$$pH = -\log(H^+) \tag{1.6}$$

or equivalently:

$$(H^+) = 10^{-pH} \tag{1.7}$$

Similarly:

$$pOH = -\log(OH^-) \tag{1.8}$$

$$pK_a = -\log(K_a) \tag{1.9}$$

Note that using a logarithm on a unit of fixed dimensions (such as molar concentration) is mathematically awkward, since it is permitted to use the logarithm only on dimensionless units such as activity and equilibrium constants (this issue is explained in Section 1.15).

1.14 Chemical equilibrium

Chemical equilibrium is achieved when the net concentrations of the products and reactants do not change overtime. In most cases, this situation occurs when the rate of the reaction toward the products ("forward") is equal to the rate of the opposite reaction, that is, in the direction of the reactants ("backward"). This is in fact a dynamic equilibrium.

Certain chemical reactions are carried out almost in their entirety in the direction of the products (denoted →). Converting the products of this type of reaction back to reactants is a very difficult process even under extreme conditions, and for this reason these types of reactions are defined as irreversible. Many other chemical reactions can easily be reversed in direction. In these reactions, the forward (toward the products) and backward (toward the reactants) reactions occur simultaneously (noted ↔) as described by eq. (1.10):

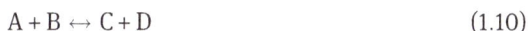

$$A + B \leftrightarrow C + D \tag{1.10}$$

1.15 The kinetic approach for describing chemical equilibrium

As stated, chemical equilibrium is achieved when the rate of the reaction toward the products (forward) is equal to the rate of the reaction toward the reactants (backward). The rate of a chemical reaction is dependent upon: (a) the concentration* of the reactants and products, (b) temperature and (c) the presence of catalysts. Therefore, the rate of the reaction (forward and backward) of eq. (1.10) can be expressed as follows:

$$r_f = k_1[A][B] = \text{rate of the forward reaction} \tag{1.11}$$

$$r_r = k_2[C][D] = \text{rate of the backward reaction} \tag{1.12}$$

k_1 and k_2 are rate constants.

   * In section 1.15, the chemical equilibrium for non-ideal systems is explained with detail and in a more precise way than the description appearing here. The

adjustment for equilibrium of non-ideal systems was done using definitions such as "activity" and "apparent equilibrium constant". However, when the kinetic approach (described here) was developed, it was not yet acknowledged that the concentration which actually takes place in the reaction (termed the "active concentration" or the "activity") differs from the measured (analytical) concentration. Thus, the writing shown in this section suits the phrasing used when the kinetic approach was formed, but does not suit the phrasing that is used today, which is described in the next section (1.15).

Since chemical equilibrium describes a dynamic situation in which at least two opposing chemical reactions occur at the same time and rate, it is obtained: $r_f = r_r$, or:

$$k_1[A][B] = k_2[C][D] \tag{1.13}$$

From eq. (1.13), a single apparent equilibrium constant can be defined, which expresses the ratio between the two rate constants and is denoted K_{eq}:

$$\frac{k_1}{k_2} = \frac{[C][D]}{[A][B]} = K_{eq} \tag{1.14}$$

And for the following general reaction:

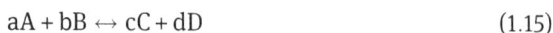

$$aA + bB \leftrightarrow cC + dD \tag{1.15}$$

The general expression for the apparent equilibrium constant is obtained as follows:

$$\frac{[C]^c [D]^d}{[A]^a [B]^b} = K_{eq} \tag{1.16}$$

It is important to remember that:

a. [A], [B], [C] and [D] are the molar concentrations of the products and reactants in equilibrium, but there is an infinite amount of possible combinations of concentrations.

b. By convention, the numerical value of the equilibrium constant describes the ratio between the products and reactants. Therefore, a weak electrolyte will have a low K_{eq} value and a strong electrolyte will have a high K_{eq} value.

c. Knowledge of the equilibrium constant alone does not provide any information regarding the amount of time needed to reach equilibrium, which can be a fraction of a second or a very long time.

d. Only with the *combined* knowledge of both the value of the equilibrium constant and the initial concentrations of the products and reactants, the direction in which equilibrium will be reached can be determined.

1.16 Adjusting the equilibrium constant to a non-ideal system

Equation (1.16) was developed by Guldberg and Waage in 1879 using molar concentrations. However, as shown further, equilibrium equations that are based on thermodynamics, the equilibrium constant is a function of *activity* rather than analytic concentration. The concept of activity was developed by J.W. Gibbs in the USA as early as the 1870s.

Guldberg and Waage assumed that all the solutions are ideal, that is, any ion in the solution behaves chemically in a way that is independent of any other ion. Gibbs proved that this assumption only holds true for highly diluted solutions. In the solution, there are interactions amongst the ions (electrical attraction, collisions), despite the fact that the ions are hydrated. As the solute concentration rises, so does the intensity of the interactions. These interactions have various effects on the solution's characteristics (freezing point, vapor pressure, boiling point, etc.). Considering this and in order to describe these systems thermodynamically, the term concentration activity or "activity" was proposed. Activity, sometimes referred to as "effective concentration" and denoted with round parenthesis, allows accurate description of reaction rates and equilibrium constants for non-ideal solutions, which are in essence all the practical solutions.

In order to find the relationship between activity and analytic concentration (molar), a proportional constant termed the "activity coefficient" is used. The relationship is described by the following equation:

$$a = \gamma \frac{C}{C^{\circ}} \qquad (1.17)$$

where a = activity (unitless), C = analytic concentration (M = moles per liter), C° = standard concentration: equals 1.0 by definition (mole per liter). The standard concentration can be real or hypothetical and γ = activity coefficient (unitless).

The activity of a chemical species (i) is unitless. The activity of pure phase states (solid, liquid, ideal gas composed of one species) is 1.0 by definition. If one wishes to develop a more meticulous definition for activity, it is preferable to use molality (moles per kg) instead of molar concentration (moles per liter), so that the activity will not be affected by the change in volume which occurs in solutions with high solute concentrations (brine, seawater, etc.). Since this book aims to deal solely with solutions of limited ionic strength (lower than 0.5 M), the activity equation here is developed with molar concentrations.

The closer a solution is to being ideal, the closer the activity coefficient is to unity, and the closer the activity value is to the molar concentration. The non-ideal behavior of electrolytic solutions arises from a number of factors that affect the interactions amongst the ions themselves and between the ions and the solvent. Two of the most significant factors are the ion concentration and their electric charge (valence). The term "ionic strength" of a solution was thus created, which expresses the effect of these two factors, and which in turn can be used to calculate the activity coefficients with relative ease. The ionic strength, I, is defined as follows:

$$I = \frac{1}{2}\sum_{i=1}^{i=i} C_i Z_i^2 \tag{1.18}$$

where I = ionic strength (M), C_i = analytic concentration of the ion i (M) and Z_i = charge (valence) of ion i.

Example 1.7 What is the ionic strength of a solution containing 0.005 M of Na_2SO_4 and 0.002 M of NaCl?

Solution

$$I = \frac{1}{2}[C_{Na^+}Z_{Na^+}^2 + C_{SO_4^{-2}}Z_{SO_4^{-2}}^2 + C_{Cl^-}Z_{Cl^-}^2] =$$

$$= \frac{1}{2}[(2\cdot0.005 + 0.002)(+1)^2 + 0.005(-2)^2 + 0.002(-1)^2] = 0.017M$$

For solutions in which the ionic makeup is unknown, the ionic strength can be approximated based on the direct relationship of the ionic makeup with the electrical conductivity or with the total dissolved solids (TDS) in the solution – two parameters that are easy to measure. Kemp (1971) [1] suggested the following approximations:

$$I = (2.5\cdot10^{-5})\cdot(TDS - 20) \tag{1.19}$$

$$I = (2.5\cdot10^{-5})\cdot(EC)\cdot670 \tag{1.20}$$

where TDS = total dissolved solids (mg/L), which is calculated based on the weight of the solids after drying a water sample of a known volume at a temperature of 105 °C (after filtering through a screen with pore sizes of 2 microns or less to separate the suspended solids).

EC = electrical conductivity at 20 °C (units: mS/cm or dS/m).

Equation (1.19) is a later addition to the equation proposed by Langelier (1946) [2]:

$$I = (2.5\cdot10^{-5})\cdot TDS$$

Langelier's equation, which was developed by linear regression of many measured samples, was defined by its author to be applicable at TDS concentrations lower than 1000 mg/L. On the other hand, Kemp (1971) (eq. (1.19)) did not limit the concentrations to which this equation is applicable. It should be noted that in practice for processes that rely on calculating the ionic strength, it is common to use the Langelier/Kemp equation for very high TDS concentrations. For example, in the Stasoft software (see Chapter 7), the TDS limitation for this type of calculation is 20,000 mg/L.

The number 20 subtracted from the TDS value in Kemp's equation was put in place to represent the presence of non-ionic silica which contributes to the TDS value but not to ionic strength or EC. It is important to note that eq. (1.19) does not consider the presence of dissolved organic matter (non-ionic), which can have a significant contribution to the TDS value, especially in wastewater. Nonetheless, the activity coefficient is not very sensitive to small changes in the ionic strength and therefore for engineering purposes the approximation is sufficient. In addition, it should be noted that in eq. (1.20), the factor 670 is generally accepted, however its value ranges between 550 and 700, and can sometimes even deviate from this range. Multiplying this factor by the EC value should yield the TDS value in eq. (1.19). Therefore, use of these two equations and comparing their results can give an indication of the accuracy of the approximations. Table 1.3 shows typical values of TDS, electrical conductivity and ionic strength for different types of water in Israel.

Table 1.3: Typical values (Range) of TDS, EC and I, for drinking, wastewater and treated water in Israel.

Water type	TDS (mg/L)	EC (dS/m)	I (M)
Surface water	~600	0.9	0.015
Ground water	200–1500	0.3–2.2	0.005–0.0375
	(Typical = 1000)	(1.5)	(0.025)
Sanitary sewage and Treated water	1000–2000	1.5–3.0	0.025–0.05
Desalinated water	150–250	0.22–0.37	0.0038–0.0063

The ionic strength calculated from eqs. (1.19) or (1.20) can be used in order to calculate activity coefficients. The literature provides a number of equations that describe the relationship between ionic strength and the activity coefficients, where the Davies equation is the most suitable for the broadest range of ionic strength values (up to $I = 0.5$ molar):

$$\log \gamma_{m,d,t} = -AZ^2 \left(\sqrt{I} / \left(1 + \sqrt{I}\right) - 0.2 \cdot I \right) \tag{1.21}$$

where $y =$ activity coefficient (unitless), m, d, $t =$ describe the absolute value of the valence of the ions: y_m the activity coefficient of monovalent ions, y_d is the activity coefficient of divalent ions, y_t is the activity coefficient of trivalent ions, etc.

where $A = 1.82 \cdot 10^6 (DT)^{-3/2}$, $D =$ dielectric coefficient of water, whose value is usually 78.3, $T =$ temperature of the solution (K), $Z =$ charge of the ions (1 for monovalent ions, 2 for divalent ions, etc.) and $I =$ ionic strength of the solution (M).

After learning how to calculate/estimate the ionic strength of a solution and in turn the corresponding activity coefficients, let us now see the utility of these coefficients in adjusting the equilibrium equation (eq. (1.16)) so that it will be expressed in terms of the apparent equilibrium constant (rather than the thermodynamic equilibrium constants, which is the figure provided in the literature). For the general reaction (eq. 1.15), the expression describing the thermodynamic equilibrium is:

$$K_{eq} = \frac{(C)^c (D)^d}{(A)^a (B)^b} \tag{1.22}$$

Substituting the appropriate expression for the activity of each species according to eq. (1.17) will allow displaying the equation in terms of analytic concentrations instead of activity concentrations and finding its relationship to the apparent equilibrium constant, K'$_{eq}$:

$$K_{eq} = \frac{(C)^c(D)^d}{(A)^a(B)^b} = \frac{(\gamma_C[C])^c(\gamma_D[D])^d}{(\gamma_A[A])^a(\gamma_B[B])^b} = \frac{[C]^c[D]^d}{[A]^a[B]^b} \cdot \frac{\gamma_C{}^c\gamma_D{}^d}{\gamma_A{}^a\gamma_B{}^b} = K'_{eq} \cdot \frac{\gamma_C{}^c\gamma_D{}^d}{\gamma_A{}^a\gamma_B{}^b} \qquad (1.23)$$

Thus,

$$K'_{eq} = \frac{K_{eq}}{\frac{\gamma_C{}^c\gamma_D{}^d}{\gamma_A{}^a\gamma_B{}^b}}$$

The thermodynamic equilibrium constants of most reactions used in environmental engineering appear in the literature. Notice that the thermodynamic equilibrium constant is expressed with activity concentrations which are dimensionless, thus it itself is dimensionless.

1.17 Acid–base equilibrium in the aqueous phase

Equation (1.24) describes the dissociation of a water molecule:

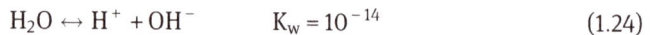

$$H_2O \leftrightarrow H^+ + OH^- \qquad K_w = 10^{-14} \qquad (1.24)$$

The expression for the equilibrium constant of reaction (1.24) is:

$$K_w = \frac{(H^+)(OH^-)}{(H_2O)}$$

The value of the equilibrium constant of reaction (1.24) was based on the convention by which the activity of water is 1.0, thus $K_w = (H^+)(OH^-)$. Still, it is sometimes common (e.g., [3]) to use $K_{eq} = 1.8 \cdot 10^{-16}$. In this case, the equilibrium constant is divided by the molar concentration of the water (as opposed to the activity which is 1.0). When calculating the concentration (moles of water per liter), the activity of protons and hydroxide ions can be neglected:

$$(H_2O) = \frac{\text{density of water}}{\text{molecular weight of water}} = \frac{1000\,g/L}{18\,g/mol} = 55.5\,mol/L$$

After substituting the water concentration in the equation, the same value for the equilibrium constant of the reaction is obtained:

$$K_{eq}(H_2O) = 1.8 \times 10^{-16} \times 55.5 = 10^{-14} = K_W$$

Recall that the value of K_W is a function of temperature and ionic strength. The value written above, 10^{-14}, is true only for a solution with a temperature of 25 °C and negligible ionic strength.

The activity of the hydroxide ions:

$$(OH^-) = \frac{K_W}{(H^+)} = \frac{10^{-14}}{(H^+)} \tag{1.25}$$

Taking log of both sides of eq. (1.25) and multiplying by –1, gives:

$$pOH = pK_W - pH = 14 - pH \tag{1.26}$$

Or

$$pH + pOH = 14 \tag{1.27}$$

From eq. (1.27), it can be seen that there is a linear relationship between the logarithm of the concentration of the hydroxide ion and the pH. This relationship can be shown graphically (see Fig. 1.3). This relationship is of great use in the following sections for understanding concepts such as equivalence points and alkalinity, as well as for solving complex problems.

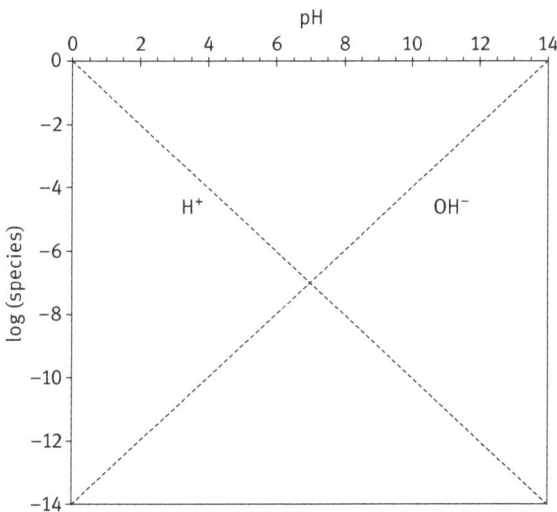

Fig. 1.3: Graph log (species) activity vs. pH – for the species H^+ and OH^-.

Example 1.8 Given groundwater with a TDS concentration of 100 mg/L and ferrous iron (i.e., Fe^{2+}) concentration of 50 mg/L. It is required to reduce the iron concentration to 4 mg/L by adding NaOH and precipitating $Fe(OH)_{2(s)}$. Water temperature is 25 °C.

What should be the target pH (neglect changes in TDS as a result of the addition of the strong base) to precipitate the iron?

Solution Calculation of the ionic strength according to eq. (1.19):

$$I = 2.5 \cdot 10^{-5} (TDS\text{-}20) = 2.5 \cdot 10^{-5} (100-20) = 0.002 \text{ M}$$

Find the activity coefficients using eq. (1.21):

$$\log \gamma_m = -1.82 \cdot 10^6 (78.3 \cdot 298)^{-\frac{3}{2}} (1)^2 \left(\frac{\sqrt{0.002}}{1+\sqrt{0.002}} - 0.2 \cdot 0.002 \right)$$

$$\gamma_m = 0.951$$

$$\log \gamma_d = -1.82 \cdot 10^6 (78.3 \cdot 298)^{-\frac{3}{2}} (2)^2 \left(\frac{\sqrt{0.002}}{1+\sqrt{0.002}} - 0.2 \cdot 0.002 \right)$$

$$\gamma_d = 0.818$$

From a table of thermodynamic constants, one can find the equilibrium constant for the precipitation reaction:

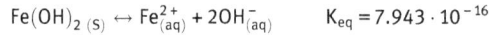

$$Fe(OH)_{2 (S)} \leftrightarrow Fe^{2+}_{(aq)} + 2OH^-_{(aq)} \qquad K_{eq} = 7.943 \cdot 10^{-16}$$

And thus (since the activity of solids is equal to unity):

$$K_{eq} = 7.943 \cdot 10^{-16} = \frac{(Fe^{2+}_{(aq)})(OH^-_{(aq)})^2}{(Fe(OH)_{2 (S)})} = \frac{[Fe^{2+}_{(aq)}]\gamma_d [OH^-_{(aq)}]^2 \gamma_m^2}{1}$$

The required remaining concentration of the divalent iron, after equilibrium is reached is (atomic weight of iron is 55.8 g per mole):

$$[Fe^{2+}_{(aq)}] = \frac{4(mg/L)}{55.8(mg/mmol)} = 0.07 \cdot 10^{-3} M$$

Now the activity of the hydroxide ion can be found:

$$[OH^-_{(aq)}]^2 = \frac{7.943 \cdot 10^{-16}}{[Fe^{2+}_{(aq)}]\gamma_d \gamma_m^2} = \frac{7.943 \cdot 10^{-16}}{0.07 \cdot 10^{-3} \cdot 0.818 \cdot 0.951^2} = 1.534 \cdot 10^{-11}$$

$$[OH^-_{(aq)}] = 3.92 \cdot 10^{-6} M$$

$$(OH^-_{(aq)}) = [OH^-_{(aq)}]\gamma_m = 3.92 \cdot 10^{-6} \cdot 0.951 = 3.728 \cdot 10^{-6}$$

pH calculation from the equilibrium constant of the water:

$$(H^+) = 10^{-pH} = \frac{K_w}{(OH^-_{(aq)})} = \frac{10^{-14}}{3.728 \cdot 10^{-6}} = 2.68 \cdot 10^{-9} M$$

$$pH = 8.57$$

Note that the pH value measured by a pH electrode describes the activity of the H^+ ions and not their concentration. Therefore, to find the theoretical pH to be reached, the activity of H^+ ions should be calculated rather than their analytic concentration. If the pH were to be calculated using the analytic concentration, the error would have been small in this case (the pH would be 8.59), yet in solutions with a high ionic strength, the error could be substantial.

In conclusion, the pH must be raised to pH 8.57, at which 46 mg/L of ferrous iron will precipitate.

1.18 The thermodynamic approach for describing chemical equilibrium

The equivalence of the rates of a reaction forward and backward is a necessary condition for the existence of chemical equilibrium, but it does not explain why chemical equilibrium occurs. The necessary and sufficient condition for a chemical equilibrium to occur is that the Gibbs ("available") free energy of a particular system would be at a minimum (for a particular constant pressure). If a certain mixture is not in equilibrium, Gibbs free energy will be released and it will be the "driving force" that changes the composition of the mixture. Only once the system arrives at equilibrium, energy will stop being released.

The significance of the above condition is that as the reaction progresses toward equilibrium, Gibbs free energy is reduced, and is null when the system reaches equilibrium. Mathematically, equilibrium at constant pressure is expressed as:

$$\Delta G_{reaction} = 0 \tag{1.28}$$

Reactions in which energy is released from the system to the environment are called exergonic reactions. Since the system loses energy, Gibbs energy in the final state is lower than in the initial state, and therefore $\Delta G_{reaction} < 0$ in these reactions. According to the laws of thermodynamics, every system tends naturally to minimal free energy. Therefore, exergonic reactions often occur spontaneously. Similarly, endergonic reactions are reactions that require an input of energy from their environment. In these reactions, $\Delta G_{reaction} > 0$ and is not spontaneous.

Gibbs free energy is defined as:

$$\Delta G = U + pV - TS \tag{1.29}$$

where G = Gibbs free energy, U = Internal energy of the system, p = pressure, V = Volume of the system, T = Absolute Temperature and S = Entropy of the system.

The development of eq. (1.30), which describes the change in the Gibbs free energy, will not be shown here. The development is based on taking the complete derivative of eq. (1.29) and substituting the derivative of U from the first law of thermodynamics. The result is as follows (neglecting external forces acting on the system):

$$\Delta G_{\text{reaction}} = VdP + SdT + \sum_{i=1}^{k} \mu_i dN_i \qquad (1.30)$$

where μ_i = Chemical potential of chemical component i. Chemical potential, for simplicity's sake, will be defined here as the specific Gibbs energy for each component in the solution and N_i = number of particles (moles) of chemical component i.

At constant pressure and temperature, eq. (1.30) becomes:

$$\Delta G_{\text{reaction}} = \sum_{i=1}^{k} \mu_i N_i \qquad (1.31)$$

It can be shown that $\Delta G_{\text{reaction}}$ is in essence the difference between the sum of the chemical potentials of the products and the sum of the chemical potentials of the reactants as described by the following equation which relies on eq. (1.15):

$$\Delta G_{\text{reaction}} = c\mu_C + d\mu_D - (a\mu_A + b\mu_B) \qquad (1.32)$$

One of the fundamental equations in thermodynamics describes the chemical potential in ideal solutions as follows:

$$\mu_i = \mu_i^O + RT \ln(i) \qquad (1.33)$$

where R = the universal gas constant $\left(R = 1.98 \frac{\text{cal}}{\text{mol} \cdot \text{K}}\right)$, T = the absolute temperature (K), I = the activity of species i and ^{O} in superscript indicates standard conditions (1 M, 101.325 kPa, 298 K).

Using eq. (1.33) on each component of eq. (1.32), yields:

$$\Delta G_{\text{reaction}} = (c\mu_C^O + d\mu_D^O) - (a\mu_A^O + b\mu_B^O) \\ + (cRT \ln(C) + dRT \ln(D)) - (aRT \ln(A) + bRT \ln(B)) \qquad (1.34)$$

which can be abbreviated as

$$\Delta G_{\text{reaction}} = \sum_{i=1}^{k} \mu_i^O N_i + RT \ln \frac{(C)^c (D)^d}{(A)^a (B)^b} \qquad (1.35)$$

Eq. (1.35) can be written in a more general form as follows:

$$\Delta G_{\text{reaction}} = \Delta G_{\text{reaction}}^O + RT \ln Q \qquad (1.36)$$

where $\Delta G_{\text{reaction}}^O$ = the change in Gibbs free energy under standard conditions (298 K, 101.325 kPa), and $Q = \frac{(C)^c (D)^d}{(A)^a (B)^b}$.

In equilibrium $\Delta G_{\text{reaction}}^O = 0$ and $Q = \frac{(C)^c (D)^d}{(A)^a (B)^b} = K_{\text{eq}}$, therefore:

$$\Delta G^O_{reaction} = - RT(\ln K_{eq}) \tag{1.37}$$

or:

$$K_{eq} = e^{\left(\frac{-\Delta G^O_{reaction}}{RT}\right)} \tag{1.38}$$

Combining eqs. (1.37) and (1.36) yields the following relationship:

$$\Delta G_{reaction} = - RT(\ln K_{eq}) + RT \ln Q = RT \ln \frac{Q}{K_{eq}} \tag{1.39}$$

1.19 Standard free energy of formation

Free energy of formation ($\Delta G^O_{formation}$) is the change in the Gibbs free energy that accompanies the formation of one mole of a certain substance from its comprising elements under their standard conditions. $\Delta G^O_{formation}$ of O_2, for example, is zero since O_2 is the standard form of oxygen and there is no change in its state under standard conditions. Similarly, $\Delta G^O_{formation}$ of all chemical elements in their standard form is zero.

The change in the free energy of a chemical reaction is defined as the difference between the free energy in the final state to that in the initial state:

$$\Delta G_{reaction} = G_{final} - G_{initial} \tag{1.40}$$

Therefore, for a reaction occurring under standard conditions, it can be written as follows:

$$\Delta G^O_{reaction} = \sum_{\substack{\text{of products}}} \Delta G^O_{formation} - \sum_{\substack{\text{of reactants}}} \Delta G^O_{formation} \tag{1.41}$$

Knowing the values of $\Delta G^O_{formation}$ for the products and reactants, $\Delta G^O_{reaction}$ can be calculated using eq. (1.41). The values for the energies of formation under standard conditions ($\Delta G^O_{formation}$) for different compounds are given in many literature sources. When $\Delta G^O_{reaction}$ is negative, the reactants in their standard form will turn into products in their standard form (exergonic reaction). A positive $\Delta G^O_{reaction}$ indicates that the reactants in their standard form will not turn into products in their standard form (endergonic reaction). If the calculated $\Delta G^O_{reaction}$ is zero, the reactants and the products in their standard form will be at equilibrium. Since $\Delta G^O_{formation}$ is given in units of kcal/mol, one must multiply by the number of moles in the balanced equation (the coefficients) in order to obtain $\Delta G^O_{reaction}$ in kcal.

Example 1.9 Divalent iron Fe^{2+} is sometimes found in groundwater (under anaerobic conditions). After extracting the groundwater and bringing it in contact with the atmosphere, the iron is likely to be oxidized into trivalent iron ions and precipitate as a solid under the standard pH levels of groundwater. This process may interfere with the various uses of the water. In an effort to remove the divalent iron immediately after extracting the groundwater, it can be oxidized by gaseous chlorine and the oxidized iron deposits (trivalent iron oxides) can be removed by gravitational deposition.

Determine whether the suggested chemical reaction is possible thermodynamically:

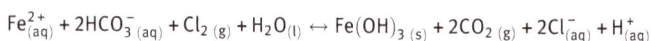

$$Fe^{2+}_{(aq)} + 2HCO_3^-{}_{(aq)} + Cl_{2\,(g)} + H_2O_{(l)} \leftrightarrow Fe(OH)_{3\,(s)} + 2CO_{2\,(g)} + 2Cl^-_{(aq)} + H^+_{(aq)}$$

Solution First, the free energy of formation for every species found in the reaction needs to be determined (Table 1.4, from the literature):

Table 1.4: Gibbs standard free energy of formation for the chemical species in Example 1.9.

Chemical species	Standard free energy of formation $(\Delta G^0_{formation})$(kcal/mol)
$Fe^{2+}_{(aq)}$	−20.3
$HCO_3^-{}_{(aq)}$	−140.31
$Cl_{2\,(g)}$	0
$H_2O_{(l)}$	−56.69
$Fe(OH)_{3\,(s)}$	−166
$CO_{2\,(g)}$	−94.25
$Cl^-_{(aq)}$	−31.35
$H^+_{(aq)}$	0 (by convention)

Using eq. (1.41), $\Delta G^0_{reaction}$ can be calculated as follows:

$$\Delta G^0_{reaction} = [-166 + 2(-94.25) + 2(-31.35)] - [-20.3 + 2(-140.31) + (-56.69)] = -59.59\,kcal/mol$$

Since $\Delta G^0_{reaction}$ is negative, the reaction from left to right occurs spontaneously under standard conditions.

1.20 Determining ΔG under non-standard conditions

In practice, only a few chemical reactions occur under standard conditions. This especially holds true for pH-dependent reactions, whose pH is almost never zero (under standard conditions the activity of H^+ is 1 M). Therefore, it is important to know how to calculate changes in free energy under non-standard conditions.

Example 1.10 Inserting gaseous chlorine into water in order to disinfect the water is a common water treatment practice. In practice, the main disinfecting agent is the hypochlorous acid (HOCl) that forms by the hydrolysis reaction of chlorine:

$$Cl_{2(g)} + H_2O_{(l)} \leftrightarrow HOCl_{(aq)} + H^+_{(aq)} + Cl^-_{(aq)}$$

a. Determine the value of the equilibrium constant of the above reaction under standard conditions.
b. Determine the change in free energy of the above reaction given that all reactants and products are in their standard state except for H^+. The pH of the solution is 7.0.

Determine in both cases whether the reaction is spontaneous.

Solution

Table 1.5: Gibbs standard free energy of formation for the chemical species in Example 1.10.

Chemical species	Standard free energy of formation ($\Delta G^0_{formation}$) (kcal/mol)
$Cl_{2\,(g)}$	0
$H_2O_{(l)}$	−56.69
$HOCl_{(aq)}$	−19.11
$Cl^-_{(aq)}$	−31.35
$H^+_{(aq)}$	0 (by convention)

$\Delta G^0_{reaction} = [(-31.35) + (0.00) + (-19.11)] - [(0.00) - (-56.69)] = 6.23\ kcal/mol$
$\Delta G^0_{reaction}$ is positive and therefore, under standard conditions, the reaction is not spontaneous.
a. To find K_{eq}, eq. (1.38) can be used to obtain:

$$K_{eq} = e^{-\frac{\Delta G^0_{reaction}}{RT}} = e^{-\frac{6230}{1.98 \cdot 298}} = 10^{-4.59}$$

b. Using eq. (1.36) yields (note that $\log Q \cdot 2.303 = \ln Q$):

$$\Delta G_{reaction} = \Delta G^0_{reaction} + 2.303\,RT \log\left(\frac{(HOCl)(H^+)(Cl^-)}{(Cl_2)(H_2O)}\right)$$

$$\Delta G_{reaction} = 6230 + 2.303 \cdot 1.98 \cdot 298 \cdot \log\left(\frac{1 \cdot 10^{-7} \cdot 1}{1 \cdot 1}\right) = -3282\ cal/mol = -3.3\ kcal/mol$$

$\Delta G^0_{reaction}$ in this case is negative, therefore it can be concluded that the reaction is spontaneous.

1.21 Temperature effect on the equilibrium constant value

In most cases, chemical reactions occur at a temperature other than the standard temperature of 25 °C. At temperatures other than 25 °C, the value of the equilibrium constant is different. Why is that?

If the forward reaction (i.e., from left to right) of a chemical system in equilibrium releases heat to the environment (exothermic), then the reaction in the opposite direction absorbs heat from the environment (endothermic). If an external source of heat is added to this system, when it is at equilibrium, the point of equilibrium will be shifted toward the reactants (the endothermic reaction is preferred) according to Le Chatelier's principle. As a result, the value of the equilibrium constant will decrease (the concentration of reactants in the final state will increase, and the concentration of products will decrease).

The *enthalpy*, (denoted H) describes the heat released/absorbed by the reaction at constant pressure (1 atm). The enthalpy of a closed system will approach minimum when pressure and entropy are constant. In other words, the change in enthalpy will be zero when the reaction reaches equilibrium at a constant pressure and entropy.

The method for calculating the enthalpy of a reaction at standard conditions, $\Delta H^O_{reaction}$, is similar to the calculation of the Gibbs's free energy:

$$\Delta H^O_{reaction} = \sum \Delta H^O_{\substack{formation \\ of\ products}} - \sum \Delta H^O_{\substack{formation \\ of\ reactants}} \tag{1.42}$$

One of the forms to describe the change in the Gibbs energy (development not shown here):

$$\Delta G^O_{reaction} = \Delta H^O_{reaction} - T\Delta S^O \tag{1.43}$$

where ΔH^O = change in enthalpy of the reaction under standard conditions and ΔS^O = change in entropy under standard conditions.

Substituting eq. (1.37) into eq. (1.43) yields:

$$\ln K_{eq} = -\frac{\Delta H^O_{reaction}}{RT} + \frac{\Delta S^O}{R} \tag{1.44}$$

Differentiation of eq. (1.44) with respect to T yields Van't Hoff's equation [3]:

$$\frac{d(\ln K_{eq})}{dT} = \frac{\Delta H^O_{reaction}}{RT^2} \tag{1.45}$$

where the assumption is that the change in enthalpy of the reaction remains constant in the range of the temperatures T_1 and T_2, integrating eq. (1.45) with limits of integration T_1 to T_2, yields the following expression:

$$\ln \frac{(K_{eq})_2}{(K_{eq})_1} = \frac{-\Delta H^{0}_{reaction}}{R} \left[\frac{1}{T_2} - \frac{1}{T_1} \right] \tag{1.46}$$

where $(K_{eq})_1$ and $(K_{eq})_2$ represent the equilibrium constants at temperatures T_1 and T_2, respectively. It is possible to use eq. (1.46) in many cases (in which the assumption that the change in enthalpy of the reaction remains constant throughout the range of the given temperatures) to estimate the change of the equilibrium constant at a different temperature. Usually, it is convenient to use the equilibrium constant found in the literature $(K_{eq})_1$ together with the standard temperature $(T_1 = 298\ K)$ and the temperature under which the reaction occurs (T_2) in order to find the value of the equilibrium constant at this temperature $(K_{eq})_2$.

Example 1.11 The precipitation of gypsum is described by the following reaction:

$$CaSO_{4(s)} \Leftrightarrow Ca^{2+} + SO_4^{2-} \quad K_{sp}=2.5 \cdot 10^{-5}$$

The equilibrium constant is defined at 25 °C. What is the value of the equilibrium constant at 15 °C?

Solution

$$\Delta H^{0}_{reaction} = \sum \Delta H^{0}_{formation \atop of\ products} - \sum \Delta H^{0}_{formation \atop of\ reactants} = -129.77 + (-216.90) - (-342.42)$$

$$= -4.25 \frac{kcal}{mol}$$

Conclusion: since $K_{15\ °C} > K_{25\ °C}$, the gypsum is more soluble at lower temperatures.

Alternatively, when the enthalpy balance results in a positive value, the solid is more soluble at a higher temperature.

Note: In case an equilibrium constant is to be converted for both temperature and ionic strength, the calculations should be carried out in series, were the first conversion is substituted into the second. The order of conversion is not important.

---

2 Acids and bases

2.1 Introduction

Acids and bases fill a most important role in drinking and industrial water quality, natural processes and water and wastewater treatment processes. The composition and molar ratios between weak-acid and base systems determine the pH of the water and its buffer capacity (defined later in this section). The pH value has both direct and indirect effects on reactions occurring in the aqueous phase: for example, it affects directly the precipitation of solids, the dissolution or stripping of gasses, adsorption processes, corrosion, flocculation, disinfection reactions and more. The pH also affects the distribution of weak acid species (e.g., NH_3, versus $NH_4{}^+$), thus it is involved indirectly in possible inhibition/promotion of biological processes (e.g., oxidation of ammonia to nitrite, termed nitrification). Moreover, biological processes such as photosynthesis and respiration and physical processes such as aeration (resulting from turbulent flow) affect pH values by changing the carbon dioxide concentration in the water (and also other gases such as H_2S, NH_3 or volatile organic species). In addition, many biological processes, such as bacterial enzymatic activity, are affected by pH.

This section focuses on the concepts and principles related to aqueous systems that contain acids and bases, allowing us later to simplify problems related to water chemistry. As such, analytical and graphical methods for determining pH and specific species concentrations in equilibrium are presented.

2.2 Basic principles

2.2.1 Defining acids and bases

Over the years, many definitions for acids (and bases) have been proposed. According to the Arrhenius theory, published around 1884, an acid is a substance that contains a hydrogen atom in its structure and increases the concentration of H_3O^+ (hydronium) ions as it dissolves in water thereby decreasing the concentration of OH^- (hydroxide) ions. Similarly, a base is a substance contains OH^- in its structure and therefore that decreases the concentration of H_3O^+ ions and increases the concentration of OH^- ions as it dissolves in water. Equation (2.1) describes the ionization of hydrochloric acid (HCl) in water. A reaction between a weak base (non-ionic ammonia) and water is described in eq. (2.2). Explanation for the determination of arrow direction is listed below.

$$HCl + H_2O \rightarrow H_3O^+ + Cl^- \tag{2.1}$$

$$NH_3 + H_2O \leftrightarrow NH_4{}^+ + OH^- \tag{2.2}$$

https://doi.org/10.1515/9783110603958-002

It was later realized that Arrhenius's definition is limited in that it can only be used for substances dissolvable in water. The Arrhenius theory also cannot explain the fact that there are bases, such as $NH_{3(aq)}$, which do not contain OH^- and acids that do not include H^+ (e.g., Fe^{3+}) in their structure.

Bronsted and Lowry expanded the definition of acids and bases. According to their definition, on which this book is based, an acid is any substance that can donate a proton (H^+) to another substance, and a base is any substance that can receive a proton from another substance (written in a general form in eq. (2.3)). According to this definition, ammonia ($NH_{3(aq)}$) is a base, since it can react with H^+ to form NH_4^+ and by that it increases the OH^- concentration ($NH_3 + H^+ \leftrightarrow NH_4^+$). Similarly, water ($H_2O$) can act as both a base and an acid. When water reacts with a base, it acts as an acid, as shown in eq. (2.4), where CO_3^{2-} is the base (receives H^+ from the water). Similarly, in eq. (2.1), the water acts as a base as it can receive a proton from hydrochloric acid. Equation (2.5), which describes the ionization of water, shows how water can act simultaneously as both an acid and a base.

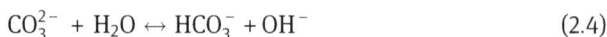

$$HAcid + Base \leftrightarrow Acid^- + HBase^+ \tag{2.3}$$

$$CO_3^{2-} + H_2O \leftrightarrow HCO_3^- + OH^- \tag{2.4}$$

$$H_2O + H_2O \leftrightarrow H_3O^+ + OH^- \tag{2.5}$$

There exists another definition for acids and bases (termed Lewis' acidity) but it will not be elaborated upon, because its practical uses differ from the practical applications discussed in this book. The definition is as follows: According to Gilbert Lewis (around 1923), an acid is a chemical species that can receive a pair of electrons, and a base is a species that can donate a pair of electrons to produce a covalent bond. It is important to note that this is not the same as oxidation–reduction reactions where electrons leave an atom entirely and are transferred to a different atom; rather, it is the sharing of a pair of electrons by two molecules in order to form a chemical bond. This definition includes reactions that do not involve the release or acceptance of a proton (H^+), and therefore this definition is more general than Bronsted and Lowry's definition. A proton can receive a pair of electrons, and therefore it follows that every base according to Lewis (in other words, all substances with a pair of unbound electrons) can receive a proton and thus are a base according to Bronsted and Lowry. On the other hand, not every Lewis acid is an acid according to Bronsted and Lowry. Take, for example, multivalent metal cations (e.g., Al^{3+}). This ion reacts with water (by accepting 2 electron from the water into the covalent bond) to form the complex (ion pair) $Al(OH)_2^+$, while releasing a proton to solution. $Al^{3+} + H_2O \leftrightarrow Al(OH)_2^+ + H^+$; $K = 1.8 \cdot 10^{-5}$. From a purely chemical standpoint, the electron density is drawn away from the O-H bond in the H_2O molecule toward the positively charged cation and H^+ is released to the solution. Al^{3+} is therefore a Lewis acid. Lewis's definition, although accurate in most circumstances, is not the most common; therefore, when Lewis acids (which are not acids according to

Bronsted and Lowry) are mentioned, they are declared as such. As stated earlier, this book will not deal with these types of acids.

The links between the concepts basicity, alkali and alkalinity

In a number of dictionary definitions, as well as daily use, there are contradictions and confusions between the terms "alkali" and "basic" and between the terms "alkalinity" and "basicity". The source for the confusion is probably historical. The term "alkali" or "alkaline" has no practical importance in the context of acids and bases and the chemistry of water treatment processes. So it's worth sticking to the definitions given in this book to the terms "base" and "alkalinity" for the description and calculation of reactions in water. Below is a possible explanation for the mixture of concepts that have become an entrenched practice:

The term alkali (Al-Qali) is derived from Arabic, its meaning is "the roasted". "the roasted" refers to lime (lime, burnt-lime), CaO, which was produced by man already in the prehistoric period by means of a calcination (decomposition or transformation by heat) of limestone, which is usually composed of the mineral calcite $CaCO_3$. The reaction, which takes place at a temperature of about 850 Celsius (the temperature of a strong fire) is:

$$CaCO_{3(s)} \rightarrow CaO + CO_{2(g)}$$

The burnt lime that remains after the limestone is burned causes the release of hydroxide ions when it comes in contact with water:

$$CaO + H_2O \rightarrow Ca^{+2} + 2OH^-$$

Hence, the lime ("alkali") is actually a base according to the definition of Arrhenius and has been used as such already in ancient times. For example, in order to make soap from animal fat, which today is a process performed using a hydroxide base such as KOH or NaOH.

In chemistry, the term "alkali" describes electrolytes of certain metals (alkali metals), which dissolving in water causes the release of hydroxides similar to lime. It should be noted that although all the alkaline materials are bases according to Arrhenius, not all bases are alkaline (e.g., ammonia - NH_3).

2.3 Acid-base pairs

In all the examples presented up to this point (eqs. (2.1) through (2.5)), the acid and base form acid-base pairs. In other words, hydrochloric acid (HCl) is accompanied by the chloride ion (Cl^-) basis species (eq. (2.1)). Similarly, Ammonia (NH_3) is the conjugate base of the ammonium ion (NH_4^+) in eq. (2.2). Sometimes the conjugate base of the acid is called the "acid's salt". For example, sodium acetate ($NaCH_3COO$), sometimes abbreviated as NaAc, is the conjugate base of acetic acid (CH_3COOH, abbreviated HAc) (eq. (2.6)). Since sodium acetate is a strong electrolyte, it dissociates entirely into the sodium ion and the acetate ion (see left hand side of eq. (2.6)). The sodium ion is of little importance with regard to the acid-base system (except for its indirect effect through the apparent equilibrium constants which are determined by the ionic strength). As such, the sodium ion can be ignored and the focus is solely on the anionic base (CH_3COO^-, abbreviated Ac^-). Accordingly, acetic acid's (CH_3COOH)

"salt" is sodium acetate ($NaCH_3COO$). However, the conjugate base of the acid is not always that acid's salt, for example, the ammonium ion (NH_4^+) is a weak acid whose conjugate base is ammonia (NH_3). In such a case, ammonia is not an ion and therefore cannot be referred to as the weak acid's salt (eq. (2.2)).

$$NaAc + H_2O \rightarrow Na^+ + Ac^- + H_2O \leftrightarrow Na^+ + HAc + OH^- \tag{2.6}$$

2.4 Polyprotic acids

Many acids can donate more than one proton. For example, $H_2CO_3^*$ and H_2SO_4 can donate two protons; H_3PO_4 can donate three protons. In general, acids such as these are termed polyprotic acids, as opposed to acids that can donate only one proton and are termed monoprotic acids. Likewise, acids that are capable of releasing two protons are termed diprotic acids, and acids that possess three protons are termed triprotic acids. Similarly, bases that can receive more than one proton are termed polyprotic bases, examples include CO_3^{2-} and PO_4^{3-}.

The ionization of a diprotic acid is divided into two stages.

The first stage:

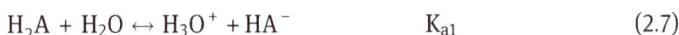

$$H_2A + H_2O \leftrightarrow H_3O^+ + HA^- \qquad K_{a1} \tag{2.7}$$

$$K_{a1} = \frac{(H^+)(HA^-)}{(H_2A)} \tag{2.8}$$

where K_{a1} denotes the equilibrium constant for the first ionization stage, or the equilibrium constant where the first proton is donated.

The second stage of ionization:

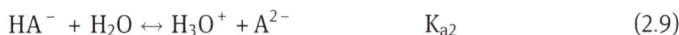

$$HA^- + H_2O \leftrightarrow H_3O^+ + A^{2-} \qquad K_{a2} \tag{2.9}$$

$$K_{a2} = \frac{(H^+)(A^{2-})}{(HA^-)} \tag{2.10}$$

where K_{a2} denotes the equilibrium constant for the second ionization stage, or the equilibrium constant where the second proton is donated.

2.5 The carbonate system

The carbonate system is a weak acid system that dominates acid-base equilibria in natural waters (surface water, groundwater and seawater) and also in most types of wastewaters. The great prevalence of this system is what leads to its importance in environmental chemistry and biology, as well as in industrial/drinking water treatment processes. The carbonate system is a weak, diprotic acid system (three species and two protonation constants). The three species are $H_2CO_3^*$, HCO_3^- and CO_3^{2-}. In practice, the carbonate system comprises of four species since $H_2CO_3^*$ is a fictitious species representing the concentration sum of the two real species: $H_2CO_{3(aq)}$ and $CO_{2(aq)}$, as explained later.

The prevalence of the carbonate system in almost every natural water body is the result of two processes: the dissolution of atmospheric carbon dioxide ($CO_{2(g)}$) and the dissolution of carbonate solids in water. The atmosphere is made up of approximately 0.041% carbon dioxide by weight (true for 2018). Despite this relatively low concentration, enough CO_2 can dissolve in natural waters to ensure (together with dissolution of carbonate rocks that is also induced by $CO_{2(g)}$ dissolving in rain water) that carbonate system species are at a higher concentration than all other acid-base system species found in natural bodies of water. Dissolution is carried out according to the following equilibrium reaction: (further discussion of this reaction can be found in Chapter 5, which focuses on *aqueous-gas equilibrium*):

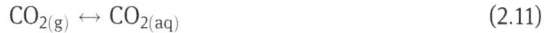

$$CO_{2(g)} \leftrightarrow CO_{2(aq)} \tag{2.11}$$

Groundwater that is found in regions made up of carbonate rocks (e.g., limestone, $CaCO_{3(s)}$ and dolomite $CaMg(CO_3)_{2(s)}$) can contain relatively high concentrations of the carbonate system due to reactions that promote dissolution of the stone. These reactions raise the concentration of carbonate ions (CO_3^{2-}) in the water (see eqs. (2.12) and (2.13)).

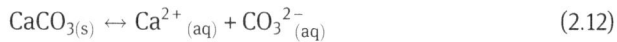

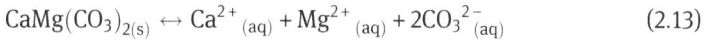

$$CaCO_{3(s)} \leftrightarrow Ca^{2+}_{(aq)} + CO_3^{2-}_{(aq)} \tag{2.12}$$

$$CaMg(CO_3)_{2(s)} \leftrightarrow Ca^{2+}_{(aq)} + Mg^{2+}_{(aq)} + 2CO_3^{2-}_{(aq)} \tag{2.13}$$

The carbonate system can also be found in groundwater that has not been in contact with carbonic rock; this is possible because groundwater originates from rainwater which was exposed for sufficient contact time with the atmosphere to ensure that carbon dioxide is dissolved in it. The dissolved carbon dioxide ($CO_{2(aq)}$) tends toward equilibrium with carbonic acid according to eq. (2.14):

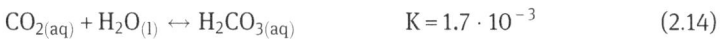

$$CO_{2(aq)} + H_2O_{(1)} \leftrightarrow H_2CO_{3(aq)} \qquad K = 1.7 \cdot 10^{-3} \tag{2.14}$$

Note that the equilibrium constant of eq. (2.14) is $K = 1.7 \cdot 10^{-3}$. Therefore, once equilibrium is reached the concentration of dissolved carbon dioxide is three orders of magnitude greater than the concentration of carbonic acid.

Carbon dioxide

Despite carbon dioxide's low concentration in the atmosphere many researchers have claimed that it is no less important to the existence of life than oxygen. When carbon dioxide dissolves in water, a weak acid is formed (as stated earlier, the acid along with its conjugate base make up the carbonate system). Life on earth originated in water and in general, can only be maintained in a relatively narrow scope of physical and chemical conditions in this phase. An especially important characteristic of natural waters and physiological solutions (e.g., blood) is the ability to maintain a neutral pH range (pH 5-pH 9). In this regard, the carbonate system is of great importance, as elaborated upon in Chapter 3 (Section 3.9) which focuses on the buffering capacity of water.

Both of the aforementioned species can donate a proton to the water according to the following reactions:

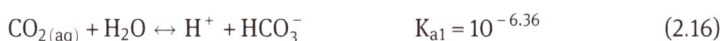

$$H_2CO_{3\,(aq)} \leftrightarrow H^+ + HCO_3^- \qquad K_{a1} = 10^{-3.6} \qquad (2.15)$$

$$CO_{2\,(aq)} + H_2O \leftrightarrow H^+ + HCO_3^- \qquad K_{a1} = 10^{-6.36} \qquad (2.16)$$

For the sake of simplifying mathematical computations, it is uncommon to refer to dissolved carbon dioxide and carbonic acid separately, but rather as the sum of the two species, with the symbol $H_2CO_3^*$:

$$[H_2CO_{3\,aq}] + [CO_{2\,aq}] = [H_2CO_3^*] \qquad (2.17)$$

As such, eqs. (2.15) and (2.16) can be written as one reaction, which describes the loss of the first proton of $H_2CO_3^*$:

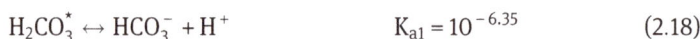

$$H_2CO_3^* \leftrightarrow HCO_3^- + H^+ \qquad K_{a1} = 10^{-6.35} \qquad (2.18)$$

Note that the equilibrium constant of the above reaction is based on the equilibrium constants of the protonation reactions of the two species (eqs. (2.15) and (2.16)) and on the distribution amongst the species; note that it is clearly much closer to the equilibrium constant of dissolved carbon dioxide (eq. (2.14)).

The second protonation reaction of carbonate weak acid system is as follows:

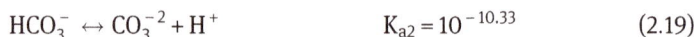

$$HCO_3^- \leftrightarrow CO_3^{-2} + H^+ \qquad K_{a2} = 10^{-10.33} \qquad (2.19)$$

Equations (2.18) and (2.19) describe the carbonate system for all practical purposes. Because of the carbonate system's importance in the chemistry of natural water bodies and wastewaters, as well as in the chemistry of the beverage industry, the biotechnological industry and more, the use of these two reactions will be abundant throughout this book.

As was stated previously, the carbonate system is a weak acid system. As such, to understand this system we must first learn about weak acid systems in general. This will be the topic of the remainder of this chapter.

2.6 The strength of an acid or base

The strength of an acid or base is determined by its tendency to donate or receive a proton, respectively. A weak acid is an acid with a weak tendency to donate a proton.

In other words, a weak acid is a weak electrolyte. However, it is difficult to define the acid's ("acid" as in eq. (2.3)) tendency to donate a proton since this tendency is also dictated by the conjugate base's ("base") tendency to receive a proton (or donate a hydroxide ion). Therefore, when in the aqueous phase, the strength of the acid-base pair, A^- – HA, is measured relative to the acid-base strength of the water system, that is, OH^- – H_2O. Similarly, the relative strength of a base acid pair, B – HB^+, is measured relative to the base-acid pair of the water system, that is, H_2O – H_3O^+.

According to this definition, the strength of an acid can be defined (in aqueous solution) with the help of the equilibrium constant for the protonation reaction. Let us focus again on eq. (2.7), which describes the first protonation reaction of a diprotic acid:

$$H_2A + H_2O \leftrightarrow H_3O^+ + HA^- \quad K_{a1} \tag{2.7}$$

The equilibrium constant for eq. (2.7) is defined as follows:

$$K_a = \frac{(H^+)(HA^-)}{(H_2A)} \tag{2.20}$$

Thus, it is clear that the stronger the acid, the larger the value of the equilibrium constant (K_a), and the smaller the value of pK_a. A strong acid is usually defined as one where $pK_a<0$. This means that adding a strong acid to water will cause (almost) complete protonation within the pH range 0 to 14, and eq. (2.7) will proceed almost entirely toward creation of the products. Therefore, when dealing with strong acids and bases, one can draw a one-way arrow from the reactants to the products (as can be seen in eq. (2.1)), as is the case for strong electrolytes which dissolve (almost) entirely in water. However, when dealing with weak-acids and bases, the solution within the pH range 0–14 will contain both the products (A^-) and the reactants (HA) in a non-negligible quantity. Accordingly, a two-way arrow will be used between the reactants and products in a reaction describing weak-acids and bases (e.g., eq. (2.7)).

It follows from the explanation above that the pH of a given solution can be calculated directly from the concentration of the strong acid (or base) that is in it (see e.g., 2.1); however, the pH of a solution cannot be directly calculated when dealing with weak acids. The method for calculating the pH in such situations is elaborated upon in the section focusing on Solving Weak-Acid/Base systems using the proton balance equation and another which focuses on Equivalent solutions and Equivalence points. Using a similar definition, the strength of a base can be determined (in aqueous solution) with the help of the equilibrium constant for the protonation reaction. For example, for a monoprotic base:

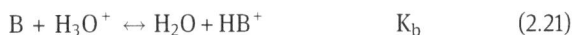

$$B + H_3O^+ \leftrightarrow H_2O + HB^+ \qquad K_b \tag{2.21}$$

The equilibrium constant of eq. (2.21) is defined as follows:

$$K_b = \frac{(HB^+)}{(H^+)(B)}$$

It is thus clear that the stronger the base, the larger the value of K_b and the smaller the value of pK_b.

Note that "weak acid" and "weak base" share the same definition and can be the same system. The difference between the two terms is only a matter of the direction of the pH range from which one defines them. Take, for example, the ammonia weak base (acid) system.

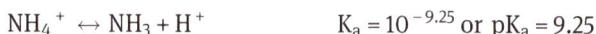

$$NH_4^+ \leftrightarrow NH_3 + H^+ \qquad K_a = 10^{-9.25} \text{ or } pK_a = 9.25$$

When the equation is written in this fashion, the ammonia system can be defined as a weak acid with a relative low constant. However, the same system can also be written as:

$$NH_3 + H_2O \leftrightarrow NH_4^+ + OH^- \qquad K_b = 10^{-4.75} \text{ or } pK_b = 4.75.$$

In this case, one can refer to the system as being a weak base with a relatively high constant. Note that in such cases always $pK_a + pK_b = pK_w = 14$ ($K_a \cdot K_b = K_w$) because the difference between the two manifestations of the system is the incorporation of the water dissociation equation. Regardless of the choice to work with K_a or K_b, the results of solving problems should be identical, and therefore there is no importance to whether the system is regarded as a weak acid or a weak base.

Example 2.1
A. Three bases are shown in Table 2.1 below. Using the equilibrium constants, arrange the bases according to their strength. What is the pH value of the strongest base?
B. Three acids of differing concentrations are also given in Table 2.1. Note that each acid has a different equilibrium constant and that not all acids are monoprotic. Arrange the acids according to their pH values. What is the pH value of the most acidic solution?

Table 2.1: Acids and bases and their equilibrium constants.

		pK_{a1}	pK_{b1}	pK_{a2}	Concentration (M)
Base	$H_2PO_4^-$	–	11.9		
	NaOH	–	0.2		
	HS^-	–	6.9		
Acid	HCl	–3	–	–	1
	HOCl	7.6	–	–	1
	H_2SO_4	–3	–	1.9	0.5
	HNO_3	–1	–	–	1

Solution

A. As mentioned earlier, the strength of a base is measured by the value of its equilibrium constant, with a stronger base having a larger equilibrium constant (and a smaller pK_b value). Thus, based on the pK_b values for the given bases, they can be arranged by descending strength, for example as follows:

$$NaOH > HS^- > H_2PO_4^-$$

B. Let us look at the pK_{a1} values for the acids shown in Table 2.1. It can immediately be seen that there are three acids with a strong tendency toward donating one proton (strong acids) and one acid (hypochloric acid, HOCl) with a lower tendency to donate its proton. HOCl is thus a weak acid. It is now noted that all four acids, except for sulfuric acid, H_2SO_4, are monoprotic. In addition, note that the concentration of the diprotic acid is 0.5 M while the concentration of the remaining three monoprotic acids is double, that is, 1 M. Therefore, the normal concentrations (see Chapter 1) of these four acids are identical. Since the concentrations of all four acids are identical, by taking a second look at the pK_a values, the pH of each solution can be determined. One might raise the question which solution would have a lower pH: HCl or H_2SO_4, where the concentration of both is 1N. The tendency of the sulfuric acid to donate its first proton is equal to the tendency of the hydrochloric acid to donate all its protons (because it is a monoprotic acid). However, the tendency of the sulfuric acid to donate its second proton is lower, and therefore the pH of the HCl 1N solution will be the lowest (the most acidic). Accordingly, the list of the acids according to their pH values (in descending order) is: $HOCl > HNO_3 > H_2SO_4 > HCl$.

Let us now calculate the pH of the HCl 1N solution. As stated, this is a strong acid which dissolves in its entirety according to eq. (2.20). The pH is therefore:

$$pH = -\log(H^+) = -\log(1) = 0 \text{ (excluding ionic strength effects)}$$

Note that if the concentration of this acid had been larger, we would have gotten a result of pH < 0 (i.e., negative pH value). For example, a hydrochloric acid with a concentration of 37% (37.0 g HCl/100 g solution) is commonplace in chemical laboratories. The molar concentration of this solution is ~12 M, and therefore:

$$pH = -\log(H^+) = -\log(12) = -1.08$$

Likewise, it is possible to achieve pH > 14 with the use of strong bases.

Notice the pH value which is calculated this way will be more correct than the value measured using a pH electrode, since normal pH electrodes show low accuracy at high H^+ activities.

Polyprotic acids can be weak or strong. There are polyprotic acids whose tendency to release their first proton is very strong (i.e., $pK_{a1} < 0$), while the tendency to release the following proton (or protons) is much weaker (pK_{a2}, $pK_{a3} > 0$), therefore they cannot be referred to as weak or strong acids but rather a combination of the two. In any case, $K_{a1} > K_{a2} > K_{a3}$ because the tendency of an acid to release its first proton is always stronger than its tendency to release the next proton.

Table 2.2 shows the equilibrium constants for common acids. Again, the K_a value is determined by eq. (2.7). The acids are presented in descending order, from the strongest acid to the weakest. In addition, the conjugate bases of the acids are presented along with their pK values.

Table 2.2: Acids and bases and their corresponding equilibrium constants. From Stumm
and Morgan [4].

Acid		pK_a*	Base	pK_b**
$HClO_4$	Perchloric acid	−7	ClO_4^-	21
HCl	Hydrogen chloride	~−3	Cl^-	17
H_2SO_4	Sulfuric acid	~−3	HSO_4^-	17
HNO_3	Nitric acid	−1.3	NO_3^-	15
HNO_2	Nitrous acid	3.7	NO_2^-	10.3
H_3O^+	Hydronium ion	0	H_2O	14
HSO_4^-	Bisulfate	1.9	SO_4^{-2}	12.1
H_3PO_4	Phosphoric acid	2.1	$H_2PO_4^-$	11.9
CH_3COOH	Acetic acid	4.7	CH_3COO^-	9.3
H_2CO_3*	Carbon dioxide	6.3	HCO_3^-	7.7
H_2S	Hydrogen sulfide	7.1	HS^-	6.9
$H_2PO_4^-$	Dihydrogen phosphate	7.2	HPO_4^{-2}	6.8
HOCl	Hypochlorous acid	7.6	OCl^-	6.4
HCN	Hydrogen cyanide	9.2	CN^-	4.8
H_3BO_3	Boric acid	9.3	$B(OH)_4^-$	4.7
NH_4^+	Ammonium ion	9.3	NH_3	4.7
HCO_3^-	Bicarbonate	10.3	CO_3^{-2}	3.7
HPO_4^{-2}	Hydrogen phosphate	12.7	PO_4^{-3}	1.3
HS^-	Bisulfide	14	S^{-2}	0
H_2O	Water	14	OH^-	0

* According to eq. (2.7), **According to eq. (2.21)

2.7 The Henderson–Hasselbalch equation

Let us now deal with special cases in which the acid-base system's concentration is significantly higher than that of H^+ and OH^-. In such cases, it can be assumed that the concentration of the protons and bases in the solution are negligible relative to the concentration of the acid-base system species. In such cases, and only in these, the Henderson–Hasselbalch equation presented below can be used:

Using the same equation that was previously used for calculating equilibrium constants, eq. (2.20) one can isolate, for example, the proton activity term:

$$(H^+) = \frac{K_a(HA)}{(A^-)} \tag{2.22}$$

Let us now take the log of each side of this equation and obtain:

$$\log(H^+) = \log K_a + \log \frac{(HA)}{(A^-)} \tag{2.23}$$

After multiplying by negative one:

$$-\log(H^+) = -\log K_a + \log \frac{(A^-)}{(HA)} \tag{2.24}$$

Or according to the "p" notation:

$$pH = pK_a + \log \frac{(A^-)}{(HA)} \tag{2.25}$$

In a more general form, this can be written as:

$$pH = pK_a + \log \frac{(\text{proton acceptor})}{(\text{proton doner})} \tag{2.26}$$

Equation (2.26) is termed as the Henderson–Hasselbalch equation. The thermodynamic equilibrium constant of different acids is known and can be calculated from data found in the literature. Therefore, the Henderson–Hasselbalch equation shows the relationship between the pH of a solution and the ratio of the concentrations of the acid and its conjugate base. Several conclusions can be drawn from this equation:

A. When the pH of a solution is equal to the pK of the acid, the concentrations of the proton accepting species and the proton donating species are equal.

$$@pH = pK: \quad 0 = \log \frac{(\text{proton acceptor})}{(\text{proton doner})} \Rightarrow \frac{(\text{proton acceptor})}{(\text{proton doner})} = 1 \tag{2.27}$$

B. When the value of pH is greater than that of pK, the concentration of the proton donor is lower than the concentration of the proton acceptor. Conversely, when the pH is lower than pK of the acid, the concentration of the proton acceptor is lower than that of the proton donor.

$$@pH > pK: \quad 0 > \log \frac{(\text{proton acceptor})}{(\text{proton donor})} \Rightarrow \frac{(\text{proton acceptor})}{(\text{proton donor})} > 1$$

$$@pH < pK: \quad 0 < \log \frac{(\text{proton acceptor})}{(\text{proton donor})} \Rightarrow \frac{(\text{proton acceptor})}{(\text{proton donor})} < 1$$

C. A low pK value (e.g., pK = 2.0) means that at pH > 2, the concentration of the proton donor is lower than that of the proton acceptor. In other words, acids with a low pK value will have a greater tendency to donate a proton than acids with a high pK value or put differently, the lower the pK_a value, the stronger the acid.

Buffer solutions are made up of a combination of proton donors (a weak acid) and proton acceptors (the weak-acid's salt). Buffer solutions are named so because of their ability to strive to maintain a steady range of pH values (see elaboration on buffer capacity in section (2.12) and in Chapter 3: alkainity and acidity). For example, phosphoric acid is a weak, triprotic acid that is commonly used to produce buffer solutions around the neutral pH range.

Example 2.2 You are given a buffer solution composed of 0.2 N sodium acetate (NaCH$_3$COO) and 0.1 N acetic acid (HCH$_3$COO). The water temperature is 25 °C.
The equilibrium constant of the acetic acid is K$_a$ = 1.8 · 10^{-5}. Calculate the pH of the solution.

Solution First calculate the ionic strength. Note that the concentration of the acid does not affect the ionic strength since it is not an ion.

$$I = \frac{1}{2}[C_{Na^+}Z_{Na^+}^2 + C_{Ac^-}Z_{Ac^-}^2] = \frac{1}{2}[0.2(+1)^2 + 0.2(-1)^2] = 0.2M$$

Now calculate the monoprotic activity coefficient:

$$\log\gamma_M = -1.82 \times 10^6 (78.3 \times 298)^{-3/2}(1)^2\left(\sqrt{0.2}/(1+\sqrt{0.2}) - 0.2 \times 0.2\right) \Rightarrow \gamma_M = 0.729$$

The pH can now be calculated using the Henderson–Hasselbalch equation:
*Reminder: pK$_a$ = −log(K$_a$). Therefore, K$_a$ = 1.8 × 10^{-5} is equivalent to pK$_a$ = 4.74

$$pH = pK_a + \log\frac{(Ac^-)}{(HAc)} = 4.74 + \log\frac{\gamma_M[Ac^-]}{(HAc)} = 4.74 + \log\frac{0.729 \times 0.2}{0.1} = 4.91$$

It is important to note that the Henderson–Hasselbalch equation can be used to calculate pH only when the concentrations of the weak-acid/base system's species are known. In the case of a strong acid (or base), ionization/dissociation is complete, and we do not need to use this equation (see Example 2.1).

One of the purposes of this book is to give the reader tools on how to characterize water-based solutions without using approximations, because approximations can be problematic and misleading in certain scenarios. In the latter of this book, explanations and examples will be provided for different ways of calculating chemical species distributions from pH and vice versa. Most of the examples in the following chapters, as opposed to those that made use of the Henderson–Hasselbalch equation, are not based upon any approximations. Therefore, in this book, the use of the Henderson–Hasselbalch equation is very limited.

2.8 Species distribution of weak-acids/bases as a function of pH

The concentrations of species in a weak-acid/base system are affected by pH and the equilibrium constants of the system. Let us begin by looking at the concentrations of the water species, OH$^-$ and H$^+$, as a function of pH. In the introductory chapter, it was defined pH = −log(H$^+$). We will look again at the reaction describing the protonation of water:

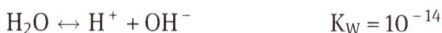

$$H_2O \leftrightarrow H^+ + OH^- \qquad\qquad K_W = 10^{-14}$$

The equilibrium constant of eq. (2.5) is very low. In other words, the degree of ionization of water is low. This is why, in calculating the amount of moles of water per liter, one can neglect the concentration of protons and hydroxides:

$$(H_2O) = \frac{\text{mass of 1 liter of water}}{\text{g molecular weight of water}} = \frac{1000}{18} = 55.5\,\frac{\text{mol}}{l}$$

An expression for the equilibrium constant of eq. (2.5):

$$K = \frac{(H^+)(OH^-)}{(H_2O)}$$

Keeping in mind that the activity of water is 1.0, let us calculate the value of the product of the proton and hydroxide concentrations:

$$(H^+)(OH^-) = K(H_2O) = K_W = 10^{-14}$$

It is important to note that the value of K_W (as of any other equilibrium constant) is a function of temperature and ionic strength. The value provided here, 10^{-14}, is only correct for water at a temperature of 25 °C and having a negligible ionic strength.

Let us now define the activity of the hydroxide ions:

$$(OH^-) = \frac{K_W}{(H^+)} = \frac{10^{-14}}{(H^+)} \tag{2.28}$$

We will use the log operator on both sides of eq. (2.28) and multiply by –1 (as per the "p" notation, see Chapter 1):

$$-\log(OH^-) = pK_W - pH = 14 - pH \tag{2.29}$$

From eq. (2.29), it can be seen that a linear relationship exists between the log of the activity of the hydroxide ion and the pH. The relationship between the activity of the protons and pH is given in eq. (1.6). These relationships are shown graphically in Fig. 2.2.

Let us now turn to calculating the different species concentrations in aquatic solutions that contain weak-acid/base systems in addition to the water system. The following example shows how the concentration of every chemical species in a weak, diprotic acid system can be calculated as a function of pH and equilibrium constants. In a similar way, we can develop expressions for the concentrations of chemical species in mono- or triprotic systems, or any other weak acid system. The expressions for mono-, di- and triprotic systems are presented in Table 2.3. The steps in developing an expression for diprotic systems are given below, however they are general and can be applied to any weak-acid/base system.

Table 2.3: Species in various weak-acid systems as a function of pH. A_T stands for the total concentration of the weak acid/base system, i.e. the sum of the concentrations of all the system's species.

System	Species	Expression
Monoprotic	HA	$\dfrac{(H^+)A_T}{K'_a + (H^+)}$
	A^-	$\dfrac{K'_a A_T}{K'_a + (H^+)}$
Diprotic $(K_{a1} > K_{a2})$	H_2A	$\dfrac{(H^+)^2 A_T}{(H^+)^2 + K'_{a1}(H^+) + K'_{a1}K'_{a2}}$
	HA^-	$\dfrac{(H^+)K'_{a1}A_T}{(H^+)^2 + K'_{a1}(H^+) + K'_{a1}K'_{a2}}$
	A^{-2}	$\dfrac{K'_{a1}K'_{a2}A_T}{(H^+)^2 + K'_{a1}(H^+) + K'_{a1}K'_{a2}}$
Triprotic $(K_{a1} > K_{a2} > K_{a3})$	H_3A	$\dfrac{(H^+)^3 A_T}{(H^+)^3 + K'_{a1}(H^+)^2 + K'_{a1}K'_{a2}(H^+) + K'_{a1}K'_{a2}K'_{a3}}$
	H_2A^-	$\dfrac{K'_{a1}(H^+)^2 A_T}{(H^+)^3 + K'_{a1}(H^+)^2 + K'_{a1}K'_{a2}(H^+) + K'_{a1}K'_{a2}K'_{a3}}$
	HA^{-2}	$\dfrac{K'_{a1}K'_{a2}(H^+)A_T}{(H^+)^3 + K'_{a1}(H^+)^2 + K'_{a1}K'_{a2}(H^+) + K'_{a1}K'_{a2}K'_{a3}}$
	A^{-3}	$\dfrac{K'_{a1}K'_{a2}K'_{a3}A_T}{(H^+)^3 + K'_{a1}(H^+)^2 + K'_{a1}K'_{a2}(H^+) + K'_{a1}K'_{a2}K'_{a3}}$

A. Write the relevant chemical equations and balance them.

The general symbol for a weak, diprotic acid, H_2A, is used. As stated, two protonation reactions occur in a diprotic system and they are described by eqs. (2.7) and (2.9):

$$H_2A + H_2O \leftrightarrow H_3O^+ + HA^- \tag{2.7}$$

$$HA^- + H_2O \leftrightarrow H_3O^+ + A^{2-} \tag{2.9}$$

B. Write expressions for the equilibrium constants.

Write the equilibrium equations that correspond to the reactions mentioned in part A:

Equation (2.8): $K_{a1} = \dfrac{(H^+)(HA^-)}{(H_2A)}$ and eq. (2.10): $K_{a2} = \dfrac{(H^+)(A^{2-})}{(HA^-)}$. Note that in the explanations and examples that follow, the thermodynamic equilibrium constants have been exchanged for the apparent equilibrium constants and the activity has accordingly been exchanged for concentration (except for the activity of the protons which were not exchanged for concentration because pH is a direct measure of activity). Therefore, the appropriate equilibrium equations are:

$$K'_{a1} = \dfrac{(H^+)[HA^-]}{[H_2A]} \text{ and } K'_{a2} = \dfrac{(H^+)[A^{-2}]}{[HA^-]}.$$

C. Write the mass balance equation.
    This equation is based on the mass balance of the acid-base system. It means that the concentration of all the species of the system can vary (depending on pH), but as long as there is no entry or exit of species into or from the solution (e.g., as a result of volatilization, sedimentation, salt dissolution, etc.), the sum of the species remains constant. In other words, the distribution of species is dependent upon pH, but the sum of the species is not.
    The mass balance equation for the current example is:

$$A_T = [H_2A] + [HA^-] + [A^{-2}]$$

D. Solve for each species separately.
    After writing the equations in parts B and C, the number of equations equals the number of unknowns (there are two unknowns in a monoprotic system and three and four unknowns in a di- and triprotic system, respectively).

As an example, let us develop an expression for the concentration of the species HA. To start, isolate the species H_2A in eq. (2.8) and the species A^{2-} in eq. (2.10):

$$[H_2A] = \frac{(H^+)[HA^-]}{K'_{a1}}$$

$$[A^{2-}] = \frac{K'_{a2}[HA^-]}{(H^+)}$$

Now substitute these expressions into the mass balance equation:

$$A_T = \frac{(H^+)[HA^-]}{K'_{a1}} + [HA^-] + \frac{K'_{a2}[HA^-]}{(H^+)} = [HA^-]\left(\frac{(H^+)}{K'_{a1}} + 1 + \frac{K'_{a2}}{(H^+)}\right)$$

After finding a common denominator:

$$A_T = [HA^-]\left(\frac{(H^+)^2 + K'_{a1}(H^+) + K'_{a1}K'_{a2}}{(H^+)K'_{a1}}\right)$$

And isolating the species HA^-:

$$[HA^-] = \frac{(H^+)K'_{a1}A_T}{(H^+)^2 + K'_{a1}(H^+) + K'_{a1}K'_{a2}}$$

As stated, a similar procedure can be used for finding expressions for all the species in different systems, these general expressions are presented in Table 2.3.

2.8.1 Species concentration as a function of pH – the graphical approach

The dependence of species concentration on pH, which is presented in Table 2.3, can also be presented graphically using a simple manual technique, from which it is possible to understand visually the behavior of the acid-base system as a whole. The technique for drawing the distribution of species is an important tool for solving various problems in aquatic chemistry and before the computer era it was actually the main solution method for simple acid-base problems.

For the purpose of explaining the general procedure, the carbonate system is sketched, given that the sum of all species in the system is $C_T = 10^{-3}$ M. A bit further in the section, a quick and simple technique for the graphical approach is provided without the explanations which are now discussed. Let us start by writing the relevant chemical equations. The given system is a diprotic weak acid. Therefore, two protonation reactions are carried out (eqs. (2.18) and (2.19)):

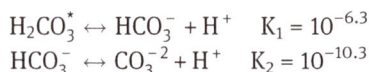

$$H_2CO_3^* \leftrightarrow HCO_3^- + H^+ \quad K_1 = 10^{-6.3}$$
$$HCO_3^- \leftrightarrow CO_3^{-2} + H^+ \quad K_2 = 10^{-10.3}$$

The equilibrium constant equations:

$$K_1 = \frac{(H^+)(HCO_3^-)}{(H_2CO_3^*)}; \quad K_2 = \frac{(H^+)(CO_3^{-2})}{(HCO_3^-)}$$

Following is the mass balance equation, which means that the total sum of species in the system does not change:

$$C_T = [H_2CO_3^*] + [HCO_3^-] + [CO_3^{-2}] = 10^{-3}M$$

where C_T is the sum of the concentrations of species in the carbonate system. At this point, we have three equations (two expressions for the equilibrium constants and one mass balance equation) and three unknowns (the three species of the carbonate system). Therefore, we can express each species with the help of the equilibrium constants, the total concentration of the system (C_T) and the concentration of H^+ and get the following expressions (see also Table 2.3):

$$[H_2CO_3^*] = \frac{(H^+)^2C_T}{(H^+)^2 + K'_{a1}(H^+) + K'_{a1}K'_{a2}} \tag{2.30}$$

$$[HCO_3^-] = \frac{(H^+)K'_{a1}C_T}{(H^+)^2 + K'_{a1}(H^+) + K'_{a1}K'_{a2}} \tag{2.31}$$

$$[CO_3^{-2}] = \frac{K'_{a1}K'_{a2}C_T}{(H^+)^2 + K'_{a1}(H^+) + K'_{a1}K'_{a2}} \tag{2.32}$$

With the help of the expressions that were developed and with the knowledge of C_T, one can sketch each species' concentration as a function of pH. As an example, let us now demonstrate how to sketch, by hand, the curve of the most acidic species, $H_2CO_3^*$.

Sketch the concentration as a function of pH in the range $0 < pH < 14$. In this example, which deals with a diprotic system, the region is divided into five separate sub-regions. Dividing the region into five sub-regions simplifies the process in that different assumptions can be made in each region, as is elaborated upon shortly. In three out of the five sub-regions marked I, II and III in Fig. 2.1A, the pH value is distant from the pK values by at least one pH unit. Between these three sub-regions lie in the two remaining sub-regions, each of which is in the vicinity of one of the system's pK values (marked as sub-regions IV and V in Fig. 2.1A). In other words, these two sub-regions span a length of two pH units (from $pK_i - 1$ to $pK_i + 1$).

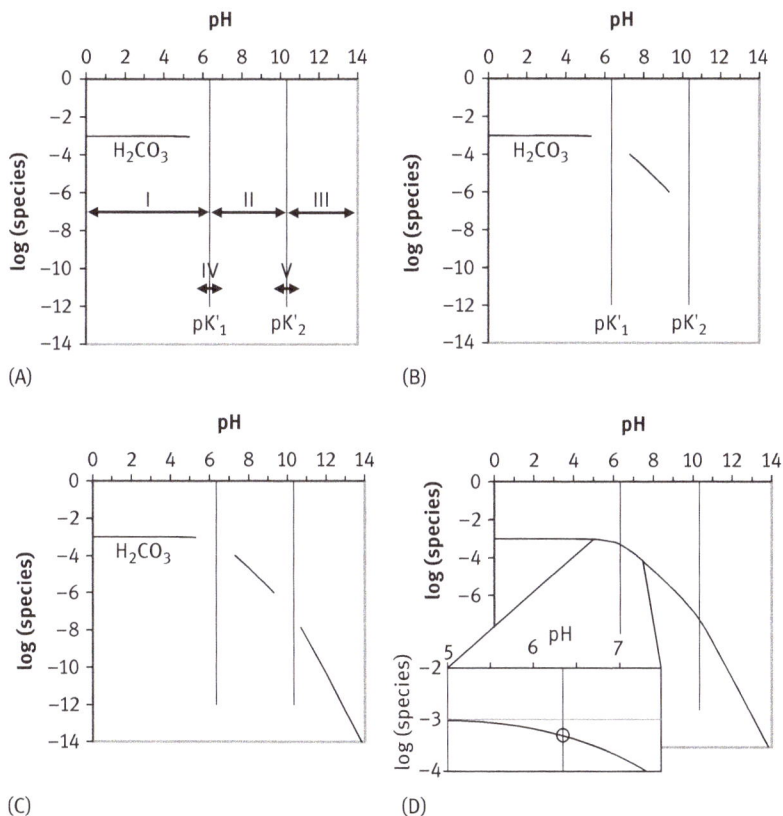

Fig. 2.1: The log-concentration of $H_2CO_3^*$ as a function of pH in sub-region I (A), sub-regions I and II (B), sub-regions I, II and III (C) and throughout the entire pH range (D).

To start, let us characterize the change in concentration of $H_2CO_3^*$ as a function of pH in the first three sub-regions:

Sub-region I: pH < pK'$_1$−1. This sub-region is marked as I in Fig. 2.1A. Recall that pH and pK'$_1$ are the negative log of (H^+) and K'_1, respectively. Therefore for pH = pK'$_1$−1 we get that$(H^+) = 10 \cdot K'_1$. Therefore in sub-region I, where pH is at least one unit smaller than pK'$_1$, it follows that $(H^+) \gg K'_1$. In general, regardless of pH $K_1 \gg K_2$ and therefore in sub-region I, $(H^+) \gg K'_1 \gg K'_2$. Turn now to the expression for the species concentration of $H_2CO_3^*$ (eq. (2.30)). Since $(H^+)^2 \gg K'_1(H^+) \gg K'_1K'_2$, negligible terms in the denominator can be omitted to obtain:

$$[H_2CO_3^*]_{@pH < pK'_1} = \frac{(H^+)^2 C_T}{(H^+)^2}$$

Canceling out yields:

$$[H_2CO_3^*]_{@pH < pK'_1} = C_T \tag{2.33}$$

From eq. (2.33), it is clear that in sub-region I of the pH values, the concentration of the species is dependent solely on C_T and is equal to it. Accordingly, we can sketch a horizontal line representing the concentration of $H_2CO_3^*$ in sub-region I, see Fig. 2.1A.

Sub-region II: pK'$_1$+1<pH<pK'$_2$−1. This sub-region is marked as II in Fig. 2.1A. In this sub-region, since the pH value is smaller than pK'$_2$ by at least one unit, it follows that $(H^+) \gg K'_2$. In addition, since the pH is greater than pK'$_1$ by at least one unit, it follows that $(H^+) \ll K'_1$. Take another look at our expression for $[H_2CO_3^*]$ (eq. (2.30)). Note that the term $(H^+)K'_1$ is the only non-negligible term in the denominator. Eliminating identical and negligible terms and get:

$$[H_2CO_3^*] = \frac{(H^+)^2 C_T}{(H^+)^2 + K'_1(H^+) + K'_1K'_2} \Rightarrow [H_2CO_3^*]_{@pK'_1 < pH < pK'_2} = \frac{(H^+)^2 C_T}{(H^+)K'_1} = \frac{(H^+)C_T}{K'_1} \tag{2.34}$$

Taking the logarithm of each side yields:

$$\log(H_2CO_3^*)_{@pK'_1 < pH < pK'_{21}} = \log\left(\frac{(H^+)C_T}{K'_1}\right) = \log\left(\frac{C_T}{K'_1}\right) - pH \tag{2.35}$$

In this region, thus, the species' concentration is inversely and linearly related to pH. Therefore, the line representing the species' concentration as a function of pH in this intermediate region will have a slope of −1, that is, to say a 45° slope (see Fig. 2.1B). It is important to note that to ensure a 45° slope, it is important that the x and y axes use the same scale.

Sub-region III: pH>pK'$_2$+1. This sub-region is marked as III in Fig. 2.1A. In this sub-region, since the pH is greater than pK'$_2$ by at least one unit, it follows that (H$^+$)≪K'$_2$. As was already mentioned, K'$_1$≫K'$_2$ regardless of the pH, so that (H$^+$)≪K'$_2$≪ K'$_1$. Looking again at the expression for [H$_2$CO$_3^*$], it transpires that K'$_1$K'$_2$ is the only non-negligible term in the denominator. Omitting the negligible terms yields:

$$\left[H_2CO_3^*\right] = \frac{(H^+)^2 C_T}{(H^+)^2 + K_1'(H^+) + K_1'K_2'} \Rightarrow \left[H_2CO_3^*\right]_{@pH > pK_2' > pK_1'} = \frac{(H^+)^2 C_T}{K_1'K_2'} \qquad (2.36)$$

Taking the log of each side of the equation results in:

$$\log\left[H_2CO_3^*\right]_{@pH > pK_2' > pK_1'} = \log\left(\frac{(H^+)^2 C_T}{K_1'K_2'}\right) = \log\left(\frac{C_T}{K_1'K_2'}\right) - 2pH \qquad (2.37)$$

Hence, in this sub-region, the species concentration is linearly related to pH as well, but multiplied by two. Therefore, the line representing the concentration of the species in this sub-region is a straight line with a slope of −2 (see Fig. 2.1C). The curves of the remaining two species of the carbonate system can be sketched similarly in these three sub-regions (I, II and III).

The next step in plotting the line is to connect the three sub-regions, in other words plotting the concentration curve in the two sub-regions adjacent to pK'$_{1,2}$ (sub-regions IV and V in Fig. 2.1A). Drawing the concentration curve in the sub-region adjacent to pK'$_1$ (sub-region IV in Fig. 2.1A) is similar in principle to drawing the curve in sub-region V, therefore only the procedure for drawing sub-region IV is explained.

As is discussed shortly, the concentration of the species CO$_3^{2-}$ is negligible at pK'$_1$ and therefore:

$$C_T \cong \left[H_2CO_3^*\right] + \left[HCO_3^-\right] = 10^{-3}M$$

As stated, the concentration of H$_2$CO$_3^*$ and HCO$_3^-$ are equal at pK'$_1$ (eq. (2.27)). Therefore:

$$C_T \cong 2\left[H_2CO_3^*\right] = 2\left[HCO_3^-\right] \Rightarrow \left[H_2CO_3^*\right] = \left[HCO_3^-\right] = 0.5 \cdot C_T$$

Convert the concentrations to fit a logarithmic scale (as they are presented in the graph):

$$\log\left[H_2CO_3^*\right] = \log\left[HCO_3^-\right] = \log(0.5 \cdot C_T) = \log 0.5 + \log C_T = -0.3 + \log C_T$$

In other words, the H$_2$CO$_3^*$ concentration curve (as well as the HCO$_3^-$ curve) passes through the point (pK'$_1$, logC$_T$−0.3). The same conclusion can be drawn for every weak-acid/base pair and their corresponding pK. After marking this point, draw an arc connecting the line in sub-region I with the line in sub-region II, making sure that the arc passes through the marked point. See enlarged sketch detailing this on the bottom left corner of Fig. 2.1D.

As stated, the same calculations can be performed for the remaining two species to get a full graph describing the concentration of each species as a function of pH, as presented in Fig. 2.2. Figure 2.2 also shows the concentration of protons (H^+) and hydroxide ions (OH^-) as a function of pH. These concentrations can be found directly from eqs. (1.4) and (2.29), respectively.

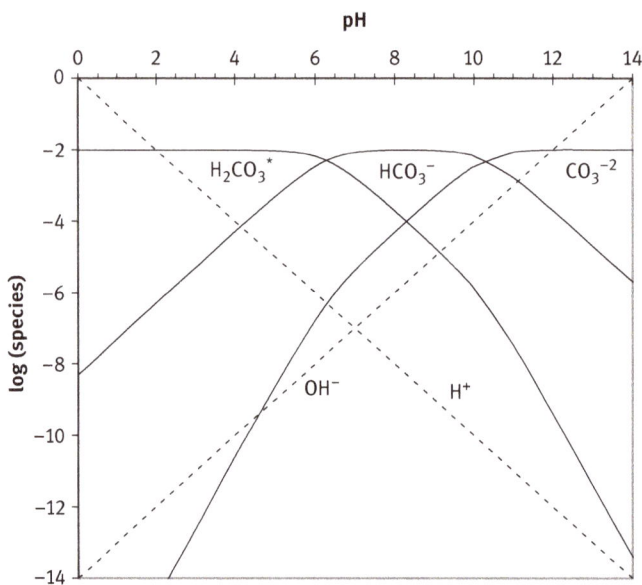

Fig. 2.2: pH vs. log (species) graph of the carbonate system. $C_T = 10^{-2}$ M.

2.8.2 Sketching the log (species) curve as a function of pH – a quick procedure

To plot the log species curve as a function of pH, there is no need to perform all the calculations shown above. Once a firm understanding of the stages outlined above is gained, the process of sketching these graphs can be simplified to merely carrying out the following steps:

1. Draw the x and y axes for pH and log species, respectively. Make sure to plot the two axes using the same scale. Start at pH = 0 and end at pH = 14. For the y-axis, start at log(species) = 0 and continue until at least log(species) = −7.
2. Mark the total species concentration (C_T in the case of the carbonate system) on the y-axis. Draw a horizontal line as a reference going through this point (portions of this line will later be erased).
3. Mark the system's pK values on the x-axis. Draw vertical lines going through the pK values as references.

4. Mark the intersection points between the vertical and horizontal reference lines. Next to each intersection point, mark two additional points on the horizontal line (the C_T line), each at a distance of one unit (to the left and to the right) from the intersection point. These points will mark the beginning and end of the lines representing the concentration of each species in the area where it is most dominant. We can now erase the horizontal reference line in the vicinity of the pK values. It is advised not to erase the intersection points quite yet.

5. Draw a 45° line downward from the intersection point. Draw this line starting in the area that is one pH unit away from the pK.

6. Mark the points where the curve passes for pH = pK. These points, as stated, are found 0.3 units underneath (vertically) the intersection points, which is at (pK, C_T-0.3). These are the actual pK points, not points used for reference.

7. Draw arched curves representing the species concentration in the pH close to the pK value. Each arc should start from the horizontal line, pass through the marked pK point, and connect with the sloped line drawn. At this point, we have drawn the concentration curves for each of the species within the pH range.

8. Write the name of the species next to its appropriate curve. The most acidic species is represented by the curve whose horizontal portion is found at the leftmost part of the graph, that is, to the left of the lowest pK. After undergoing one protonation reaction, one get a species which is less acidic, whose horizontal portion is found to the right of the lowest pK value.

9. Add the line representing the H^+ concentration. This line starts at the origin and slopes downward at 45° (slope of −1).

10. Add the line representing the OH^- concentration. This line starts at the point (14, 0) and slopes downward at 45° (slope of +1). Ensure that the H^+ and OH^- lines meet at point (7, −7).

2.8.3 Sketching the log (species) curve as a function of pH – important points

These type of graphs provide information regarding the distribution of species at all pH values. It can be seen that for almost every pH value (except for those that are close to the pK values), only one dominant species is expressed, whose concentration is significantly higher than the other species'. Therefore, in practice, the concentrations of the other species can be ignored and the concentration of the dominant species is practically equal to C_T. For example, in the second pH range (pK$_1$ < pH < pK$_2$) one can say, with considerable accuracy, that [HCO$_3$$^-$] = C_T. Recall that the y-axis uses a logarithmic scale, and therefore a one unit drop means a drop of one order of magnitude of the species' concentration. Observe, for example, that in Fig. 2.2 at pH = 5 the concentration of $H_2CO_3^*$ is more than one order of magnitude greater than the concentration of HCO$_3$$^-$.

Let us now turn to the case in which the pH changes (C_T remains constant), for example, as a result of adding a base to our solution. In this case, as long as the pH remains in the same sub-region (i.e., to say, the same species remains dominant) in the practical sense, the species percentage (out of C_T) will not change significantly. For example, if the pH of a solution that includes the carbonate system changes from pK – 4 (i.e., the pK value minus four pH units) to pK – 2, in both cases, the concentrations of HCO_3^- and CO_3^{2-} are negligible and $H_2CO_3^* = C_T$.

In the case where pH = pK, the concentrations of the two species relevant to the pK are equal, while the concentration of the third species is negligible (note that this is correct when the difference between the pK values is significant, i.e., > 3 pH units). For example, for pH = pK_2 we get $\left[HCO_3^-\right] = \left[CO_3^{-2}\right] \cong 0.5C_T$ and the concentration of $H_2CO_3^*$ is smaller by many orders of magnitude. As we move one pH unit away from the pK the concentration of one of the species falls significantly and makes up only 9% of the C_T, while the concentration of the other species grows to the point where it makes up 91% of the C_T. This can be seen, for example, in the carbonate system at pH = 7.3, where we have $[HCO_3^-] = 0.91C_T$ and $[H_2CO_3^*] = 0.09C_T$.

The last points mentioned are important in understanding the buffer capacity of solutions, which will be elaborated upon in the next chapter.

2.9 The effect of a change in the total species concentration, C_T

For any system, for a given pK value, the shapes of the species concentration curves do not change because of a change in the total species concentration, C_T (or A_T). From the graph sketching procedure described above, we see that a change in C_T will only affect step 2, and therefore the vertical position of the curves change with varying values of C_T, but their shapes remain the same (Fig. 2.3). It can be seen that the distribution among the species remains the same for a given pH, for every pH value, the dominant species remains the same regardless of the value of A_T. Recall that a rise in C_T must lead to a rise in the ionic strength of the solution. Therefore, the statement that a change in C_T leads solely to a vertical change in the curves is contingent upon an assumption that the effects of ionic strength on the pK values may be ignored.

The log (species) – pH graphs help in developing and understanding the theoretical terms such as weak-acids and bases and their mixtures, alkalinity and acidity, buffer capacity and pH titration. These concepts are elaborated upon in the next chapter.

2.10 Interpreting acid/base systems using proton balance equations

It is possible to use pH-log (species) graphs to solve various problems; one such example is finding the pH of different solutions or mixes of different solutions. To

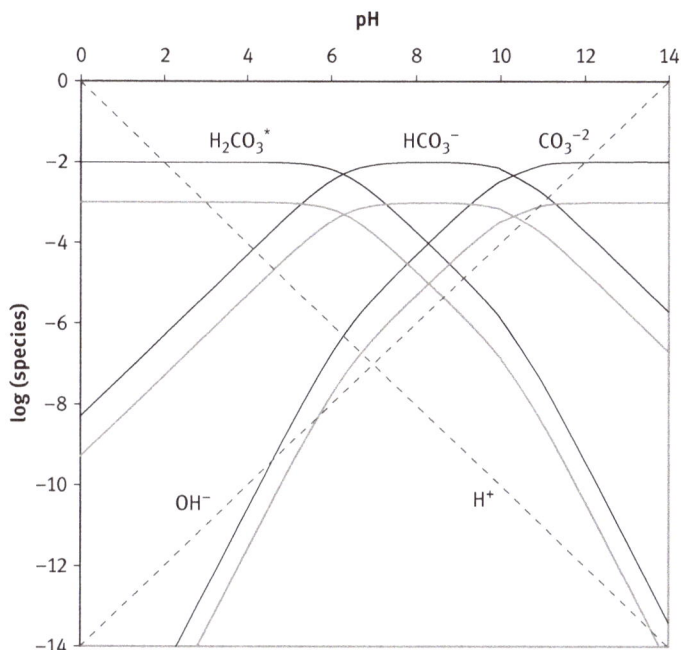

Fig. 2.3: pH-log (species) graph of the carbonate system for the cases $C_T = 10^{-2}$M and $C_T = 10^{-3}$ M using black and grey lines, respectively. The concentration of protons and hydroxides are represented by thinner lines.

fully characterize an acid-base system (i.e., calculating the concentration of each and every species in the water), the following must be defined: (1) the equilibrium equations of the system, including the values of the constants; and (2) two additional independent parameters. The two parameters can be, for example, pH and the total species concentration, A_T. As was already shown and explained, knowing the pH value, the system's equilibrium constants and A_T is sufficient for determining the concentration of each of the species (see Table 2.3), and thus to define the system in its entirety. We will now show how knowledge of the equilibrium constants and the concentration of the species put into the solution is sufficient to determine the pH value on which the solution will stabilize, and therefore also the distribution of its species' concentrations at equilibrium. The solution to this problem can be done either graphically or analytically.

The solution relies on the concept of proton balance (the equation is termed proton balance equation or PBE), at the time the solution is prepared (dissolution of a chemical in the water), according to [5]. The simplest example is of distilled water, without the addition of any chemicals. The "elementary state" is defined as the state of each component prior to the preparation of the solution. The elementary state of the water will be defined as H_2O. A water molecule can donate a proton, H^+, and become OH^-. The proton that separates from the water molecule is not stable and must be accepted by another molecule in the solution. Since there are no other molecules present in the water except for water, the proton will be accepted by a different water molecule yielding the molecule H_3O^+, but symbolized by convention as H^+ (see also eq. (2.5)). Figure 2.4 presents the elementary species (H_2O) as well as the species that will result if it donates a proton or receives a proton.

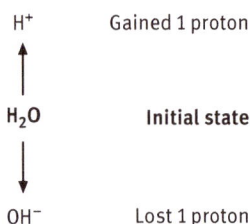

H^+ Gained 1 proton

$\uparrow$

H_2O Initial state

$\downarrow$

Fig. 2.4: Water and the species that result from it donating or

OH^- Lost 1 proton   accepting a proton.

It is clear that every OH^- molecule that was formed as a result of a water molecule donating a proton, necessitates the creation of an H^+ molecule (H_3O^+) as a result of a different water molecule receiving the donated proton. Therefore, the proton balance equation, PBE, for distilled water is:

$$[H^+] = [OH^-] \tag{2.38}$$

Now, if we would like to know the pH of the water and the concentration of the two species (H^+, OH^-), we require an additional equation, that is, the water equilibrium equation:

$$[OH^-] = \frac{K_w}{[H^+]} = \frac{10^{-14}}{[H^+]} \tag{2.28}$$

Substituting eq. (2.28) into the proton balance equation (eq. (2.38)) yields:

$$[H^+] = \frac{10^{-14}}{[H^+]} \Rightarrow [H^+] = 10^{-7}M \Rightarrow pH = 7.0$$

It is also possible to find the pH and the species concentrations of the water system in a quick fashion by employing the graphical approach. The intersection point between the $\log(H^+)$ line and the $\log(OH^-)$ line is in fact the solution to the proton balance eq. (2.38) since this is the only point in the diagram where the equation holds, that is, to say the concentrations of both species are equal. This intersection point can be seen in Figs. 2.2 and 2.3.

When some chemical is added to the water, the proton balance occurs in the same way. Let us take the example of a weak acid, HA, added to distilled water. The elementary state of the water is H_2O and the elementary state of the acid is

HA. The elementary state of the acid HA can only donate a proton and become the base species A⁻. The water, as stated, can donate a proton and become OH⁻ or it can receive a proton and become H⁺. Figure 2.5 presents the relevant elementary species and the species that result from their donating or receiving a proton.

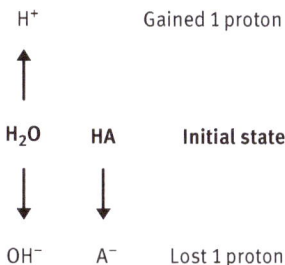

H⁺   Gained 1 proton

↑

H₂O   HA   Initial state

↓   ↓

OH⁻   A⁻   Lost 1 proton

Fig. 2.5: The elementary species and the species that result from their donating or receiving a proton in a solution mixture of HA and water.

The PBE is based on the necessary comparison between concentrations of the products that form as a result of the elementary species donating a proton, and between the concentrations of products that form as a result of receiving a proton from species (usually others) in their elementary state. Thus, the proton balance equation in this case is:

$$[H^+] = [A^-] + [OH^-] \qquad (2.39)$$

To complete the example, let us assume the addition of 10^{-3} M of a weak acid, HA, with pK = 6.0 to distilled water. With the help of eq. (2.39) and the equilibrium constants of the weak acid system and the water system, it is possible to solve the system (of equations) analytically and find the concentrations of all species and the pH on which the system will stabilize. For didactic and convenience reasons, the solution to the problem is presented graphically, since in more complex systems the graphical approach is simpler and there is little, if any, loss of accuracy. Figure 2.6 presents a pH-log (species) graph describing the current example. Equation (2.39) presents an additional constraint to Fig. 2.6 and has only one solution on the graph. This solution is the point at which the left side of eq. (2.39) equals the right side. Let us now explain how to locate this point on the graph.

   In Fig. 2.7, the curves representing each side of eq. (2.39) (the relevant PBE) are labeled. The left-hand side of the PBE is comprised solely of the concentration of protons, and therefore it is represented by line "a" in Fig. 2.7. The right-hand side is comprised of the sum of two species ([OH⁻] and [A⁻]) and is represented by the line labeled "b" in Fig. 2.7. The intersection point between the two lines, "a" and "b", satisfies eq. (2.39) and is therefore the solution to the problem (finding the concentrations of the species and the final pH). In order to sketch line "b", we must

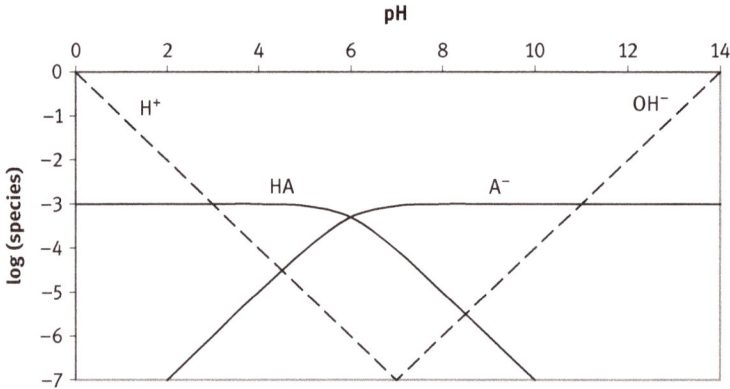

Fig. 2.6: pH-log (species) graph of aqueous solution with $A_T = 10^{-3}$ M. Proton and hydroxide concentrations are represented by dashed lines.

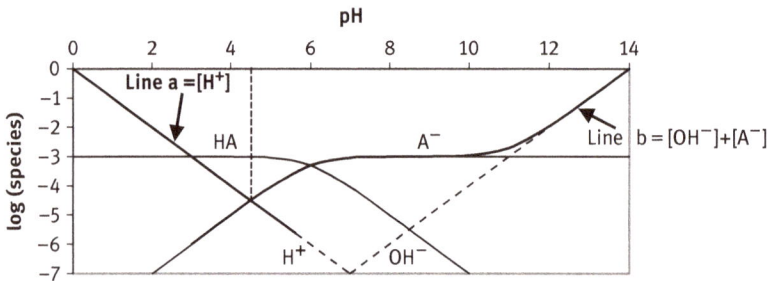

Fig. 2.7: PBE Sketched upon a pH-log (species) graph of a system with $A_T = 10^{-3}$ M. The highlighted lines represent the left and right-hand sides of the equation (lines "a" and "b", respectively). Proton and hydroxide concentrations are represented by dashed lines.

understand the behavior of the species at the different pH ranges: For high values of pH (pH > 12), the sum of the two species is essentially equal to the concentration of OH^-, since the concentration of A^- (at that pH range) is negligible.

$$@pH > 12: \qquad [A^-] + [OH^-] \cong [OH^-] \qquad (2.40)$$

Similarly, for pH values lower than 10, the sum is equal to A^-.

$$@pH < 10: \qquad [A^-] + [OH^-] \cong [A^-] \qquad (2.41)$$

Determining the position of line "b" (i.e., the sum of $[OH^-]$ and $[A^-]$) at pH = 11 is based on the same principle that determined the position of the pK points when sketching the pH-log species graph. At pH = 11, the curve representing the concentration of OH^- and the curve representing the concentration of A^- meet. This means that at this pH, the concentrations become equal, and therefore their sum is equal to twice the concentration of either of them alone:

$$@pH = 11: \quad \log\left([A^-] + [OH^-]\right) = \log\left(2[OH^-]\right) = 0.3 + \log[OH^-] \quad (2.42)$$

As stated, the location on the graph where the left-hand side of the PBE equals the right-hand side is at the intersection of the two lines representing the two sides of the equation.

As can be seen from Fig. 2.7, the addition of HA to distilled water until $A_T = 10^{-3}$ M is reached, results in a pH of 4.5.

As stated, we can solve this problem analytically as well. To do so, we must insert into the PBE (eq. (2.39)) the expression for A^- as a function of A_T and $[H^+]$, as shown in Table 2.3, and the expression for the concentration of hydroxides as a function of $[H^+]$, as shown in eq. (2.28). The insertion yields:

$$[H^+] = \frac{K_a A_T}{K_a + [H^+]} + \frac{K_W}{[H^+]} \quad (2.43)$$

This is a single equation with a single unknown, $[H^+]$. Solving this equation, we find that the solution stabilizes on pH = 4.502 (note that the effects of ionic strength were ignored in this example). In conclusion, the analytic method is more cumbersome than the graphic method, while the results of the two methods are equal in a practical sense.

One can, obviously, use the graphic method described above for finding the pH that results from the addition of a strong acid (or base) to distilled water. In this case, the analytic method is simpler but in the interest of practice we will herein provide an example of the graphic solution:

Let us assume that a monoprotic strong acid, HCl (pK~−3) at a concentration of $10^{-2.5}$ M, is added to distilled water. Figure 2.8 presents the proton balance for this system. Aided by Fig. 2.8, one can write the appropriate PBE:

$$[H^+] = [Cl^-] + [OH^-] \quad (2.44)$$

H⁺   Gained 1 proton

↑

H₂O   HCl   Initial state

↓   ↓

OH⁻   Cl⁻   Lost 1 proton

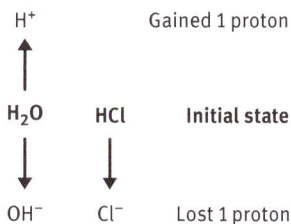

Fig. 2.8: The protonation in terms of elementary states of a system comprising HCl and water.

Figure 2.9 presents a pH-log (species) graph for this system. Note that for strong acids, with pK<0, the x-axis begins at pH < 0. We can see that for pH > 0, the resulting concentration of the species of the strong acid is uniform since the concentration of the strong acid, HCl, is negligible, while the concentration of Cl⁻ is equal to A_T. Figure 2.9

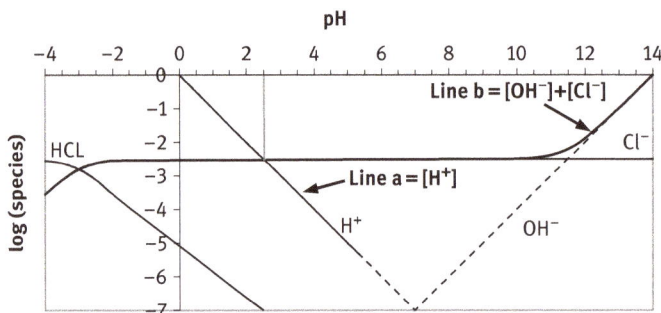

Fig. 2.9: PBE, shown on pH-log (Species) graph of a system containing $10^{-2.5}$ M of strong acid. Highlighted lines represent the left and right-hand sides of the equation (lines "a" and "b", respectively). Proton and hydroxide concentrations are represented by dashed lines.

presents the lines representing the two sides of the appropriate PBE, eq. (2.44). The left-hand side is simply the concentration of the protons. The right-hand side is the sum of $[Cl^-] + [OH^-]$. From Fig. 2.9, it can be seen that the intersection between the two sides of the PBE occurs at pH = 2.5, as expected $(pH = -\log(10^{-2.5}M) = 2.5)$.

The behavior of strong acids at pH < 0

Contrary to popular belief, negative pH values, as well as pH values greater than 14, do exist. For example, nitric acid is a strong acid with pK = −1. Therefore, in a solution with 1 M HNO_3 the theoretic pH is zero. However, a solution of 10 M HNO_3 results in pH = −0.8 (and not pH = −1). A theoretical solution of 100 M HNO_3 will stabilize on pH = −1.43 (not −2 which might have been anticipated). The reason for this is because for pH values close to pK values, strong acids act similarly to weak acids. However, contrary to the weak acids, strong acids have negative pK values and therefore pH values rarely stabilize close to the pK values. The above also holds true for strong bases, where for high concentrations (above 1 M) pH>14 will be observed. It is important to remember that for these pH values the strong base will also act as a weak base (as was the case for the strong acids).

The aforementioned can be tested by using the graphical method to determine the pH as was done for acid salt (see Figs. 2.8 and 2.9). For a better understanding of this topic, the reader is referred to Section 2.11: *Equivalent solutions and Equivalence points*.

2.11 Equivalent solutions and equivalence points

The terms *equivalence point* and *equivalent solution* are important as they are extensively used further in this book to provide a detailed definition of the terms alkalinity and acidity.

An *equivalent solution of the species i* is a solution that has reached equilibrium and contains water (distilled) to which the species *i* of a weak-acid or base is added.

It follows that for every acid-base system, we can set a number of equivalent solutions equal to the number of species in that system.

An *equivalence point of species i* is defined as the pH value of the equivalent solution of the species i. The accepted notation is i_{EP}. An important point to mention now is that, for the most part, the equivalence point is not a set pH value for every species i, but rather is based on the mass of the species i that was added to the water. A more detailed explanation is provided later.

We can determine the pH value of an equivalence point graphically (with the help of the pH-log (species) graph), by solving the appropriate PBE. An example of this is now provided using the general diprotic system, H_2A. The diprotic system has three possible species: the most acidic species H_2A, the result of the first protonation, HA^-, and the result of the second protonation, A^{-2}. We can therefore define three equivalent solutions and three matching equivalence points. The process for finding the equivalence point in relation to the acidic species, H_2A, is described in detail herein. The process for finding the remaining two equivalence points is similar and therefore only an outline of the process will be provided.

2.11.1 H_2A_{EP} – the equivalence point of H_2A

As stated, the equivalent solution of a species i is obtained by dissolution of the same species in distilled water. In the beginning of the chapter, a number of calculations were performed to determine the pH of a solution that results from the dissolution of a strong acid or base in water. These calculations were based on the fact that strong acids and bases ionize completely; however, a weak acid dissociates incompletely (or partially), and therefore the pH that results from its addition to distilled water is not solely a function of the weak acid's concentration.

Figure 2.10 presents the proton balance for dissolution of a diprotic acid H_2A in distilled water. Using Fig. 2.10, let us now write the appropriate proton balance equation, eq. (2.45):

$$[H^+] = [OH^-] + [HA^-] + 2[A^{-2}] \tag{2.45}$$

Note that the concentration of A^{-2} is multiplied by two in the proton balance equation. The reason is as follows: The origin of the protons in the solution can either be from the water or from the acid. For the case where the water donates a proton, it follows that a hydroxide ion must be donated as well. For the case where the acid donates one proton, it follows that the species HA^- will also be donated to the solution. For the case where the acid donates two protons, it follows that the basic species A^{-2} will be donated to the solution. So, every A^{-2} species found in our solution indicates that two protons were released/donated for it to form. Therefore,

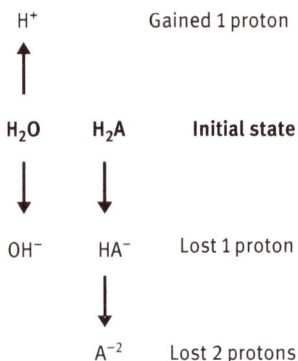

H^+ Gained 1 proton

H_2O H_2A Initial state

OH^- HA^- Lost 1 proton

A^{-2} Lost 2 protons

Fig. 2.10: Level of protonation in terms of the elementary states in an equivalent solution of the species H_2A.

calculating how many protons were released in the solution requires the summation of the hydroxide concentration, the HA^- concentration and twice the A^{-2} concentration. In general, when writing a proton balance equation, one needs to multiply the concentration of the species by the number of protons donated (or received, depending on whether the elementary species is an acid or a base) by the elementary species for that species to have formed.

Let us assume that the diprotic system is the carbonate system, and therefore $pK_1 = 6.3$ and $pK_2 = 10.3$. Also, we will assume the total species concentration to be $C_T = 10^{-2}$ M. Figure 2.11 presents the pH-log (species) graph for this system. The left-hand side of the PBE is marked as line "a", and is, in effect, the proton (H^+) concentration. The right-hand side of the PBE is divided by several pH ranges: the right-most side, defined by pH > 13, is where the concentration of OH^- ions is dominant. According to eq. (2.46), we can see that in this region, the line must overlap the line representing the OH^- concentration:

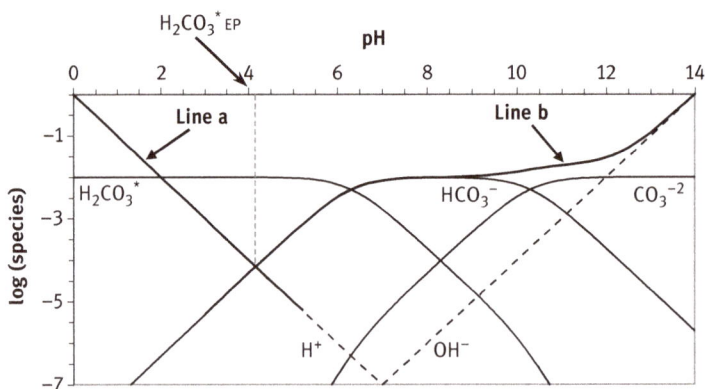

Fig. 2.11: Carbonate system species distribution and presentation of the equivalence point of $H_2CO_3^*$. The concentrations of protons and hydroxide ions are represented by dashed lines.

$$@pH > 13: \qquad [OH^-] + [HA^-] + 2[A^{-2}] \cong [OH^-] \qquad (2.46)$$

The next region is where the most basic species, A^{-2}, is dominant, in our case ($C_T = 10^{-2}$ M) this is at pH ~ 11:

$$@pH \approx 11: \qquad [OH^-] + [HA^-] + 2[A^{-2}] \cong 2[A^{-2}] \qquad (2.47)$$

In this region, line "b" is parallel to the line representing the concentration of A^{-2}, and is 0.3 units above it, since:

$$\log(2[A^{2-}]) = \log 2 + \log[A^{2-}] = 0.3 + \log[A^{2-}]$$

The left most region is where HA^- is dominant, which is when pH < 9.5, at which:

$$@pH < 9.5: \qquad [OH^-] + [HA^-] + 2[A^{-2}] \cong [HA^-] \qquad (2.48)$$

At the transition regions between the dominant species, there are two dominant species rather than one. For example, at pH = 12, the concentrations of OH^- and CO_3^{-2} are equal to each other (see Fig. 2.11). According to eq. (2.49), line "b" is found 0.47 units above this intersection point.

$$@pH = 12: \qquad [OH^-] = [A^{-2}]$$
$$\Rightarrow \log([OH^-] + 2[A^{-2}]) = \log(3[A^{-2}]) = \log 3 + \log[A^{-2}] = 0.47 + \log[A^{-2}]$$
$$(2.49)$$

Similarly, in the transition between the next two regions (meaning at pH = 10.4), the concentrations of CO_3^{-2} and HCO_3^- are equal. Therefore:

$$@pH = 10.4: \qquad [A^{-2}] = [HA^-] \Rightarrow \quad \log([HA^-] + 2[A^{-2}]) = 0.47 + \log[HA^-] \quad (2.50)$$

Figure 2.11 shows that lines "a" and "b" meet at approximately pH = 4.1. In other words, the equivalence point of $H_2CO_3{}^*$ for $C_T = 10^{-2}$ M is located at pH ~ 4.1. Note that this point occurs at the intersection of the H^+ concentration line and the HCO_3^- concentration line. Had the total concentration, C_T, been different, the curves representing the concentration of the carbonate system's species would shift vertically, while the line representing proton concentration would remain the same. Therefore, it is clear that the intersection between H^+ and HCO_3^- concentration lines (i.e., the equivalence point) would be different.

2.11.2 HA^-_{EP} – The equivalence point of HA^-

We will now find the equivalence point under the same conditions: we will assume the diprotic system to be the carbonate system and the total species concentration is $C_T = 10^{-2}$ M.

As mentioned, an equivalent solution for a particular species is attained by the species' dissolution in distilled water. Figure 2.12 demonstrates the protonation relative to the elementary states in an equivalent solution of HA^-. Using Fig. 2.12, we can write the following PBE:

$$[H^+] + [H_2A] = [OH^-] + [A^{-2}] \tag{2.51}$$

H^+	H_2A	Gained 1 proton
$\uparrow$	$\uparrow$	
H_2O	HA^-	**Initial state**
$\downarrow$	$\downarrow$	
OH^-	A^{-2}	Lost 1 proton

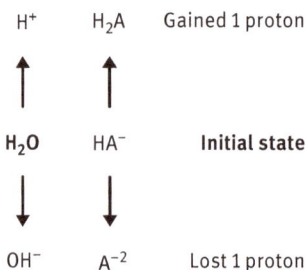

Fig. 2.12: Level of protonation in terms of the elementary states in an equivalence solution of the species HA^-.

The left-hand side of the PBE is represented by line "a" in Fig. 2.13 and is actually the sum of the concentrations of protons and $H_2CO_3^*$. The right-hand side of the PBE is represented by line "b" in Fig. 2.13 and is actually the sum of the hydroxide ion concentration and the CO_3^{-2} concentration.

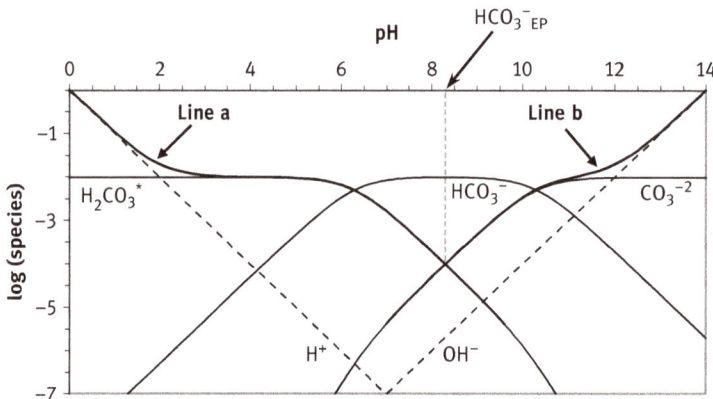

Fig. 2.13: Carbonate system species distribution for finding the equivalence point of HCO_3^-. The concentrations of protons and hydroxide ions are represented by dashed lines.

In Fig. 2.13, we can see that lines "a" and "b" intersect at pH = 8.3. In other words, the equivalence point of HCO_3^- when $C_T = 10^{-2}$ M is at pH = 8.3. Note that this point is reached at the intersection between the lines representing the concentrations of CO_3^{2-} and $H_2CO_3^*$. As stated, a change in the value of C_T will result in a vertical shift in the curves representing the species concentrations of the carbonate system. As the equivalence point of HCO_3^- is found at the intersection between two species of the system (and not at an intersection with the concentration of protons or hydroxide ions), we can see that a change in C_T will not alter the pH on which the point is found (assuming solutions with equal ionic strength values).

Natural pH of rain water

A known environmental problem is acid rain. However, clean rain water also has acidic pH. Clean rain water is in essence pure water that comes in contact with the atmosphere and thus carbon dioxide dissolves in it, as described by eq. (2.11). In other words, using the terminology developed in this chapter, the obtained solution is at the equivalent point of $H_2CO_3^*$ (the sum of dissolved carbon dioxide and carbonic acid with which it is in equilibrium). Therefore it is clear that rainwater is acidic (pH of approximately 5.5).

In order to calculate the pH of rainwater, one must first calculate the concentration of the dissolved carbon dioxide in the water at equilibrium with the atmosphere (for more detail see Chapter 5 gas liquid equilibrium). In the second stage, the pH of the equivalence point of $H_2CO_3^*$ can be determined given the concentration that was found.

2.11.3 A^{-2}_{EP} – the equivalence point of A^{-2}

Let us now find the equivalence point of A^{-2} under the same conditions: we will assume the diprotic system to be the carbonate system and that the total species concentration is $C_T=10^{-2}$ M.

As already mentioned, an equivalent solution of a given species is attained by dissolution of the given species in pure water. Figure 2.14 demonstrates the level of protonation relative to the elementary state in an equivalent solution of A^{-2}. With the help of Fig. 2.14, we arrive at the following PBE:

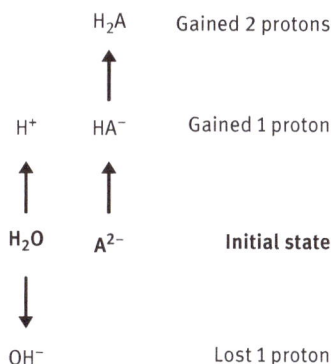

$$[H^+] + 2[H_2A] + [HA^-] = [OH^-] \qquad (2.52)$$

Fig. 2.14: Level of protonation in terms of the elementary states in an equivalent solution of the species A^{-2}.

As was already explained, when writing a PBE, one must multiply the concentration of every species by the number of protons that were received by the elementary species for that species to have formed. Therefore, as was the case with the PBE of H_2A (eq. (2.45)), the concentration of one of the species of the A^{-2} PBE is also multiplied by a factor of two (the unit of this factor is eq/mol).

The right-hand side of the PBE is represented by line "b" in Fig. 2.15, which in this case becomes the line representing the hydroxide ion concentration. The left-hand side of the PBE is represented by line "a" in Fig. 2.15, which in this case is the sum of the concentration of the protons, the concentration of HCO_3^- ions and twice the concentration of the species $H_2CO_3^*$.

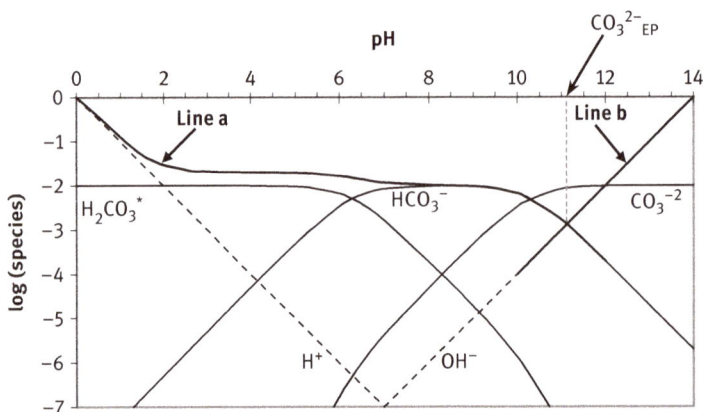

Fig. 2.15: Carbonate system species distribution for finding the equivalence point of CO_3^{-2}. The concentrations of protons and hydroxide ions are represented by dashed lines.

In Fig. 2.15, we can see that lines "a" and "b" intersect at pH = 11.1. In other words, the equivalence point of CO_3^{-2} when $C_T = 10^{-2}$ is at pH = 11.1. Note that this point is obtained at the intersection between the concentrations of the lines representing the hydroxide ions and the species HCO_3^-.

Notice that in general at the equivalence points, the dominant species in the solution is the species that was added to water. In other words, in an equivalence solution of the species i, one will find a predominance of the species i. Thus, it can be seen for example, that at the equivalence point of $H_2CO_3^*$ (see Fig. 2.11), the concentration of $H_2CO_3^*$ is practically equal to C_T, the concentration of HCO_3^- is smaller by two orders of magnitude and the concentration of CO_3^{-2} (which cannot even be seen on the given graph) is smaller by eight orders of magnitude. This observation is true in cases where the pK values are neither excessively high nor low, in other words not too close to the extremes of the natural pH range (0 < pH < 14). Another condition is that the concentration of the acid-base system be relatively large. An example where the observation (of the resulting dominant species being the species that was added to the solution) does not hold is for the equivalence point of CO_3^{-2} for C_T values that are lower than 10^{-3} M. This equivalence point is found at pH > 10.5, in other words close to the pK value. Recall that for pH values that are close to the pK, the concentrations of the two species that are relevant to that pK are equal thereby at this equivalence point, the concentration of HCO_3^- is far from being negligible.

It is important to note that the pH of the equivalence points of the most basic species and the most acidic species is dependent on the total species concentration (C_T in our case). If the value of C_T should rise, it would follow that the equivalence point of $H_2CO_3^*$ would stabilize on a lower pH value; observe the intersection point between the proton concentration line and the HCO_3^- concentration in Fig. 2.3, note that this is not the same point when $C_T = 10^{-2}$ M as it is when $C_T = 10^{-3}$ M. This result is logical in that a rise in C_T means that more $H_2CO_3^*$ was added to the water, in other words more acid, and therefore it is clear that the pH will drop. In the same way in terms of the equivalence point for the basic species (CO_3^{-2} in the case of the carbonate system): a rise in C_T means that more basic species, CO_3^{-2}, was added to the water. Therefore, the equivalence point will have a higher pH value, see the intersection point between the hydroxide ion concentration line and the HCO_3^- concentration in Fig. 2.3, for each case.

Finally, we have already seen that adding HCO_3^- stabilizes our solution at a pH of 8.3 regardless of C_T. To understand this better, we will look at the relevant PBE (eq. (2.51)) and at Fig. 2.11.

$$[H^+] + [H_2A] = [OH^-] + [A^{-2}] \tag{2.51}$$

Equation (2.51) is the general equation for a diprotic system. For the special case of the carbonate system, we write:

$$[H^+] + [H_2CO_3^*] = [OH^-] + [CO_3^{2-}] \tag{2.52}$$

In most practical cases, the value of C_T is greater than or equal to 10^{-5} M (its value in aqueous-gas equilibrium with atmospheric CO_2). Therefore, we can assume that the equivalence point of HCO_3^- must be found in the region where HCO_3^- is the dominant species, that is, between the pH values pH = 7.3 and pH = 9.3. Figure 2.13 shows the species distribution for the carbonate system when $C_T = 10^{-2}$ M. From Fig. 2.13, we can conclude that the concentration of protons and hydroxide ions at the equivalence point must be negligible with respect to the concentration of the carbonate system species. Therefore, we can write eq. (2.52) as follows:

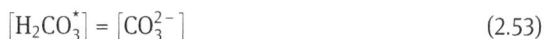

$$[H_2CO_3^*] = [CO_3^{2-}] \tag{2.53}$$

Now insert expressions for the species, according to Table 2.3, to yield:

$$\frac{(H^+)^2 C_T}{(H^+)^2 + K'_1(H^+) + K'_1 K'_2} = \frac{K'_1 K'_2 C_T}{(H^+)^2 + K'_1(H^+) + K'_1 K'_2} \tag{2.54}$$

The denominators cancel out to yield:

$$(H^+)^2 C_T = K'_1 K'_2 C_T \Rightarrow (H^+)^2 = K'_1 K'_2 \tag{2.55}$$

One can see that C_T cancels out from both sides of eq. (2.55). The solution to this equation is therefore [H⁺] = $10^{-8.3}$, regardless of a specific value for C_T, except for its effect on the ionic strength. A higher value of C_T causes an increase in the ionic strength, which in turn affects the value of the constants and therefore the solution of eq. (2.53) as well. Note that for the concentrations of the protons and hydroxide ions to be considered negligible, we need to make sure that $C_T > 10^{-3}$ M. Figure 2.16 presents the equivalence point of HCO_3^- when $C_T = 10^{-4}$ M. Line "b" in Fig. 2.16, which represents the right-hand side of the PBE equation (eq. (2.51)) is very close to the line representing (OH⁻); this is because the concentration of CO_3^{-2} is lower than (OH⁻) for the entire pH range. Thus, it is clear than in such a case we cannot ignore the (OH⁻) concentration in eq. (2.51), and the equivalence point is therefore shifted left and attained at pH = 8.06.

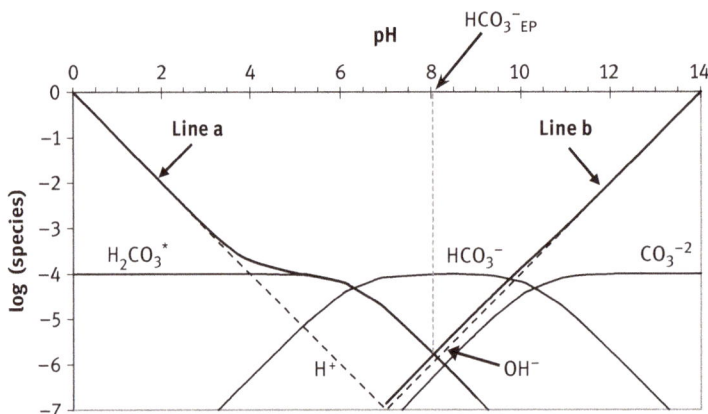

Fig. 2.16: Effect of a low C_T value on the equivalence point of HCO_3^-.

2.12 Buffer capacity

As mentioned previously, weak-acid/base solutions have what is known as buffer capacity, the ability of the solution to resist to changes to its pH as a result of the addition of a strong acid or base. Following is a qualitative explanation of this concept here, further elaboration can be found in Chapter 3.

For simplicity, let us discuss buffer capacity only with regards to a given solution's resistance to lowering its pH as a result of the addition of an acid. The following also pertains, in principle, to the resistance of a solution's pH to rise as a result of the addition of a base.

Let us assume a given monoprotic weak-acid solution which we will refer to as HA. This solution is in equilibrium with all relevant weak-acid species, as described in eq. (2.7). Let us now add some strong acid to this solution. The addition of a strong acid is equivalent to the addition of protons to the solution. If the solution would not have a buffer capacity, it is clear that the pH value would drop according to the concentration of protons that were added to it. In practice, in solutions with a buffer capacity, there exist species of the weak base (A^- in our example), that reacts with the protons to form a weak-acid species; in other words eq. (2.7) will be carried out from the right to left. As a result, some of the protons that were added to the solution will not affect the pH value. Recall that pH is a measure for the activity of protons. In conclusion, while the pH does drop, the drop is less substantial than if the basic species (A^- in our case) had not been present.

Two parameters affect buffer capacity: the total species concentration, A_T, and the pH value. As A_T is higher, there are more species in the solution that can react with the protons and therefore to curb the pH drop, and as a result the buffer capacity is higher. The pH value determines the species distribution; for pH values far from the pK, there is only one dominant species (see the section *Sketching the log (species) curve as a function of pH – a quick procedure*). The change in the concentration of this species is negligible, that is, the concentration of protons with which this species reacts is negligible. In a polyprotic system, aside from the dominant species, there can be other species that react with protons to halt the pH drop, however, in these pH ranges (those far from the pK), the concentrations of these species are negligible, and therefore the buffer capacity is relatively low. Conversely, for pH values that are close to a pK value, there are two species that are dominant. A change in the concentration of each of these as a function of pH is significant, since the tendency of the more basic species of the two to react with protons and form the acidic species is high. In these pH ranges, the buffer capacity is high. Again, the above explanation pertains to buffer capacity when adding an acid, but the same holds true for the addition of a base.

2.13 Graphical method for solving problems

As was demonstrated earlier in the section, sketching a pH-log (species) graph allows drawing conclusions about the pH value of equivalent solutions and their species distribution. Let us now see how to solve more complex problems using the same basic procedure.

Example 2.3 1.36 g of the strong electrolyte KH_2PO_4 and 0.2 g of the strong base NaOH are added to 1 L of distilled water. Calculate the resulting pH. Disregard the effects of ionic strength and assume a water temperature of 25 °C.

Solution As explained in Chapter 1, compounds containing potassium or sodium ions are usually soluble. Therefore, addition of the two mentioned compounds will result in their full dissociation as described by the following reactions:

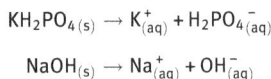

$$KH_2PO_{4(s)} \rightarrow K^+_{(aq)} + H_2PO^-_{4(aq)}$$

$$NaOH_{(s)} \rightarrow Na^+_{(aq)} + OH^-_{(aq)}$$

The dissolution of the salt KH_2PO_4 in the water does not have a direct effect on the pH of the solution. Nevertheless, its dissolution contributes a species of the triprotic, weak-acid system (the orthophosphate system termed for simplicity the "phosphate weak acid system"), and therefore through its dissociation it does have an effect on the pH. The three protonation reactions of the phosphate system are written below:

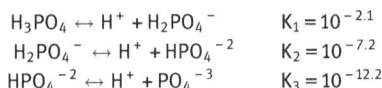

$$
\begin{array}{ll}
H_3PO_4 \leftrightarrow H^+ + H_2PO_4^- & K_1 = 10^{-2.1} \\
H_2PO_4^- \leftrightarrow H^+ + HPO_4^{-2} & K_2 = 10^{-7.2} \\
HPO_4^{-2} \leftrightarrow H^+ + PO_4^{-3} & K_3 = 10^{-12.2}
\end{array}
$$

Let us now perform unit conversion to express the concentration of the chemicals that were added to the water in molar. The salt concentration added to the water:

$$\frac{1.36g\ KH_2PO_4}{Mw(NaH_2PO_4) \cdot l} = \frac{1.36g}{136\frac{g}{mol} \cdot l} = 0.01\frac{mol}{l}$$

Therefore, the total species concentration for the phosphate system is $P_T = 10^{-2}$ M.
The concentration of the base in the water:

$$\frac{0.2g\ NaOH}{Mw(NaOH) \cdot l} = \frac{0.2g}{40\frac{g}{mol} \cdot l} = 0.005\frac{mol}{l}$$

Let us sketch a pH-log (species) graph based on the equilibrium constants of the phosphate system, P_T, and the concentration of the base. See Fig. 2.17. Note that regarding the strong base, as with a strong acid (see Fig. 2.7), the concentration of the sodium ion is constant within the relevant pH range. The reason for this lies in the pK of the base, which is above 14, in other words for pH < 14 the concentration of sodium ions is equal to the concentration of the base (in molar terms) that was added to the water. On this graph, we would like to draw the lines which correspond to the two sides of the PBE equation. We will therefore develop the appropriate PBE. Figure 2.18 shows the protonation states in relation to the elementary states in our example. Note that while the elementary state of the salt is KH_2PO_4, as stated, its dissolution in the water does not affect the pH. Therefore, its elementary state is the ion that results from its dissolution, which is $H_2PO_4^-$.
Using Fig. 2.18, we can write the appropriate PBE:

$$[Na^+] + [H^+] + [H_3PO_4] = [OH^-] + \left[HPO_4^{-2}\right] + 2\left[PO_4^{-3}\right]$$

As noted, Fig. 2.19 presents the two sides of the PBE equation. It shows that the intersection between the two lines is obtained at pH = 7.2. That is, precisely at pK_2 of the phosphate system.
To solve this problem (and others) analytically as opposed to graphically, we must insert into the PBE expressions for the species as functions of constants, total species concentration and pH, equilibrium constants and total species concentration (P_T alone in this case). These expressions can be found

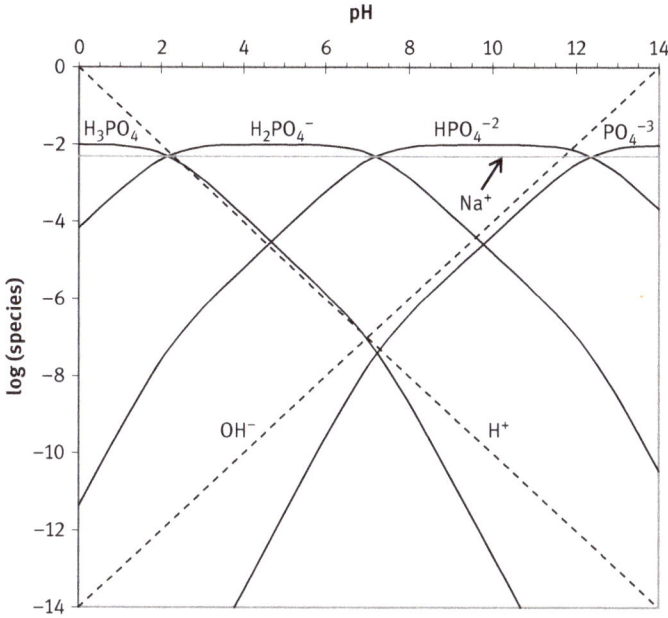

Fig. 2.17: Species distribution of the phosphate system, water system (dashed lines) and the strong base (represented by [Na⁺], grey line) according to the given parameters in Example 2.3.

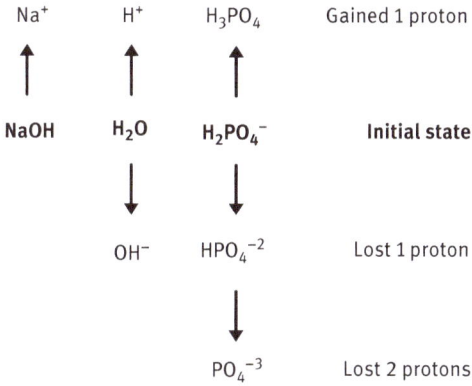

Fig. 2.18: Level of protonation in terms of the elementary states for Example 2.3.

in Table 2.3. After substituting the expressions and shifting around the terms, we get an equation with one unknown (pH):

$$\frac{(H^+)^3 P_T - K'_{a1}(H^+)^2 P_T - 2K'_{a1}K'_{a2}(H^+)P_T}{(H^+)^3 + K'_{a1}(H^+)^2 + K'_{a1}K'_{a2}(H^+) + K'_{a1}K'_{a2}K'_{a3}} = \frac{K_w}{(H^+)} - [Na^+] - (H^+)$$

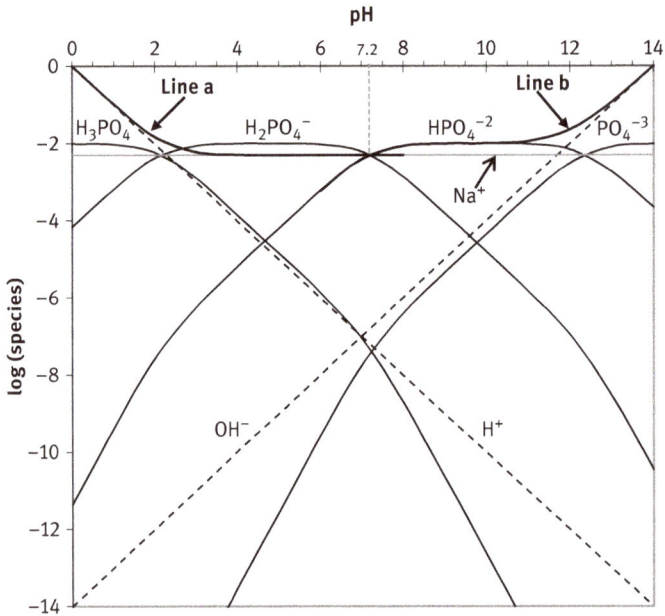

Fig. 2.19: The two sides of the PBE equation for Example 2.3.

We insert the values $[Na^+] = 0.005$ M and $P_T = 10^{-2}$ M and the values of the constants and get that the pH value on which the solution in this example will stabilize is pH = 7.20.

We shall now discuss the case in which NaH_2PO_4 is dissolved in the water instead of KH_2PO_4. In this case, the concentration of sodium ions in the solution would be a result of dissolution of the salt as well as the dissociation of the base. However, the concentration of sodium to be sketched on the graph would not change in this case, since the purpose of the Na^+ line is to represent the concentration of protons that were received by the base.

3 Alkalinity and acidity as tools for quantifying acid-base equilibrium and designing water and wastewater treatment processes

3.1 Introduction

In this chapter, the quantitative parameters "acidity" and "alkalinity" are developed from the basis. These parameters are widely used on a daily basis in the chemical, water and environmental engineering fields. Experience has shown that these terms are not sufficiently understood, often even by the professionals who make frequent use of them, and as a result errors are made in the interpretation of lab results as well as in the design and operation of water and wastewater treatment processes, and also in industrial processes involving aqueous solutions. The great importance of understanding the alkalinity and acidity concepts lies in two of their basic properties (which are elaborated upon later in this chapter):

a. Alkalinity and acidity values may usually be regarded as having "conservative" properties in aqueous solutions, a fact which allows performing simple arithmetical operations on them, such as addition, subtraction or weighted average calculations; as opposed to other measurable parameters such as pH or a specific chemical species' concentration in a weak-acid system, on which such operations cannot be performed.

b. Some alkalinity and acidity values (defined below) are easily and cheaply measured in the lab.

As stated, experience shows that correct laboratory analysis of these values, as well as adequate interpretation of laboratory results are deficient in many cases. The aim of this chapter is to convey the basic knowledge on this theory and demonstrate its many practical uses, both in the case of natural waters (dominated, in terms of acid-base relationships, by the carbonate system) and in wastewaters (which in certain cases contain a reasonably large number of weak-acid systems at significant concentrations).

3.2 Alkalinity and acidity – definitions

Defining alkalinity (or alternatively, acidity) in simple terms is not a trivial task. Perhaps, one should start by noting that these are mathematical/empirical values rather than true "chemical" values in the accepted sense of the word. In other words, they do not refer to the concentration of an ion or molecule that can directly take part in a chemical reaction, but rather in a certain combination of species

https://doi.org/10.1515/9783110603958-003

concentrations derived from a mathematical definition. Having said this, if we insist on verbal definitions, it can be provided in two main ways:

3.2.1 First verbal definition

One way of defining "alkalinity" is by saying that it expresses "the total concentration of dissolved species in water that are capable of neutralizing H^+ ions, relative to a specific reference point(s)". The reference points are the equivalence point(s) as defined in Chapter 2 *Acids and Bases*. Similarly, a verbal definition of the term "acidity" is "the total concentration of dissolved species in water with the ability to neutralize OH^- ions, relative to a specified equivalence point(s)".

3.2.2 Second verbal definition

When a strong base (NaOH for example) is added to a solution that has stabilized on one of its equivalence points (prior to dosage of the base), the alkalinity concentration (with respect to that same equivalence point) equals the concentration of the strong base that was added to the solution (in eq/L units). Acidity is defined in a similar fashion: When a strong acid (HCl for example) is added to a solution which has stabilized on a certain equivalence point, the acidity concentration with respect to that same equivalence point is equal to the concentration of the strong acid that was added to the solution (in eq/L units).

Three scenarios are described below to illustrate and implement the above terms:

1. 0.01 mol of a weak acid, HA, were added to one liter of distilled water. What is the alkalinity of the solution with respect to the HA equivalence point?
   Solution: According to the second definition of alkalinity above, we can immediately see that the alkalinity value of this solution is zero, since the solution stabilizes on the HA equivalence point. We can also draw this conclusion in the following manner: to bring the solution to the equivalence point of HA, we will need to add 0.0 mol of a strong acid. The reader is reminded that the equivalence point of HA is, in practice, the pH value on which the solution stabilizes following HA addition to distilled water (see Chapter 2).
2. Assume that an additional 0.02 mol/L of the acid HA are added to the previous solution. What will the resulting alkalinity be with respect to the species HA?
   Solution: Here too, the system stabilizes on HA's equivalence point, therefore the alkalinity with respect to this point does not change and remains zero.
3. Now we add 0.01 mol of NaOH to the water. What is the alkalinity concentration with respect to HA?

Solution: the alkalinity will be 0.01 eq/L since this is the concentration of the strong base that was added to a solution that was at the HA equivalence point. If we add a strong acid, HCl for example, to the solution, it will require a concentration of 0.01 mol/L to neutralize all the base that was added, so that the solution will return to the equivalence point of HA. Note that it was not specified whether the base was added to the first solution or the second solution (0.01 mol/L of HA or 0.03 mol/L) since, in terms of alkalinity, it has no significance (in both cases the alkalinity value with respect to the HA equivalence point would be 0.01 eq/L).

Two fundamental insights arise from the above definitions and examples, which can be utilized for solving complex problems involving physical and chemical processes in water:
a. Alkalinity and acidity are conservative parameters, since they describe the final (concentration) amount of chemical species whose sum (sum of their concentrations) can only change due to mass transfer out of or into the solution.
b. Mass transfer out of or into the solution of the species with which the alkalinity or acidity is in relation to, does not affect the alkalinity or acidity value.

One can see from these descriptions that it is possible to define, quantitatively, quite a few types of alkalinity and acidity values, since (a) there are many weak-acid species, each of which has a different equivalence point against which alkalinity or acidity terms can be defined, and (b) for the case where more than one weak-acid system exists in a solution, there are many possible combinations of equivalence points for the various weak-acid systems present.

Despite these limitations, an attempt is made to elucidate in this chapter that the mass balance method that is based on simple laboratory measurements of alkalinity and acidity values and a number of simple mathematical relationships, is the most efficient method for characterizing acid-base interactions in aqueous solutions and constitutes a very accurate and to a large extent a very necessary tool for efficient design and monitoring of water and wastewater treatment processes. Beyond that, the principles developed in this chapter enable an understanding of the algorithms underlying commercial software used for simulation and calculation of chemical doses in water treatment processes (such as STA-SOFT4 described in detail in Chapter 7, or the RTW Model by the American Water Works Association, AWWA).

The remainder of the chapter is structured as follows: First, equations for alkalinity and acidity are developed with respect to a given equivalence point for weak, monoprotic acid systems based on proton balance equations. Next, a similar method is developed for equations of diprotic and triprotic systems. Afterwards, a generalized method of developing alkalinity and acidity equations for any weak-acid system is described, followed by a method for developing equations where more than one weak-acid system is involved, that is, alkalinity/acidity values measured in respect to a number of equivalence points simultaneously. Chapter 4 presents practical examples from different fields for the use of this theory for the control and design of water supply, water treatment systems and more.

3.3 Development of alkalinity and acidity equations for monoprotic, weak-acid (weak-base) systems

Development of the equations is done for a solution containing a fictitious monoprotic weak-acid system composed of the species HA and A^- (HA $\Leftrightarrow$ A^- + H^+) with an apparent equilibrium constant pK'_a = 6.0 and total concentration A_T = 10^{-3} M. For this system, we can write the following equations:

The equilibrium equation between the systems chemical species:

$$K'_a = \frac{[A^-](H^+)}{[HA]} = 10^{-6} \tag{3.1}$$

A mass balance equation:

$$A_T = [A^-] + [HA] \tag{3.2}$$

The equation for dissociation of water:

$$K'_w = [OH^-](H^+) \tag{3.3}$$

Equations (3.1), (3.2) and (3.3) yield (see Table 2.3 in Chapter 2):

$$[HA] = \frac{(H^+)A_T}{K'_a + (H^+)} \tag{3.4}$$

$$[A^-] = \frac{K'_a A_T}{K'_a + (H^+)} \tag{3.5}$$

3.3.1 Developing an equation for the alkalinity of a monoprotic, weak acid with HA as the reference species

In Chapter 2 *Acids and Bases*, it was clarified that when an acidic species of a monoprotic weak acid (HA) is added to water, it can be shown through a simple proton balance that the pH stabilizes on the value defined as the equivalence point of HA.

Now assume that NaOH (a strong base) is dosed into this solution at a concentration of 10^{-4} M (final concentration in the receiving solution). Upon contact with the water, NaOH, which is a strong electrolyte, will dissociate completely to form the ions Na^+ and OH^-. The hydroxide ions will react with HA to form A^- and water and the pH value will rise. It is possible to calculate the resulting pH value by substituting appropriate values into the proton balance equation using the method shown in Chapter 2 (Fig. 3.1).

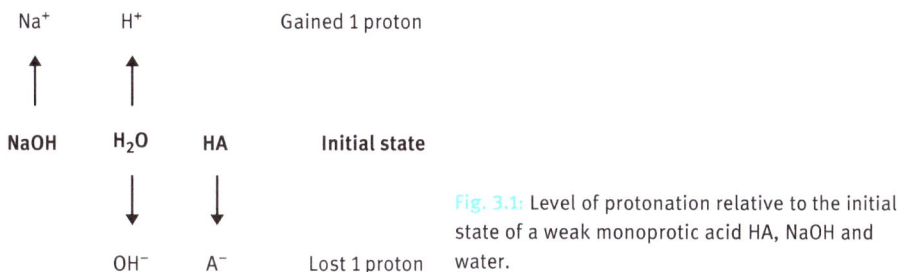

Na⁺   H⁺   Gained 1 proton

↑   ↑

NaOH   H₂0   HA   Initial state

↓   ↓

OH⁻   A⁻   Lost 1 proton

Fig. 3.1: Level of protonation relative to the initial state of a weak monoprotic acid HA, NaOH and water.

The relevant proton balance equation is:

$$[Na^+] + [H^+] = [A^-] + [OH^-] \tag{3.6}$$

where the concentration of Na^+ is known and is equal to the concentration of NaOH that was dosed, that is, 10^{-4} M.

Note: In proton balance equations (such as eq. (3.6)) all the species (including H^+), appear in analytic concentrations (units of M), rather than active concentrations. However, the term $[H^+]$ is represented in eq. (3.7) based on the pH measurement, which is, in fact, a measure of the activity of protons, i.e. (H^+). Thus, to obtain better accuracy, $[H^+]$ should have been divided by the activity coefficient of monovalent ions. However, in order to keep things relatively simple, the activity of H^+ is used in this text rather than the concentration, in the depiction of alkalinity and acidity equations.

To analytically find the resulting pH on which the solution in the current example will stabilize, one must substitute, in place of each species' concentration, an equation as a function of pH, equilibrium constants and total concentration of the system ($A_T = 10^{-3}$ M), (use Table 2.3 in Chapter 2). After substituting, we are left with an equation with only one unknown (pH):

$$10^{-4} + 10^{-pH} = \frac{K'_a A_T}{K'_a + 10^{-pH}} + \frac{K'_w}{10^{-pH}} \tag{3.7}$$

Solving eq. (3.7) with $K'_a = 10^{-6}$ yields that in the current example, the solution will stabilize at pH = 5.08.

With regard to the example above, another important question may be raised: "what is the alkalinity value in the given solution with respect to the equivalence point of HA?" According to the second verbal definition of alkalinity, it is clear that the alkalinity is equal to the concentration of the strong base (NaOH) that was added to the water, which in this case is 10^{-4} molar, or to be more precise by using the acceptable units for alkalinity, 10^{-4} eq/L. We can therefore use the (second) verbal definition in developing a mathematical equation that describes the alkalinity with HA as a reference species; this is done by isolating the species Na^+ in eq. (3.6) (since the concentration of this species is precisely equal to the concentration of NaOH that was added, that is, to the alkalinity value with respect to the equivalence point):

$$[Na^+] = \text{Alkalinity}_{HA}\,(eq/l) = HA_{alkalinity} = [A^-] + [OH^-] - [H^+] \qquad (3.8)$$

Note the two interchangeable ways of noting the alkalinity value with respect to reference species HA.

Equation (3.8) provides a general mathematical expression of the alkalinity value for a monoprotic, weak acid in which the acidic form constitutes the reference point. Equation (3.8) can also be written more explicitly (as a function of the total concentration of the weak-acid system (A_T) and the pH value):

$$HA_{alkalinity} = \frac{K_a' A_T}{K_a' + 10^{-pH}} + \frac{K_w'}{10^{-pH}} - 10^{-pH} \qquad (3.9)$$

3.3.2 Deriving an equation for the acidity of a monoprotic, weak acid with A^- as the reference species

In Chapter 2 (Acids and Bases), it was explained that when the basic species (A^-) of a monoprotic, weak acid is added to water, it can be shown by a simple proton balance equation that the pH stabilizes on a point defined as the equivalence point of A^-. Suppose now that to a solution which has stabilized on the equivalence point of A^- a strong acid (HCl) at a concentration of 10^{-4} M is added. Upon contact with the water, HCl will dissociate into Cl^- and H^+ ions. The hydrogen ions will react with A^- to form HA to maintain equilibrium between A^- and HA, as a result the pH will drop. As in the previous example, one can calculate the resulting pH using the following protonation level diagram (Fig. 3.2):

Accordingly, the proton balance equation in this case would be:

$$[Cl^-] + [OH^-] = [HA] + [H^+] \qquad (3.10)$$

According to the second verbal definition of the concept acidity, the concentration of the strong acid that is added to the solution (which is found at some equivalence point), is equal to the acidity value with respect to the reference species of that equivalence point. Since the concentration of the Cl^- ion is equal to the concentration of HCl

H+   HA   Gained 1 proton

↑   ↑

HCl H_2O A⁻   **Initial state**

↓   ↓

Cl⁻   OH⁻   Lost 1 proton

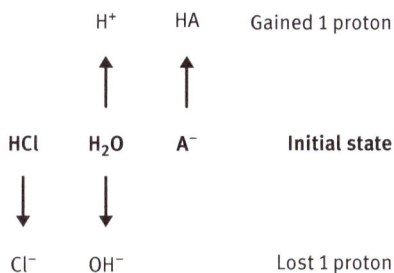

Fig. 3.2: Level of protonation relative to the initial state of the basic species of a weak monoprotic system (A⁻), HCl and water.

that was added to the water, it follows that isolation of this species allows for a mathematical definition of acidity with A⁻ as the reference species:

$$[Cl^-] = A^-_{acidity} = Acidity_{A^-} \ (eq/L) = [HA] + [H^+] - [OH^-] \tag{3.11}$$

And in a more explicit form:

$$A^-_{acidity} = \frac{10^{-pH} A_T}{K'_a + 10^{-pH}} + 10^{-pH} - \frac{K'_w}{10^{-pH}} \tag{3.12}$$

3.3.3 Generalization and elaboration of the concepts alkalinity and acidity and the relationships between them for a monoprotic weak acid

Up to this point, we have defined and derived equations for alkalinity with respect to an acidic reference species and acidity with respect to a basic reference species; however, it is also possible to define an alkalinity equation with a basic reference species (i.e., Alkalinity $_{A-}$), or an acidity equation with HA as the reference species. In fact, an alkalinity or acidity value can be defined with respect to any equivalence point (in any weak-acid system), with the procedure identical to what has been demonstrated so far, namely substitution of the proton balance equation and isolating the species which makes up the concentration of the strong acid (for acidity equations) or strong base (for alkalinity equations).

We can therefore write the following two equations for a monoprotic, weak-acid system (it is recommended that the reader develop these equations on his own by placing the appropriate proton balance equation):

$$Alkalinity_{A^-} = [OH^-] - [H^+] - [HA] \tag{3.13}$$

$$Acidity_{HA} = [H^+] - [OH^-] - [A^-] \tag{3.14}$$

From looking at eqs. (3.11) and (3.13), we can immediately see that Alkalinity $_{A^-}$ = - Acidity $_{A^-}$ Similarly, by looking at eqs. (3.8) and (3.14), it can be seen immediately that Alkalinity $_{HA}$ = -Acidity $_{HA}$. This result will always be obtained when developing alkalinity and acidity equations around the same equivalence point.

Rule: the alkalinity and acidity values around the same equivalence point are always the additive inverse of the second value.
For example: $Alk_{HA} = -Acd_{HA}$

Another useful mathematical relationship exists for the values Alkalinity $_{HA}$ and Acidity $_{A-}$, where their summation is exactly equal to the total concentration of the acid system, A_T:

$$Alkalinity_{HA} + Acidity_{A-} = \{[OH^-] - [H^+] + [A^-]\}$$

$$+ \{[H^+] - [OH^-] + [HA]\} = [HA] + [A^-] = A_T$$

3.3.4 Introduction to measuring alkalinity and acidity in the laboratory

The main strength of the alkalinity (or acidity) value lies in the fact that it can be quantified in the laboratory, with relative ease and precision. Since the (second) verbal definition of alkalinity is that the alkalinity value with respect to some reference species equals the concentration of a strong base added to a solution found at the equivalence point corresponding to the same reference species; it is possible to determine the alkalinity value in the laboratory by adding a strong acid to the solution (where the concentration of the monoprotic, weak-acid system is not known) until the pH value of the equivalence point is reached. At this point, one may raise the question: "How do we know the pH of the equivalence point if we do not know the total concentration of the system?" Although the exact pH value of the equivalence point is usually not known, there are however ways (which are explained later in this chapter following discussion on the concept "buffer capacity" of solutions) of estimating the location of the equivalence point; or alternatively, there is also a way to bypass the need to know the specific location of the point altogether using a laboratory procedure that allows for precisely finding the alkalinity value without knowing the exact location of the equivalence point.

For the time being, assuming that the equivalence point's location is known, the alkalinity/acidity value can be calculated in the following manner (for example, for alkalinity with reference species HA):

$$Alkalinity_{HA}(meq/L) = \frac{V_e \cdot C_a}{V_s} \tag{3.15}$$

where V_e = Volume of strong acid that must be added to reach the equivalence point of HA (L), V_s = Volume of the sample (L) and C_a = Concentration of the strong acid (meq/L).

The analysis and calculations are the same for both alkalinity and acidity as well as for any equivalence point chosen, regardless of the type of weak acid (monoprotic, diprotic, etc.).

3.4 Developing equations for the description of alkalinity and acidity values of diprotic weak-acid systems

As explained in Chapter 2 (Acids and Bases), diprotic systems are composed of three species that are related to one another by two equilibrium constants. A diprotic system therefore has three equivalence points (an equivalence point for each species). In water and wastewater treatment processes, the carbonate system is the predominant weak-acid system. This system is usually described as a diprotic system, although it is essentially made up of four species (see explanation in Chapter 2 and Section 3.5). Another relevant diprotic system is the sulfide system ($H_2S/HS^-/S^{2-}$). Since the carbonate system is the most important system for issues relating to water treatment processes, equations are developed herein for diprotic systems using this weak-acid system, without compromising the generality which is true for diprotic systems in general. Equations for the carbonate system were presented in Chapter 2, however due to their importance in developing the most important alkalinity and acidity equations with regards to water treatment processes, they are presented here again, in order to maintain continuity of reading in the present chapter.

3.5 The carbonate system as an example of a diprotic system

The carbonate system is the predominant weak-acid system in natural waters (surface water, groundwater and seawater) and in almost all types of wastewaters (except for special types of industrial wastewaters). Three main equations describing the carbonate system can be defined in the aqueous phase.

The first equation describes the reaction of carbon dioxide with a water molecule to form carbonic acid:

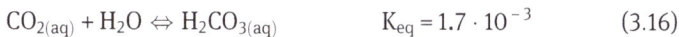

$$CO_{2(aq)} + H_2O \Leftrightarrow H_2CO_{3(aq)} \qquad K_{eq} = 1.7 \cdot 10^{-3} \qquad (3.16)$$

From the low value of the equilibrium constant in eq. (3.16), it is apparent that the reaction is hardly carried out toward the right-hand side. The significance of this is that, at equilibrium, most of the carbon dioxide (approximately 99.7%) is present as $CO_{2(aq)}$ in the water and only a small amount (0.3%) is present as carbonic acid. In order to simplify the calculations, it was decided as a convention to combine the concentrations of H_2CO_3 and $CO_{2(aq)}$ under a common name $H_2CO_3^*$:

$$[H_2CO_3] + [CO_{2(aq)}] = [H_2CO_3^*] \qquad (3.17)$$

Combining these two species results in a weak-acid system composed of three species: $H_2CO_3^*$, HCO_3^- (the bicarbonate ion) and CO_3^{2-} (the carbonate ion).

The mass balance equation for the carbonate system's species:

$$C_T = [H_2CO_3^*] + [HCO_3^-] + [CO_3^{2-}] \tag{3.18}$$

The equilibrium reactions between the carbonate system's species:

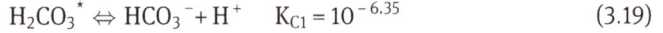

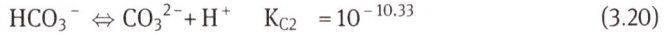

$$H_2CO_3^* \Leftrightarrow HCO_3^- + H^+ \quad K_{C1} = 10^{-6.35} \tag{3.19}$$

$$HCO_3^- \Leftrightarrow CO_3^{2-} + H^+ \quad K_{C2} = 10^{-10.33} \tag{3.20}$$

Theoretically, an additional equation describing the relationship between $H_2CO_3^*$ and CO_3^{2-} can be defined but it would be unnecessary (it would not provide any additional information).

The equilibrium equations:

$$K_{C1}' = \frac{[HCO_3^-](H^+)}{[H_2CO_3^*]} \tag{3.21}$$

$$K_{C2}' = \frac{[CO_3^{2-}](H^+)}{[HCO_3^-]} \tag{3.22}$$

From eqs. (3.18), (3.21) and (3.22), the equations describing the concentrations of the individual species of the carbonate system in terms of pH, C_T and equilibrium constants can be derived. These equations serve in solving a wide range of problems throughout this chapter (see also the general derivation in Chapter 2).

$$[H_2CO_3^*] = \frac{10^{-2pH}C_T}{K_{C1}'K_{C2}' + K_{C1}'10^{-pH} + 10^{-2pH}} \tag{3.23}$$

$$[HCO_3^-] = \frac{K_{C1}'10^{-pH}C_T}{K_{C1}'K_{C2}' + K_{C1}'10^{-pH} + 10^{-2pH}} \tag{3.24}$$

$$[CO_3^{2-}] = \frac{K_{C1}'K_{C2}'C_T}{K_{C1}'K_{C2}' + K_{C1}'10^{-pH} + 10^{-2pH}} \tag{3.25}$$

3.5.1 Developing alkalinity and acidity equations with respect to the equivalence point of $H_2CO_3^*$

To develop an equation for alkalinity ($H_2CO_3^*{}_{alk}$ or Alk $H_2CO_3^*$), let us assume that $CO_{2(aq)}$ is added to pure (distilled) water such that the total carbonate concentration,

C_T, of the solution becomes 10^{-3} molar and NaOH is added simultaneously such that its concentration is 10^{-4} molar in the solution. The order in which the chemicals are added is of no importance, but for the sake of explanation let us say that the CO_2 is added first and only afterwards the base is added.

In the first stage, after adding carbon dioxide, the solution will stabilize on some acidic value of pH which is defined as the equivalence point of $H_2CO_3^*$ (see Chapter 2). After adding the strong base, the pH will rise to a more basic value than the equivalence point. According to the second verbal definition of alkalinity (see definition at the beginning of this chapter), the concentration of the strong base added to a solution present originally at an equivalence point is equal to the alkalinity value with respect to that same equivalence point (or with respect to that reference species).

The example above can be used as a basis for discussing two issues:

One possible question, which was already discussed earlier in the chapter, is "what will be the pH on which the solution will stabilize following dosage of the carbon dioxide and NaOH?" More generally, we can also use the above example to develop a general equation for alkalinity (with $H_2CO_3^*$ as the reference species).

The pH value on which the solution will stabilize can be found by substituting the following potential proton transitions (Fig. 3.3):

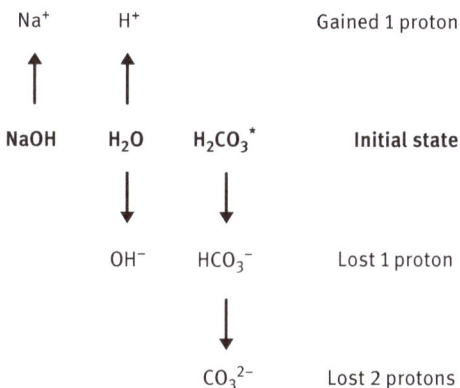

| Na⁺ | H⁺ | | Gained 1 proton |
| | | | |

(figure)

Fig. 3.3: Levels of protonation with respect to the initial state of carbon dioxide ($H_2CO_3^*$), NaOH and water.

Accordingly, the proton balance equation (PBE) is:

$$[Na^+] + [H^+] = 2[CO_3^{2-}] + [HCO_3^-] + [OH^-] \qquad (3.26)$$

To determine the pH value, eq. (3.26) can be written with one unknown (pH): the concentration of Na^+ in the solution is equal to the concentration of NaOH that was dosed (since it is a strong base which dissociates completely), that is, 10^{-4} molar. The carbonate system species are represented by eqs. (3.24) and (3.25) and in place of $[OH^-]$ we assign $K'_W/10^{-pH}$, yielding:

$$10^{-4} + 10^{-pH} = \frac{2K'_{C1}K'_{C2}C_T + K'_{C1}10^{-pH}C_T}{K'_{C1}K'_{C2} + K'_{C1}10^{-pH} + 10^{-2pH}} + \frac{\gamma_m 10^{-14}}{10^{-pH}} \tag{3.27}$$

Solving eq. (3.27), we get pH = 5.39.

From eq. (3.26), we can also develop the general alkalinity equation with $H_2CO_3{}^*$ as the reference species, under the assumption that [Na$^+$] equals the alkalinity concentration with respect to the equivalence point of this species:

$$\text{Alk} (H_2CO_3{}^*) \, (eq/L) = 2[CO_3{}^{2-}] + [HCO_3{}^-] + [OH^-] - [H^+] \tag{3.28}$$

The acidity equation with respect to the reference species $H_2CO_3{}^*$ is as stated above the negative value of the alkalinity equation (try developing this equation for yourself using the proton balance equation!):

$$\text{Acd} (H_2CO_3{}^*) = -\{2[CO_3{}^{2-}] + [HCO_3{}^-] + [OH^-] - [H^+]\} \tag{3.29}$$

Note the coefficient of the carbonate concentration in eqs. (3.28) and (3.29). The significance of this coefficient is that each carbonate mole contributes two equivalences to alkalinity with $H_2CO_3{}^*$ as a reference species (eq. (3.28)), while a mol of each of the other components contributes one equivalent for this alkalinity. It is therefore more accurate to write down the equation as follows:

$$\text{Alk} (H_2CO_3{}^*) = 2\frac{eq}{mol}[CO_3{}^{2-}] + 1\frac{eq}{mol}[HCO_3{}^-] + 1\frac{eq}{mol}[OH^-] - 1\frac{eq}{mol}[H^+] \tag{3.28a}$$

If the concentrations of the species are known, and with them we calculate the alkalinity, it is possible to understand from eq. (3.28a) that substituting the concentrations in molar units requires a doubling of the carbonate concentration by two. On the other hand, if the concentration of the carbonate placed in the equation is in normal units, no coefficient two will appear before it, since there is no need to convert units.

3.5.2 Developing alkalinity and acidity equations around the equivalence point of HCO$_3{}^-$

In order to develop an equation for acidity, (HCO$_3{}^-{}_{acd}$ or Acd $_{HCO3-}$) let us assume that the strong electrolyte NaHCO$_3$ is added to pure (distilled) water such that the concentration C_T in the solution is 10^{-3} molar and in parallel HCl is added to that solution at a concentration of 10^{-4} molar.

Since sodium bicarbonate (NaHCO$_3$) is a strong electrolyte, it will fully dissociate in the water:

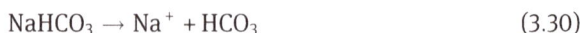

$$NaHCO_3 \rightarrow Na^+ + HCO_3 \tag{3.30}$$

To develop the proton balance equation, we will rewrite the potential protons transitions (Fig. 3.4):

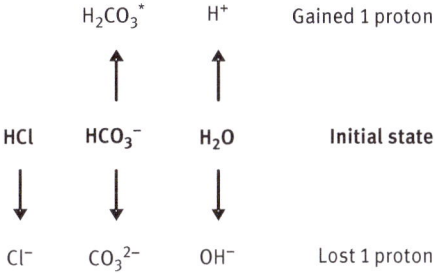

$H_2CO_3^*$ H^+ Gained 1 proton

↑   ↑

HCl HCO_3^- H_2O Initial state

↓   ↓   ↓

Cl^- CO_3^{2-} OH^- Lost 1 proton

Fig. 3.4: Levels of protonation relative to the initial state of bicarbonate (HCO_3^-), HCl and water.

The proton balance equation is thus:

$$[H^+] + [H_2CO_3^*] = [Cl^-] + [CO_3^{2-}] + [OH^-] \tag{3.31}$$

The resulting pH can be found by inputting $[Cl^-] = 10^{-4}$ M, substituting $K'_W/10^{-pH}$ for $[OH^-]$, and substituting eqs. (3.25) and (3.23) for the carbonate system species, so that one equation is obtained with a single unknown, pH.

From eq. (3.31), it is possible to develop the acidity equation with respect to the reference species HCO_3^-. Since the concentration of strong acid added to a solution found at a given equivalence point is defined as the acidity with respect to that point; therefore, isolating the concentration of Cl^- in eq. (3.31) will yield the acidity equation with respect to HCO_3^-:

$$Acd\,(HCO_3^-)\,(eq/L) = [H_2CO_3^*] + [H^+] - [OH^-] - [CO_3^{2-}] \tag{3.32}$$

The equation for alkalinity with respect to HCO_3^- (try developing the equation on your own using the proton balance equation):

$$Alk\,(HCO_3^-)\,(eq/L) = -\left\{[H_2CO_3^*] + [H^+] - [OH^-] - [CO_3^{2-}]\right\} \tag{3.33}$$

3.5.3 Developing the alkalinity and acidity equations around the equivalence point of CO_3^{2-}

In order to develop an equation for acidity with respect to the reference species CO_3^{2-} ($CO_3^{2-}{}_{acd}$ or Acd (CO_3^{2-})), let us assume that the strong electrolyte Na_2CO_3 is added to distilled water such that the concentration C_T in the solution is 10^{-3} molar and in parallel HCl is added to that solution at a concentration of 10^{-4} molar.

Potential proton transfers for this case are as follows (Fig. 3.5):

$H_2CO_3^*$ Gained 2 protons

↑

H^+ HCO_3^- Gained 1 proton

↑   ↑

HCl H_2O CO_3^{2-} Initial state

↓   ↓

Cl⁻   OH⁻   Lost 1 proton

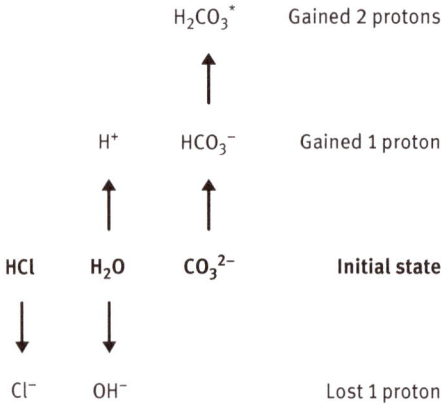

Fig. 3.5: Levels of protonation with respect to the initial state of carbonate (CO_3^{2-}), HCl and water.

The proton balance equation is:

$$2[H_2CO_3^*] + [HCO_3^-] + [H^+] = [Cl^-] + [OH^-] \tag{3.34}$$

And therefore, the acidity equation with respect to the species CO_3^{2-} is:

$$\mathrm{Acd}\,(CO_3^{2-}) = 2[H_2CO_3^*] + [HCO_3^-] + [H^+] - [OH^-] \tag{3.35}$$

The alkalinity equation with respect to the species CO_3^{2-} is the negative value of the acidity equation:

$$\mathrm{Alk}\,(CO_3^{2-}) = -\left\{2[H_2CO_3^*] + [HCO_3^-] + [H^+] - [OH^-]\right\} \tag{3.36}$$

3.5.4 Useful notes about the carbonate system and useful relationships between the values of pH, alkalinity, acidity and C_T

The links between the various parameters characterizing the carbonate system are most useful for solving quantitative problems, as demonstrated in detail further in the chapter. In practice, with the help of these connections, we can calculate any parameter of the carbonate system (every type of alkalinity or acidity, pH, individual species concentrations, C_T, etc.) through the knowledge of two independent parameters (independent parameters refer to any two parameters except for a combination of alkalinity and acidity defined with respect to the same equivalence point). Below are some simple and useful connections:

1. Relationship among $H_2CO_3^*{}_{alk}$, $CO_3^{2-}{}_{acd}$ and C_T

    Summation of the alkalinity (with $H_2CO_3^*$ as the reference species) and acidity values (with CO_3^{2-} as the reference species) is equal to twice the total carbonate system species concentration, C_T.
    Derivation:

$$\text{Acd}(CO_3^{2-}) = 2[H_2CO_3^*] + [HCO_3^-] + [H^+] - [OH^-]$$

$$+$$

$$\text{Alk}(H_2CO_3^*) = 2[CO_3^{2-}] + [HCO_3^-] + [OH^-] - [H^+]$$

- -

$$\text{Acd}(CO_3^{2-}) + \text{Alk}(H_2CO_3^*) = 2[H_2CO_3^*] + 2[HCO_3^-] + 2[CO_3^{2-}] = 2C_T$$

$$(3.37)$$

2. The relationship among $H_2CO_3^*$ $_{alk}$, HCO_3^- $_{acd}$, and C_T:

Summation of the alkalinity (with $H_2CO_3^*$ as the reference species) and acidity (with HCO_3^- as the reference species) values is equal to the total carbonate system species concentration, C_T.

Derivation:

$$\text{Alk}(H_2CO_3^*) = 2[CO_3^{2-}] + [HCO_3^-] + [OH^-] - [H^+]$$

$$+$$

$$\text{Acd}(HCO_3^-) = [H_2CO_3^*] + [H^+] - [OH^-] - [CO_3^{-2}]$$

- -

$$\text{Acd}(HCO_3^-) + \text{Alk}(H_2CO_3^*) = [H_2CO_3^*] + [HCO_3^-] + [CO_3^{2-}] = C_T$$

$$(3.38)$$

3. Relationship among CO_3^{2-} $_{acd}$, HCO_3^- $_{alk}$ and C_T:

$$\text{Alk}(HCO_3^-) = [CO_3^{2-}] + [OH^-] - [H_2CO_3^*] - [H^+]$$

$$+$$

$$\text{Acd}(CO_3^{2}) = 2[H_2CO_3^*] + [HCO_3^-] + [H^+] - [OH^-]$$

- -

$$\text{Acd}(CO_3^{2-}) + \text{Alk}(HCO_3^-) = [H_2CO_3^*] + [HCO_3^-] + [CO_3^{2-}] = C_T$$

$$(3.39)$$

4. Relationship between any alkalinity or acidity value to pH and C_T:

This relationship can be shown by writing the appropriate alkalinity or acidity equations and then writing the concentration of each of the carbonate system species as functions of C_T and pH (eqs. (3.23), (3.24) and (3.25)) and writing the concentration of the hydroxide ion as a function of K_W and pH. This yields equations relating the alkalinity or acidity to C_T and pH (examples later in the book).

5. Relationship among pH, $H_2CO_3^*{}_{alk}$, and $CO_3^{2-}{}_{acd}$

$$CO_3{}^{2-}_{acd} = \frac{1 + 2 \cdot 10^{-pH}/K'_{C1}}{1 + 2K'_{C2}/10^{-pH}} \cdot \left\langle H_2CO_3^*{}_{alk} - \frac{K'_w}{10^{-pH}} + 10^{-pH} \right\rangle + 10^{-pH} - \frac{K'_w}{10^{-pH}} \quad (3.40)$$

Mode of derivation:
Write the explicit equations for the two values (alkalinity and acidity) where all species of the carbonate system are expressed as functions of HCO_3^- (e.g., $H_2CO_3^* = 2K'_{C2} \cdot [HCO_3^-]/10^{-pH}$). Next, isolate $[HCO_3^-]$ in both equations and compare the two. Now isolate $CO_3^{2-}{}_{acd}$.

6. When a solution is found at the equivalence point of HCO_3^-, the pH value will be in the middle between the two pK values (at low ionic strength it will be around pH 8.3) regardless of C_T value.

   In principle, the pH value of equivalence points is a function of the total concentration of the weak-acid system. For example, the pH value of $H_2CO_3^*{}_{EP}$ changes drastically when $C_T = 1$ mM as compared with when $C_T = 5$ mM, both values are typical of natural waters, so we cannot say in advance (without knowledge of the C_T value) what will be the pH at the equivalence point. The exception is the pH value at the equivalence point of HCO_3^-, which will always be found equidistant from the pK' values of the carbonate system (i.e., exactly in the middle between the two values) in natural water. See detailed explanation in Chapter 2.

7. When a solution is found at the equivalence point of HCO_3^-, the values of $H_2CO_3^*{}_{alk}$ and $CO_3^{2-}{}_{acd}$ are equal (in practice; mathematically they are almost equal).

This relationship is not mathematically precise, but it is accurate enough to be mentioned here, since knowledge of this relationship enables a quick solution of certain problems. Explanation:

The proton balance at the equivalence point of HCO_3^- (eq. (2.51)) yields that the concentration of the species CO_3^{2-} is approximately equal to the concentration of the species $H_2CO_3^*$ (this is because at pH 8.3, the concentrations of H^+ and OH^- ions are typically negligible compared to the carbonate and bicarbonate ions). As a result, it follows that $H_2CO_3^*{}_{alk}$ values (eq. (3.28)) and $CO_3^{2-}{}_{acd}$ (eq. (3.35)) must also be (in practice) equal.

3.6 Determining alkalinity and acidity values in the lab: Characterization of acid–base relationships in natural waters

To characterize natural water where the carbonate system is the most prevalent system (surface water, groundwater, seawater and with good approximation most types of wastewater and agricultural facility waters), one is required to find two independent parameters from the following list. Knowledge of two parameters allows for calculation of all the system's remaining parameters (those listed below plus the concentrations of individual species) through use of the relationships detailed above.

- pH
- Alkalinity (with respect to some equivalence point)
- Acidity (with respect to some equivalence point, other than the equivalence point that was used for alkalinity determination)
- C_T

3.7 Standard laboratory alkalinity analysis

The two parameters which are usually (though not always) the easiest to determine are pH and alkalinity. The pH should normally be measured at the water source (to prevent loss of accuracy by emission/absorption of $CO_{2(g)}$ from/to the water) using an electrode attached to a portable pH-meter, following proper calibration and consideration of water temperature.

The general equation for determining the alkalinity value is $V_e \cdot C_a / V_s$ where V_e = volume of the titrant (strong acid) required to reach the chosen equivalence point (L), C_a = concentration of the titrant (meq/L) and V_s = volume of the sample (L).

Of the three types of alkalinity that can be defined for the carbonate system (with respect to each of the three species), $H_2CO_3^*$ alkalinity is the most convenient to determine in the laboratory for the following reasons:

- The first reason lies in the amount of titrant that needs to be added to the titration endpoint. Natural waters are usually found within the neutral pH range (6 < pH < 8.5), therefore a significant amount of acid must be added to reach the pH of the equivalence point of $H_2CO_3^*$ (which is usually found in the range 4.0 < pH < 5.3); adding a significant amount of titrant increases the accuracy of the analysis. A titrant concentration that is convenient to work with varies between 0.05 and 0.2 eq/L, this concentration influences the amount of titrant that must be used, see eq. (3.15).
- The second reason arises from the convenience of using strong acid for a titrant as opposed to a strong base. Since natural waters are found at pH values lower than the equivalence points of HCO_3^- and CO_3^{2-}, the alkalinity with respect to these reference species would be negative. It is possible, theoretically, to find the alkalinity values with respect to these reference species by finding their acidity (by titration with a base) and multiplying the result by −1. However, titration with a base is less precise and convenient than titration with a strong acid because the strong base is much less stable (a base tends to adsorb carbon dioxide from the atmosphere and therefore its pH and chemical composition change overtime).

- The third reason for measuring alkalinity with respect to $H_2CO_3^*$ stems from the fact that around the equivalence point of $H_2CO_3^*$, the buffer capacity of the solution is very low. A low buffer capacity around the equivalence point (i.e., significant change in pH value due to very small addition of strong acid) allows for simple and precise determination of the titration end point. Note that this characteristic is also true for the equivalence point of HCO_3^-. A mathematical and chemical description of the buffer capacity concept is discussed later in this chapter.

3.8 More on water characterization through analysis of C_T and additional forms of alkalinity and acidity

Since the knowledge of two independent parameters enables the calculation of all other parameters of water which contains the carbonate system as a sole weak-acid system, natural water can also be characterized by the knowledge of a combination of other parameter pairs (other than pH and alkalinity with respect to $H_2CO_3^*$), such as $H_2CO_3^*{}_{alk}$ and $HCO_3^-{}_{acd}$, C_T and $H_2CO_3^*{}_{alk}$, C_T and $HCO_3^-{}_{acd}$ and more. Before describing the advantages and disadvantages of the analyses that can be performed, it is worth mentioning that in laboratories it is not customary to routinely perform alkalinity/acidity analyses around the equivalence point of CO_3^{2-} (that is, $CO_3^{2-}{}_{acd}$ and $CO_3^{2-}{}_{alk}$). The equivalence point of CO_3^{2-} is found at pH > 10.3 (exact location is dependent on C_T), measuring alkalinity at such high pH values is problematic for two reasons: first and more important is that in the region of CO_3^{2-} equivalence point, the buffering capacity of the water is quite high and therefore the titration end point is difficult to determine precisely. The second reason stems from the possibility of carbonaceous solids precipitation (mainly $CaCO_{3(s)}$) at the high pH value present at this equivalence point. Precipitation of carbonaceous solids results in a change in the C_T value during titration and may introduce inaccuracy into the analysis.

Regarding possible analyses with respect to bicarbonate's equivalence point (that is $HCO_3^-{}_{acd}$ and $HCO_3^-{}_{alk}$): When the pH of the sample is below pH 8.3, one can add a strong base (such as NaOH) to determine $HCO_3^-{}_{acd}$, or alternatively when the pH of the sample is above pH 8.3, one can add a strong acid to determine $HCO_3^-{}_{alk}$. The buffer capacity of the solution at this titration end point is very low around pH 8.3 and allows for rather clear determination of the titration end point. However, there is one drawback in determining alkalinity/acidity with bicarbonate as the reference species. The drawback is the small volume of strong base (or acid) that needs to be added to natural waters to reach pH 8.3. As a result, a small error in the volume of the titrant can cause a significant error in analysis. Clearly, one can perform the titration with a lower concentration titrant (which will increase the volume of the titrant that must be added to reach the end point), but this also introduces a certain error since it makes it harder to accurately distinguish the end point. This disadvantage is not considered too significant and therefore this analysis is carried out in certain cases (usually along with $H_2CO_3^{*}{}_{alk}$) where the pH measurement proves to be inaccurate or unstable (or in solutions with a very low buffer capacity in which the pH electrode has a hard time stabilizing, or in solutions with a high supersaturated concentration of carbon dioxide, that when the $CO_{2(g)}$ is emitted into the atmosphere the pH varies considerably).

Another laboratory parameter that can aid in determining acid/base characterization of natural waters is C_T (the total inorganic carbon concentration). C_T can be measured using a "total organic carbon (TOC) analyzer". This device is originally intended to measure the concentration of dissolved organic carbon in water but can also be used for determining inorganic carbon concentration in water (which is measured anyway during the measurement of organic carbon concentration). The advantage of this analysis is its accuracy. The disadvantages are: (1) the TOC analyzer is expensive and requires a skilled operator, (2) one must ensure that no $CO_{2(g)}$ is lost/gained to/ from the atmosphere in the sampling procedure, which is not a trivial task and (3) this analysis is not readily available in simple laboratories often found at treatment facilities or on site. In summary, C_T analysis as a tool for characterizing natural waters is certainly possible, but will often be a last resort for routine analyses.

3.9 Buffer capacity of solutions

3.9.1 Buffer capacity or buffer intensity

Buffer capacity is defined as the amount of base (or acid) to be added to a solution to change the pH by a given value. The greater the amount required, the greater the buffer capacity. According to this definition, the buffer capacity can be represented by the local derivative (slope) of the titration curve, with the latter describing the pH values measured in the solution as a result of a cumulative addition of an acid or a base.

Mathematically, the "buffer capacity" value can be defined as follows:

$$\beta = \left(\frac{dC_b}{dpH}\right)_{A_T = constant} \tag{3.41}$$

Where β denotes the buffer capacity and C_b the concentration of the base added to the solution (the base concentration is defined here as the mass of the base dosed divided by the total volume of the solution). As mentioned, buffer capacity can also be defined as the change in pH as a function of the addition of acid, in which case eq. (3.41) is slightly altered to:

$$\beta = -\left(\frac{dC_a}{dpH}\right)_{A_T = constant} \tag{3.42}$$

where C_a is the concentration of acid added to the solution.

3.9.2 Deriving the buffer capacity equation for a monoprotic, weak acid

Let us assume for the derivation that a certain monoprotic system (species: HA/A$^-$) is given, located at the HA equivalence point, to which a strong base is dosed (say NaOH). An increase in the concentration of the strong base will lead to an equivalent rise of the alkalinity concentration (HA$_{alk}$) in the solution. The alkalinity concentration of the solution can be expressed using eq. (3.9). Therefore, the change in alkalinity will be expressed by the following equation:

$$\Delta C_B = \Delta HA_{Alk} = \Delta \left(\frac{A_T K'_a}{K'_a + (H^+)} + \frac{K'_W}{(H^+)} - (H^+) \right) \tag{3.43}$$

According to eq. (3.43), the explicit mathematical description of buffer capacity will be:

$$\beta = \left(\frac{dC_b}{dpH} \right)_{A_T = constant} = \frac{d \left(\frac{A_T K'_a}{K'_a + (H^+)} + \frac{K'_W}{(H^+)} - (H^+) \right)}{dpH} \tag{3.44}$$

From derivative laws it follows:

$$\frac{dC_B}{dpH} = \frac{dC_B}{d(H^+)} \cdot \frac{d(H^+)}{dpH} \tag{3.45}$$

Substituting eq. (3.9) into eq. (3.45) and taking the derivative of the first term, that is, the derivative of alkalinity with respect to (H$^+$):

$$\frac{dC_B}{d(H^+)} = \frac{- A_T K'_a}{\left[K'_a + (H^+) \right]^2} - \frac{K'_W}{(H^+)^2} - 1$$

For the second term of eq. (3.45), $\frac{d(H^+)}{dpH}$, we can take the inverse derivative, that is, $\frac{dpH}{d(H^+)}$:

$$\frac{dpH}{d(H^+)} = \left(\frac{d(-\log(H^+))}{d(H^+)} \right) = \frac{d(-\ln(H^+)/2.303)}{d(H^+)} = \frac{-1}{2.303(H^+)} \tag{3.46}$$

Accordingly:

$$\frac{d(H^+)}{dpH} = -2.303(H^+) \tag{3.47}$$

The product of the two derivatives from eq. (3.45) yields an equation for calculating the buffer capacity of a monoprotic system:

$$\beta = \frac{dC_B}{dpH} = 2.303 \frac{A_T K'_a (H^+)}{\left[K'_a + (H^+) \right]^2} + 2.303 [(OH^-) + (H^+)] \tag{3.48}$$

The first term in eq. (3.48) represents the buffer capacity of a weak-acid system. The second term in eq. (3.48) represents the buffer capacity of the water itself.

Figure (3.6) describes the buffer capacity curve of a weak monoprotic solution with a total concentration of 10^{-3} molar and $pK_a = 6.0$.

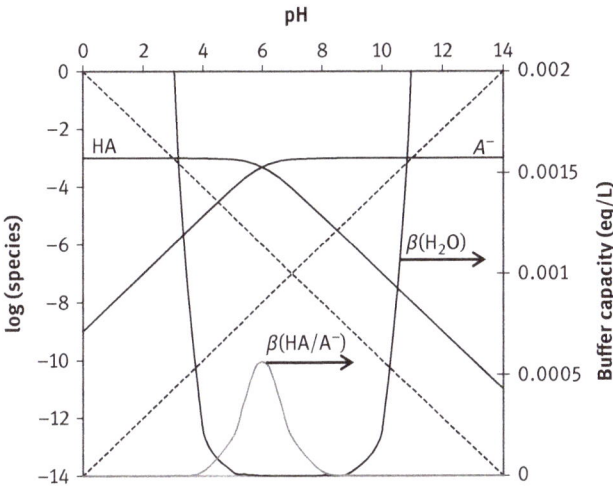

Fig. 3.6: Species distribution (thin black lines), buffer capacity curve of a monoprotic weak-acid system (bell curve) and the buffer capacity of water curve (thick black line). Proton and hydroxide ion concentrations are represented by dashed lines.

Figure 3.6 shows that the buffer capacity curve is a bell-shaped curve where the value of β is high when the pH is the same as the pKa and the buffer capacity decreases steeply when moving away from the pK in both directions. For the pH range in which the buffer capacity is mainly influenced by the weak-acid system and less by the water system, the buffer capacity value is proportional to A_T - as A_T rises, the buffer capacity rises, according to intuition. At a distance of half a pH unit from the pK (in both directions), the buffer capacity is 73% of the maximum capacity and at a distance of one pH unit it is 33%. It is generally acceptable to say that at two pH units to the left and the right of the pK, the buffer capacity becomes insignificant (it is in fact 3.9% of the buffer capacity found at the pK). On the other hand, the water system buffer capacity curve is high on the extremes of the pH scale, due to the buffer capacity of the water dissociation species, that is, H^+ and OH^-. As stated, the second term in eq. (3.48) describes the buffer capacity of water. Looking at this term, it is clear that as pH is higher, the buffer capacity arising from OH^- concentration becomes increasingly significant. At pH 10 for example, the buffer capacity of OH^- is 2.3×10^{-4} molar and at pH 13 it rises to 2.3×10^{-1} molar. The same is true for the H^+ concentration on the acidic end of the pH scale. The buffer capacity of water results from the high concentrations of hydroxide and hydrogen ions at the edges of the pH scale and is not similar in terms of the mechanism, to the buffer capacity resulting from weak acids. Despite it being dominant only at the edges of the pH scale, the buffer capacity of water is of great significance in many practical situations – for example, the reason why there is no clear titration end point at the equivalence point of CO_3^{2-}, is that the high pH at this point gives rise to a significantly high buffer capacity of the OH^- species. This problem is not apparent at the equivalence point of $H_2CO_3^*$ since the pH at this point is quite high and H^+ concentration is low and therefore does not play a significant role with respect to buffering capacity.

3.9.3 Expansion on the derivation of β to include diprotic and polyprotic systems using the carbonate system as an example for polyprotic systems

When the pK values of a polyprotic system are at a distance of at least three orders of magnitude from each other, it is generally assumed that the interplay of one pK region on another is small enough to refer to the polyprotic system as separate monoprotic systems that are interconnected (even at the distance of only two orders of magnitude between the constants the error obtained is less than 5%). This assumption is true for two central weak-acid systems in the field of water and wastewater treatment: the carbonate system ($pK_{C1} = 6.33$, $pK_{C2} = 10.32$) and the orthophosphate system ($pK_1 = 2.12$, $pK_2 = 7.21$, $pK_3 = 12.32$).

When this assumption is true, we can use the derivation that led us to eq. (3.48) to write an expression for β also for polyprotic systems. Accordingly, the approximated description of buffer capacity for the carbonate system is:

$$\beta = 2.303 \frac{C_T K'_{C1}(H^+)}{[K'_{C1} + (H^+)]^2} + 2.303 \frac{C_T K'_{C2}(H^+)}{[K'_{C2} + (H^+)]^2} + 2.303[(OH^-) + (H^+)] \tag{3.49}$$

And for the phosphate system:

$$\beta = 2.303 \left(\frac{P_T K'_1(H^+)}{[K'_1 + (H^+)]^2} + \frac{P_T K'_2(H^+)}{[K'_2 + (H^+)]^2} + \frac{P_T K'_3(H^+)}{[K'_3 + (H^+)]^2} + [(OH^-) + (H^+)] \right) \tag{3.50}$$

The buffer capacity curve of the carbonate system, given $C_T = 2 \cdot 10^{-2}$ M and the thermodynamic equilibrium constants, calculated using eq. (3.49), appear in Fig. 3.7. The graph shows separately the buffer capacity curve of the carbonate system and the water system and a common curve of both representing the "real" buffer capacity in the solution, which is always a superposition on the y-axis of the buffer capacity of each of the different systems that have buffer capacity in the solution. As with monoprotic systems, so too with the carbonate system (or any other polyprotic system), the highest buffer capacity is found precisely at the pH values equal to the pK values and this capacity diminishes significantly when moving away from pKa in the basic and acidic direction. As stated, it is accepted that the buffer capacity is insignificant once we are two pH units to the right or left of the pK. Figure 3.7 also shows why there is no

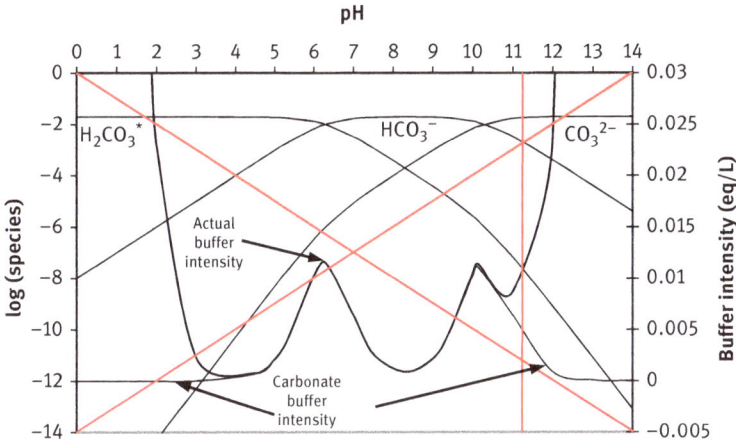

Fig. 3.7: Species distribution and the buffer capacity curve for a solution containing the carbonate system, where $C_T = 2 \cdot 10^{-2}$ M.

clear titration end point at the equivalence point of CO_3^{2-} (marked by a vertical line), and that is since this point is located where both the carbonate system and the water system have reasonably high buffer capacities. The following paragraph deals with titration curves, which will explain, among other things, why a low buffer capacity allows for a more accurate titration analysis.

3.10 Titration curves

The term "buffer capacity" was defined earlier as the derivative (local slope) of the titration curve. The titration curve describes the change in pH that results from the addition of a strong base or acid to the solution. Assuming that the y-axis of the titration curve is the volume of the titrant (cumulative volume) which is added to the water and the x-axis represents the pH, it follows that the titration curve will have a moderate slope (a large change in pH resulting from the addition of a small titrant volume) in the area where the buffer capacity is low, that is, far from the pK values and close to the equivalence points, and the titration curve will have a steep slope for pH values near the pK of the weak acid or for very basic or very acidic pH values (where the buffer capacity of the water dominates).

The upper part of Fig. 3.8 describes the distribution of the carbonate system species as a function of pH, and the corresponding three equivalence points. The bottom part of Fig. 3.8 describes the titration curve for a case in which the initial water quality is: pH 11, TDS = 500, and Alk $H_2CO_3^* = 150$ mg/L as $CaCO_3$. The added acid concentration is 0.1 N, and the volume of the sample is 0.1 L. It can be seen that when the pH value is equal to the pK values of the carbonate system, the buffer

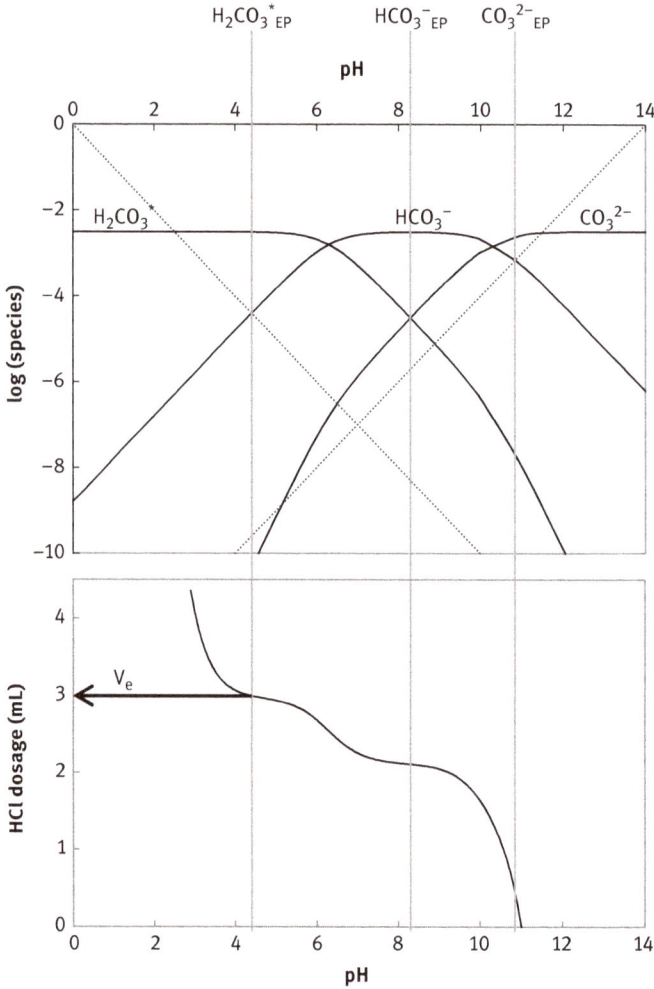

Fig. 3.8: Species distribution and the titration curve of a solution containing only the carbonate system.

capacity is large (a large slope in the titration curve), while at the equivalence points of $H_2CO_3^*$ and HCO_3^-, the buffer capacity is minimal. Note that at the equivalence point of CO_3^{-2}, the buffer capacity is large, because of the water buffering capacity at the high pH range. Due to the low buffer capacity at the $H_2CO_3^*$ equivalence point, a titration curve can easily determine the location of this point with only a small loss of accuracy and therefore the alkalinity value too.

Figure 3.9 describes the species distribution and the buffer capacity curve (both separately for each system and for the solution as a whole). The solution depicted in Fig. 3.9 includes the following weak-acid systems: the carbonate system with a total

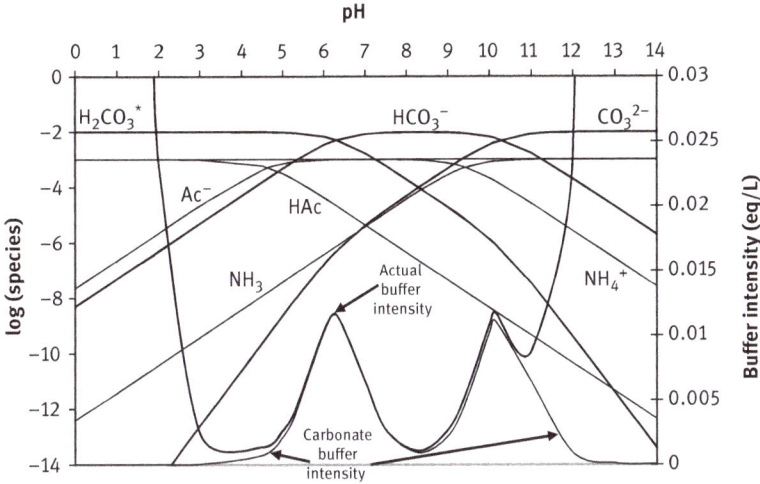

Fig. 3.9: Species distribution and buffer capacity for a system composed of several weak acids.

concentration of 10^{-2} molar, the acetic acid system (CH_3COO^-/CH_3COOH pK = $10^{-4.75}$) with a total concentration of 10^{-3} molar and the ammonia system at a total concentration of 10^{-3} molar. Equation (3.51) presents the equation used to calculate the total β of the solution, which approximates the actual buffer capacity of the solution. Note that the water system appears (logically) only once in the equation.

$$\beta = 2.303 \left(\begin{array}{l} \dfrac{10^{-2}10^{-6.35}(H^+)}{[10^{-6.35}+(H^+)]^2} + \dfrac{10^{-2}\cdot10^{-10.4}(H^+)}{[10^{-10.4}+(H^+)]^2} + \\[3mm] \dfrac{10^{-3}\cdot10^{-9.25}(H^+)}{[10^{-9.25}+(H^+)]^2} + \dfrac{10^{-3}\cdot10^{-4.7}(H^+)}{[10^{-4.7}+(H^+)]^2} + (OH^-) + (H^+) \end{array} \right) \tag{3.51}$$

Note that the area under the buffer curve between two points (resulting from the integration of the buffer capacity curve between these two points) equals (by definition) the mass of the titrant (in eq/L units) that must be added in order to change the pH from its value at one point to its value in another. Alternatively, the derivative of the titration curve is equal to the buffer capacity. Therefore (again, by definition), this area is equal to the difference in the alkalinity values between the two points (or alternatively, the difference in the acidity between the two points – dependent on which direction the titration begins, i.e., if an acid or a base is added). The alkalinity at a specific point (with respect to reference species i) equals the area under the curve between the above-mentioned point and the equivalence point of species i. Two important conclusions can be drawn from this: first, the relationships amongst C_T, pH and alkalinity should be now clearer. For a given pH, as C_T increases, the alkalinity increases (the curve will be positioned higher on the y-axis and therefore the area underneath it will be larger); on the other hand, for a given C_T, as pH increases the alkalinity will also be larger. The second conclusion is that the definition of "alkalinity" or "acidity" in a solution must consider the species of all the different weak-acid systems in the solution and cannot be defined solely for one system. For the simplest case, the alkalinity and acidity equations contain species of the carbonate and water system alone, but in the current example the general alkalinity (and acidity) equations will include species from four systems: carbonate, acetic, ammonia and water. In more complex situations (e.g., anaerobic water facilities for treatment of household or industrial wastewater), there can be additional systems (the phosphate system, different organic acids etc.). The following section explains how to write an alkalinity (or acidity) equation for a solution composed of a number of weak acids, and how to measure this value in the laboratory.

3.11 Alkalinity and acidity equations composed of several weak-acid systems

One of the main features that characterize the concepts of alkalinity and acidity that gives them such great importance in solving practical problems is the fact that the values are "additive", meaning that one can add alkalinity values of different weak-acid systems using simple arithmetic. A result of this characteristic (among other things) is that the alkalinity value of a solution containing several weak-acid systems is the sum of the alkalinity values of each of the weak-acid systems. For the sake of explanation, suppose mixing of several solutions, each containing a different acid-base system. It is clear, that the alkalinity (by mass) of the mixture will be the sum of the alkalinity values of all the solutions together. Another way to understand this feature is from the measurement method of the alkalinity: At the time of the acid titration, all the basic species (from all the acid base systems in the water) will neutralize the acid, until reaching the pH that constitutes the end of the titration. However, the question is how to define the alkalinity of a solution containing several systems and how to measure such alkalinity (in other words, what is the end point?).

At the beginning of the chapter, the first verbal definition of alkalinity (acidity) was introduced, to read "the total concentration of species able to neutralize H^+ (OH^-) ions in the solution, with respect to a specific equivalence point". Accordingly, when one defines an equation for more than one weak-acid system, one must first define the equivalence points with respect to which the alkalinity value (or acidity value) should be defined. For example, in raw municipal wastewater, there are usually three main weak-acid systems: the carbonate system (diprotic), the ammonia system (monoprotic) and the orthophosphate system (triprotic). Altogether, these three systems have nine equivalence points. Therefore, it is necessary to decide wisely against which points it is desirable to define, for example, the value of alkalinity. Equivalence point selection influences both the species that appear in the alkalinity equation and the pH at which the titration endpoint is determined by in the laboratory alkalinity analysis. Even if the selected alkalinity is not actually measured in the laboratory (that is, if the equation is used only for calculations), it is necessary to select one reference species from each system, so that the guiding rule for selecting the species is that there is a pH in which all the selected reference species are dominant. This pH value is the value of the most acidic equivalence point of the selected ones. Only in this way will all the alkalinity in the solution be measured. If the titration is done to the point where one of the species is not dominant, then the alkalinity of the acid-base system that this species represents is not measured.

In order to select the reference species, it is first necessary to determine whether this is solely for computational purposes or that the value of alkalinity should be measured (i.e., to perform titration, as explained later in this chapter). From a computing perspective, the equivalence points with which the value is taken in reference to is not of great importance, so long as we maintain the same reference point throughout our calculations. On the other hand, regarding the laboratory analysis, it is desirable to use an alkalinity value with reference points that are easily quantified in the laboratory with minimal error.

For the current example then, we can intuitively define two alkalinity equations: the first equation defines an alkalinity value with respect to the equivalence point of the most acidic species in each system, that is: $H_2CO_3^*$, NH_4^+ and H_3PO_4. This equation is obtained by summing the alkalinity expressions for each of the three species, as presented in eq. (3.52):

$$Alk(H_3PO_4, H_2CO_3^*, NH_4^+) = 3[PO_4^{3-}] + 2[HPO_4^{2-}] + [H_2PO_4^-] + 2[CO_3^{2-}]$$
$$+ [HCO_3^-] + [NH_3] + [OH^-] - [H^+]$$

(3.52)

Equation (3.52) describes the alkalinity value with respect to the most acidic equivalence point of each system, and therefore, on paper, this is the most intuitive equation. Further on, it is demonstrated that we can use eq. (3.52) for the purpose of calculating (for example) results of mixed water, or the concentration of a single species (e.g., $NH_{3(aq)}$) of importance in wastewater. However, the alkalinity value defined by eq. (3.52) cannot be directly measured in the laboratory. The reason is that in order to determine an alkalinity value in the laboratory, one must titrate the solution with a strong acid until reaching the most acidic equivalence point amongst the equivalence points chosen to define the alkalinity value (in this example this would be the equivalence point of H_3PO_4). This equivalence point is found at a very low pH value (more or less between pH 1.5 and pH 3.0 depending on the value of P_T). Therefore, the mass of the acid required to reach the most acidic equivalence point will be so large; and the relative value of the [H$^+$] term in eq. (3.52) will be so significant in relation to other species, that in practice this analysis will have no value, mainly because the mass of acid required to reach such a low pH will be, in most cases, a whole order of magnitude greater than the total concentration of all the other species, which greatly influences the accuracy of the analysis (a small error in the final pH value during titration will cause a significant error in the accuracy of the analysis).

Therefore, for the current example (raw wastewater), the alkalinity expression that appears in eq. (3.52a) is more fitting:

$$Alk\ (H_2PO_4{}^-, H_2CO_3{}^*, NH_4{}^+) = 2[PO_4{}^{3-}] + [HPO_4{}^{2-}] - [H_3PO_4] + 2[CO_3{}^{2-}] + [HCO_3{}^-]$$
$$+ [NH_3] + [OH^-] - [H^+]$$

(3.52a)

An explanation as to why eq. eq. (3.52a) is more fitting for determining the alkalinity of raw wastewater in the lab is provided in Figs. 3.8 and 3.9. We can see that in the region of the most acidic equivalence point appearing in eq. (3.52a) (i.e., near $H_2CO_3{}^*$ equivalence point, found around pH 4.5), the buffer capacity of the solution is very low, this is in contrast with the very high buffer capacity found at the most acidic equivalence point in eq. (3.52). When the buffer capacity is low, the analysis' accuracy is high because even a certain error in the pH value of the titration end point will only slightly influence the accuracy of the final analysis results. In other words, the difference between the amount of acid required to complete titration precisely at the equivalence point to the amount required to complete titration in proximity of the equivalence point (but not at the equivalence point) is small, because of the low buffer capacity. For the same reason, it is of no importance that the titration ends at a pH significantly lower than the equivalence point of $NH_4{}^+$, since in regard to the ammonia system, after the pH of the solution drops below the equivalence point of $NH_4{}^+$, the amount of acid to titrate to lower the pH more is negligible (the ammonia system's buffer capacity in this pH range is negligible).

The following discussion can also be applied to acidity equations (in all their forms). It should be noted that choosing the "appropriate" equivalence should be made according to the nature of the solution and the purpose of the analysis, and there are no defined rules, except for one rule, namely to choose a titration end point that does not fall in a range that has a significant buffer capacity of some weak-acid system; since in such a situation, the information attained from the laboratory analysis would be inaccurate. A good example of a common yet erroneous analysis is the measurement of alkalinity in anaerobic reactors by lowering the pH to around pH 4.5 (that is, measuring alkalinity with respect to the equivalence point of $H_2CO_3{}^*$, which is identical to the method used for measuring alkalinity of natural waters). Besides the fact that those who carry out this measurement often disregard the ammonia, phosphate and sulfide systems, anaerobic digester waters usually contain a non-negligible amount of volatile fatty acids (VFA's) (acetic acid, propionic acid, butyric acid, etc.), which are weak, monoprotic acids with pK values in the approximate range of 4.0 to 5.0. Therefore, titration to pH 4.5 allows the expression of the carbonate species as well as some other systems, however only a fraction of the organic acid species that are capable of receiving H^+ ions will be shown in the analysis; resulting in effect, in an analysis devoid of mathematical significance (since it will not be fully defined with respect to an equivalence point). In such cases (significant presence of VFA), if we are interested in measuring the alkalinity of the water also in respect to the volatile fatty acids, we must define an equation that also contains their relevant species with respect to the appropriate equivalence points. Laboratory titration to determine this type of alkalinity requires lowering the pH of the solution to around pH 2.7.

3.12 Elaboration on laboratory methods for measuring alkalinity and acidity

Chapter 4 describes in detail the possible uses of concepts developed in this chapter. The original reason for developing the concepts of alkalinity and acidity is the ability to measure them in the laboratory and on site in a relatively simple manner, which does not require sophisticated instrumentation beyond a calibrated pH electrode and a simple titration device (which can even be a glass burette). Nevertheless, or rather because of the relative simplicity of the analysis and as explained above in detail, when analyzing the alkalinity (or acidity), it is important to determine with respect to which equivalence points it is defined, what we wish to gain from the analysis, what are the other systems to be measured in the water to get a clearer picture of the weak acids' effects (e.g., phosphate or ammonia concentration in the water) and what accuracy is expected from the analysis. Recall that in most cases, we do not know the precise location of the equivalence points (especially with regards to $H_2CO_3^*{}_{EP}$), therefore, especially when C_T is low (for example desalinated water), a rather significant error can result from titration to an incorrect pH value.

The following discussion focuses mainly on measuring alkalinity in the laboratory for the simple reason that, in most cases, alkalinity together with pH are the parameters chosen for characterizing acid-base relations in solutions. The principles for measuring acidity values are essentially the same, except that a strong base is used instead of a strong acid for titration. This fact has practical significance mainly in relation to the shelf life of the titrant: while strong acids (HCl or H_2SO_4) can last for months or even longer, strong bases "spoil" relatively quickly as a result of $CO_{2(g)}$ adsorption from the atmosphere, and for the duration of their short life, it is important to keep the solution sealed, making it difficult for routine analysis.

Before conducting a discussion on alkalinity measurement in solutions containing a few weak acids, we should first discuss techniques used for measuring alkalinity in natural waters that are dominated solely by the carbonate system.

In most cases, when professionals use the term "alkalinity" or "general alkalinity" in the context of drinking water, they are referring to $H_2CO_3^*{}_{alk}$ (side note: the term "general alkalinity" is fundamentally incorrect since each alkalinity expression should be accompanied by its reference species, and its origins are unclear).

The standard technique used in many laboratories for measuring $H_2CO_3^*{}_{alk}$ is titration of the solution using a strong acid (usually at a concentration of 0.05–0.2 meq/L) until reaching pH 4.5, where the indication of reaching the endpoint is done either by reading a pH meter or by a color change.

After reaching the titration endpoint, the alkalinity is calculated using the following equation:

$$H_2CO_3^*{}_{alkalinity}\,(eq/L) = \frac{V_e \cdot C_a}{V_s} \tag{3.15}$$

where V_e = volume of the titrant required to reach the titration end point (L), C_a = concentration of the titrant (eq/L) and V_s = volume of the solution (L).

If one is interested in obtaining the alkalinity value in units of mg/L as $CaCO_3$, one should multiply the value from eq. (3.15) by a factor of 50,000 (the equivalence weight of $CaCO_3$ in units of mg per eq).

As shown in Fig. 3.10, the equivalence point of $H_2CO_3^*$ can vary as much as half a pH unit on the pH scale, resulting from a change in C_T of one order of magnitude (between 1 mM and 10 mM).

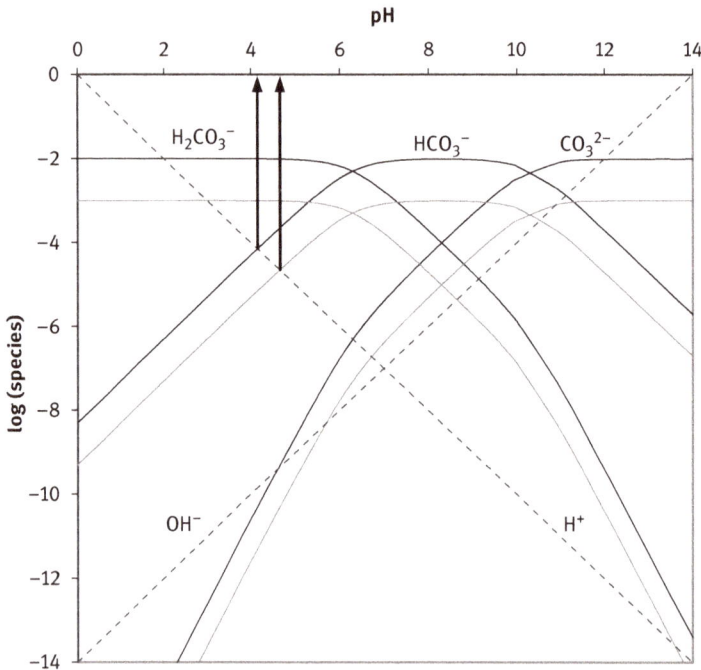

Fig. 3.10: Carbonate system species and equivalence points of $H_2CO_3^*$ for two cases: $C_T = 1$ mM and $C_T = 10$ mM. The arrows indicate the different pH values of the two equivalence points.

There are two disadvantages in using the accepted method for determining alkalinity (titration to pH 4.5): the first disadvantage, as mentioned, arises from the lack of precise knowledge of the location of the equivalence point. It is important to note that this disadvantage can cause a significant error in the analysis only in cases where the alkalinity is considerably low (less than ~50 mg/L as $CaCO_3$). When the value is greater, as is the case for most natural waters, the error becomes minimal with respect to the high background concentration. In desalinated water (where the alkalinity is sometimes low), on the other hand, a significant error may result. The second disadvantage is the fact that during titration $CO_{2(g)}$ tends to volatilize from the sample and as a result the pH value rises during the titration. $CO_{2(g)}$ stripping does not affect the alkalinity value (see explanation in following box) but does cause a drop in C_T and a rise in pH. Because of the increase in pH, the amount of acid required will increase, leading to an error in the analysis. The common solution to this problem is to perform the titration while slowly stirring the sample, to minimize carbon dioxide volatilization. This solution does reduce the error, but it complicates and slows down the analysis. The Gran method, described below, bypasses the problem of not knowing the precise location of the equivalence point.

3.12.1 Gran titration for determining alkalinity

The Gran method for determining $H_2CO_3^*{}_{alk}$ (and other alkalinity forms) was developed by a Swedish scientist in the early 1950s and is widely used in the world, mainly for determining seawater alkalinity, but also for natural waters and wastewaters [6]. Yet, for some reason, it is less common than the fixed-pH (pH 4.5) analysis, despite it being more accurate, faster and more convenient to the operator compared to the conventional method.

The Gran method is more accurate than the standard method since (1) alkalinity is measured with respect to the exact equivalence point rather than to some pH that is close to the equivalence point, without having to know the precise location of the equivalence point, and (2) as a result of titrating to pH values lower than the $H_2CO_3^*$ equivalence point, the measurement is not affected by volatilization of carbon dioxide and change in C_T during the titration.

The basis of the method is titrating to several pH values that are certainly more acidic than the equivalence point (that is below pH 4.0 for $H_2CO_3^*{}_{alk}$, as an example).

Why Stripping or Dissolution of $CO_{2(g)}$ from/ in Water Does Not Change the $H_2CO_3^*{}_{alk}$ Value

When $CO_{2(g)}$ is stripped from or dissolved in a solution, the C_T and pH values change, while the alkalinity with respect to $H_2CO_3^*$ does not change.

Explanation regarding $CO_{2(g)}$ stripping from a solution: As a result of the carbon dioxide volatilization, the equilibrium equations amongst the species will move to the right in the following manner:

$$HCO_3^- + H^+ \rightarrow H_2CO_3^*$$

$$CO_3^{2-} + H^+ \rightarrow HCO_3^-$$

Since H^+ is consumed in both equations, the pH will rise, but what happens to the value of $H_2CO_3^*{}_{alk}$?

Recall: $H_2CO_3^*{}_{alk} = 2[CO_3^{2-}] + [HCO_3^-] + [OH^-] - [H^+]$

Note that regarding the alkalinity species both equations consume and contribute identical alkalinity values:

In the first equation, bicarbonate consumption will lead to a drop in the alkalinity value, but at the same time there is an identical increase in the alkalinity value as a result of H^+ consumption. The same is true for the second equation: consumption of 1 mol of CO_3^{2-} results in a drop of 2 equivalents in the alkalinity concentration, but this is compensated by a release of 1 equivalent of bicarbonate and the consumption of 1 equivalent of H^+.

Therefore, overall, volatilization of $CO_{2(aq)}$ does not change the alkalinity. Similarly, absorption of $CO_{2(aq)}$ to the solution does not change the $H_2CO_3^*{}_{alk}$ value.

3.12.2 Mathematical derivation of the Gran method

The method is based on equating two equations that describe the mass of alkalinity in the solution at some point X throughout the titration process.

The first equation:

$$\text{Alkalinity}_x = V_e \cdot C_a - V_x \cdot C_a \qquad (3.53)$$

where Alkalinity_x = mass of alkalinity in the solution following titration of V_x mL of titrant (eq), V_e = volume of the titrant required to reach the endpoint, whose location is unknown (L), V_x = volume of the titrant required to reach some point, X, during the titration (L) and C_a = titrant concentration (eq/L).

Explanation of eq. (3.53): The value $V_e \cdot C_a$ is, by definition, the alkalinity mass in the sample prior to titration. The value $V_x \cdot C_a$ describes the mass of alkalinity consumed as a result of adding V_x L of titrant. The difference between them is the alkalinity mass remaining in the solution.

Notice that a negative alkalinity mass will result if $V_x \cdot C_a$ is greater than $V_e \cdot C_a$. This situation arises when titration is carried out to a pH value that is certainly lower than the equivalence point.

The second equation:

$$\text{Alkalinity}_x = \{2[CO_3^{2-}]_x + [HCO_3^-]_x + [OH^-]_x - [H^+]_x\}(V_s + V_x) \qquad (3.54)$$

where V_s = volume of the solution (L).

The values of the species with subscript "X" describe the concentrations of the species at point X throughout the titration (mol/L), that is, after adding V_x L of titrant.

Explanation of eq. (3.54): since Alkalinity_x describes the mass of alkalinity in the solution (and not the alkalinity concentration), one must multiply the sum of the species concentrations by the total volume of the solution. The volume of the solution following the addition of V_x L of titrant is $V_x + V_s$.

Equating eqs. (3.53) and (3.54) yields:

$$V_e \cdot C_a - V_x \cdot C_a = \{2\,[CO_3{}^{2-}]_x + [HCO_3{}^-]_x + [OH^-]_x - [H^+]_x\}(V_s + V_x) \qquad (3.55)$$

It was stated earlier that the basis of Gran titration lies in the addition of a titrant (strong acid) to reach pH points that are certainly more acidic than the corresponding equivalence point (in this case pH < 4.0). Also stated that if the pH is lower than the equivalence point, the left-hand side of eq. (3.55) becomes negative. In order for the right-hand side of the equation to be smaller than zero, the following condition must be satisfied:

Concentrations of the species $[CO_3{}^{2-}]$, $[HCO_3{}^-]$ and $[OH^-]$ must be negligible relative to the concentration of $[H^+]$. Indeed, this condition is satisfied at pH < 4.0. After neglecting these species, eq. (3.55) becomes:

$$(V_e - V_x)C_a = -[H^+]_x(V_s + V_x) \qquad (3.56)$$

Or, equivalently:

$$-(V_e - V_x)C_a = 10^{-pH_X}(V_s + V_x) \qquad (3.57)$$

where pH_x = the pH value that results after titration of V_x liters of acid.

All the values on the right-hand side of eq. (3.57) are known, therefore we can calculate the value of the right-hand side, which is denoted by F_x.

According to the analysis procedure, a strong acid should be titrated to some pH_x value below pH 4.0 and above ~pH 3.3, wait for the pH value to stabilize, and record the volume of the titrant (V_x) used to reach that pH_x. Then, the process is to be repeated two or three more times while writing the results in the left columns of the following table in an electronic spreadsheet (the values in the table below represent an example of titration). F_x is calculated automatically in the electronic spreadsheet using the right-hand side of eq. (3.57).

Equation (3.57): $(V_x - V_e)C_a = F_x$, is a linear equation. That is, if we create a graph in which the V_x variable represents the y-axis and F_x represents the x-axis, we can expect to get a straight line (if the underlying assumptions used in the derivation are correct). This line will cross the y-axis (that is at the point where $F_x = 0$) at the location where $V_e = V_x$. Thus, a linear regression of the line that results from our titration points allows us to directly extrapolate the value of V_e, without knowing the exact location of the equivalence point.

Figure 3.11 provides a graphical presentation of the method, based on the analysis results in Table 3.1. Note that for the method to be accurate, the regression line must usually have a regression coefficient higher than $R^2 = 0.99$. For water governed by the carbonate system, such precision should be obtained without any special problem.

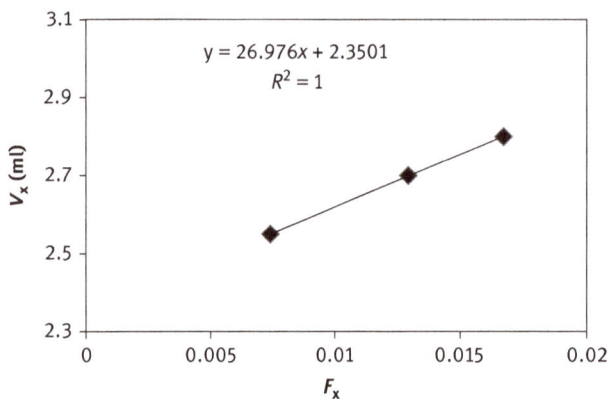

Fig. 3.11: Graphical representation of results for the Gran titration. The resulting V_e is 2.35 mL. As such, the calculated alkalinity value ($H_2CO_3^{*}$ alk) is $2.35 \cdot 0.05/50 = 2.35$ meq/L or 117.5 mg/L as $CaCO_3$.

Table 3.1: Gran titration example of a 50 mL sample using a strong acid with concentration of 0.05 N.

V_x(mL)	pH_x	F_x
2.6	3.85	0.0074
2.7	3.61	0.0129
2.8	3.5	0.0167

Note that it is common to execute the Gran titration in the pH range $3.3 < pH < 4.0$, however, for natural waters, where the C_T concentration is around 10^{-3} M, titration to the pH range $3.0 < pH < 4.0$ will also yield accurate results.

4 Use of alkalinity and acidity equations for quantifying phenomena in chemical/ environmental engineering and design of water and wastewater treatment processes

4.1 Theoretical background related to acid-base calculations in aqueous solutions from the knowledge of alkalinity and acidity parameters

The alkalinity and acidity parameters (in all their forms) have two main features that make them practical and computational tools most useful for designing and controlling of water and wastewater treatment processes, as well as control tools for processes in the aqueous phase. The first feature, mentioned in the previous chapter, is the ability to quantify them in the laboratory and also on site in a simple, inexpensive and fast manner. The second feature is that both parameters can be defined as "conservative and additive".

Fundamental rule: when using alkalinity/acidity values to quantify acid/base phenomena in an aqueous solution, once the alkalinity/acidity equation to be used in the calculations has been defined, a change in the total alkalinity/acidity value of the solution can be caused <u>only</u> by direct addition or removal of a species comprising the originally defined alkalinity/acidity equation.

As stated, alkalinity and acidity values are parameters that can be defined as having conservative and/or additive properties. This means that simple arithmetic calculations such as addition and subtraction and weighted average calculations can be performed on these parameters, as long as all the reactions involving species which make up the alkalinity/acidity equation are taken into account (this quality is derived from the fundamental rule above). The pH and calcium carbonate precipitation potential (CCPP, defined later), for example, are not conservative parameters according to this definition. C_T (as well as any total concentration of any weak-acid system) can be defined as conservative. On the other hand, individual weak-acid species are not conservative, since we must consider the equilibrium reactions in which they are involved, and the pH value in the solution should be known to determine their concentration and this value, as stated, is not conservative. Having said this, calculating the alkalinity/acidity resulting from mixing solutions and/or the addition/removal of one of the species that comprise the equation as a result of a known external dosage, does not require determining the pH and thus can be done.

https://doi.org/10.1515/9783110603958-004

Let us now discuss a set of engineering-related examples that form specific cases where the use of alkalinity and acidity values is particularly convenient. This group includes cases in which it is possible to assume that addition/detraction of species comprising the alkalinity/acidity equations does not occur as a result of interactions with the solid or gas phase; or cases where it can be assumed that all reactions occur in any case only in the aqueous phase. In practical terms, these are cases in which the alkalinity/acidity values are not affected neither by the volatilization of species that comprise the equation nor by precipitation of solids containing these species. However, there are cases that are not included in this group, for example processes in which precipitation of carbonate or phosphate solids and others occur (which affect the alkalinity value), or processes in which stripping of $CO_{2(g)}$ (which affects, for example, the $CO_3^{2-}{}_{acd}$ value) or of $NH_{3(g)}$ occur (which affect $NH_4^+{}_{alk}$ in wastewater). In many cases in natural water treatment processes, the assumption that no precipitation or stripping that affects the alkalinity or acidity values occurs is reasonable. This is especially true for the alkalinity parameter, since the alkalinity equation (with $H_2CO_3^*$ as the reference species) does not include any species that has the potential to be released to the gas phase, and also because in many cases in the aqueous phase, it is possible to assume that no precipitation occurs during the analysis. Note that even if the solids have a positive precipitation potential in the solution (the concept "precipitation potential" is described in detail in Chapter 6), in the absence of nucleation seeds, the sedimentation processes are characterized by much longer time constants (days or weeks) than the time constants characteristic of reactions occurring in the aqueous phase (often fractions of a second). Therefore, the assumption that the total alkalinity value is maintained can be well approximated even in systems open to the atmosphere, where there is sometimes a positive precipitation potential for carbonate solids. That said, these generalizations should be taken with caution, since in some cases the assumption that no precipitation occurs is incorrect (like, for example, effluents of anaerobic wastewater treatment plants, in which rapid stripping of carbon dioxide to the atmosphere occurs followed by a sharp rise in pH, which in turn may lead to precipitation of carbonate and phosphate solids).

In addition, even with respect to the acidity parameter, the assumption that there is no addition/detraction of species to/from the gas or solid phases is reasonable in certain cases. A typical example is water supply networks, which are defined as a system that is not open to the atmosphere ("closed system"). On the other hand, in open systems where gases can be exchanged with the atmosphere (mainly $CO_{2(g)}$), it is not usually possible to assume that there is no stripping of acidity species (in their various forms) to the atmosphere. Such an assumption can only be made for the case where the concentration of $CO_{2(aq)}$ is close to equilibrium with the partial pressure of carbon dioxide in the atmosphere (see Chapter 5). As explained below, the assumption that there is no reduction in the acidity species (stripping to the atmosphere) is correct in cases of closed-atmosphere systems such as water supply lines, anaerobic digesters for wastewater or sludge treatment, etc.

The fundamental rule appearing in the beginning of the chapter is not always intuitive, but deviation from it will almost always lead to an error in calculations. For example, a very common mistake is that stripping (or dissolution) of carbon dioxide ($CO_{2(g)}$) from water, changes the alkalinity value ($H_2CO_3^*{}_{alk}$) of the solution. According to the fundamental rule, stripping of $CO_{2(g)}$ from the solution or its dissolution into the solution will in no way affect the alkalinity value with $H_2CO_3^*$ as a reference species, since this mathematical term does not comprise $CO_{2(aq)}$:

$$\text{Alk}\,(H_2CO_3^*) = 2[CO_3{}^{2-}] + [HCO_3{}^-] + [OH^-] - [H^+]$$

The source of confusion is that when $CO_{2(g)}$ enters the solution or volatilizes from it, the pH value changes (pH rises in the case of carbon dioxide stripping and drops in the case of carbon dioxide dissolution). It should be understood that a change in pH is not necessarily related to whether the alkalinity value changed or not, but only to a change in H^+ concentration in the solution. Since H^+ is only one of the species comprising the alkalinity equation, if it changes while, simultaneously, a different value comprising the alkalinity equation changes equivalently but with the opposite sign, then the total alkalinity value does not change. This is exactly what happens: Take, for example, the case where $CO_{2(aq)}$ is supersaturated relative to the atmosphere and therefore will strip from the solution when exposed to the atmosphere.

As a result of the stripping, the system will veer from equilibrium, and the following reactions will occur in order to maintain the equilibrium equations in the aqueous phase:

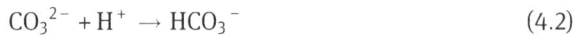

$$HCO_3{}^- + H^+ \rightarrow H_2CO_3^* \tag{4.1}$$

$$CO_3{}^{2-} + H^+ \rightarrow HCO_3{}^- \tag{4.2}$$

How will these reactions affect the total alkalinity value? Well, according to the first equation (eq. (4.1)), the concentrations of the species $HCO_3{}^-$ and H^+ in the solution will decrease (both in molar and equivalent terms) by the same proportion. However, note that these species appear with opposite signs in the alkalinity equation. The net change in the alkalinity value is therefore zero. The same phenomenon occurs in the second equation (eq. (4.2)): the concentration of the $CO_3{}^{2-}$ species that appears in the alkalinity equation with a coefficient of +2, decreases by a certain value – causing a drop in the alkalinity value. On the other hand, the concentration of H^+ (coefficient of −1) drops as well while the concentration of $HCO_3{}^-$ (coefficient of +1) rises by the same molar proportion. These two processes contribute to the increase in alkalinity. Once again, there is no net change in the alkalinity value.

What can cause a change in the above alkalinity value (Alk $H_2CO_3^*$)? – any direct addition or removal of the ions H^+, OH^-, bicarbonate and carbonate will cause a change in the alkalinity, as it was defined here, **and nothing else!** This rule is a fundamental rule that will be used later also for examples dealing with dosage of chemicals to water in order to reach a situation desirable to the engineer in terms of acid-base relations in the water. Deviation from this rule, as stated, invariably leads to errors.

Note: It can be shown, in the same manner that addition or reduction of the CO_3^{2-} ion (for example, as a result of dissolution or precipitation) does not change the $CO_3^{2-}{}_{acd}$ value (confirm this on your own!). On the other hand, stripping or dissolution of $CO_{2(g)}$ in an aqueous solution will change the alkalinity value where $CO_{2(aq)}$ comprises part of the equation – for example, the value of $HCO_3^-{}_{alk}$ is affected by changes in the concentration of $CO_{2(aq)}$ since this parameter is defined as follows:

$$HCO_3^-{}_{alk} = [CO_3^{2-}] + [OH^-] - [H_2CO_3^*] - [H^+]$$

The examples in the remainder of the chapter and their solutions present scenarios from the simple group defined above (reactions occurring solely in the aqueous phase) as well as cases in which it is not possible to assume that there is no transition of the alkalinity/acidity species out of or into the aqueous phase.

4.2 Examples of acid-base problems from chemical/ environmental engineering in which alkalinity and acidity terms can be used

The following examples are intended to deepen the understanding of the topics and concepts presented in the previous chapters. The examples are from the chemical/environmental engineering field, but similar problems (in principle) exist in many other fields. In practice, the solution methods are relevant to any field in which acid-base interactions in a solution are of importance, for example, various fields in biotechnology, food, agricultural engineering and more.

The examples demonstrate methods of solution and use of different tools, each example brings additional levels of understanding, and relies on the example/s preceding it. A summary of the topics in which the examples and tools are used is presented in Table 4.1.

Table 4.1: Summary of topics in the examples presented in the chapter and relevant fields in chemical/environmental engineering.

#	Mixed weak-acid systems	Fields in chemical/ environmental engineering that are of interest in the example	Tools/solution methods/ important points
4.1	Carbonate	Water supply system.	Mixing of water sources – weighted average.
		Treatment of drinking water.	Finding independent conservative parameters.
		Water stabilization.	
4.2	Carbonate and acetic		Defining alkalinity values for several systems. Solution with iterations. A sensible initial guess of pH value.
4.3	Carbonate and hydrogen sulfide	Laboratory experiments.	Finding alkalinity/acidity directly in an equivalence solution. Effect of H_2S on acid base ratios.
4.4	1. Carbonate with a strong base (i.e., carbonate system alone)	Mixing wastewater with river water.	Addition/reduction of alkalinity/acidity as a result of the addition of a strong acid/base.
	2. Carbonate, ammoniacal and orthophosphate	Conservation of streams. Quality of effluents of industrial and municipal wastewater.	pH correction by strong acid/base dosage.
4.5	Carbonate, orthophosphate and acetic	Biological treatment of wastewater.	Effect of the quality of raw wastewater on biological treatment processes.
		Control of biological processes.	Selecting the appropriate parameter for real-time control. Water quality change due to weak-acid/ base dosage.
4.6	Carbonate, hypochlorite and ammoniacal	Quality of drinking water: chemical stability.	Defining positive/negative precipitation potential.
		Treatment of drinking water: antiseptics, oxidation and fluoridation.	Water quality change due to weak-acid dosage.
4.7	Carbonate	Ground water quality.	Change in water quality due to weak-acid dosage (CO_2).
		Water supply.	

(continued)

#	Mixed weak-acid systems	Fields in chemical/ environmental engineering that are of interest in the example	Tools/solution methods/ important points
4.8	Carbonate and ammoniacal	Biological treatment of wastewater, removal of nitrogen compounds.	The effect of biological processes on the quality of effluents. Effect of the quality of raw wastewater on biological treatment processes.

Example 4.1: Mixing two streams of water that contain the carbonate system as a single weak acid
Figure 4.1 shows two water supply lines that converge into a single pipe. Assuming complete mixing at the intersection (a reasonable assumption), calculate the acid-base characteristics of the mixed stream (pH, alkalinity, acidity, C_T etc.) given the discharge rates and characteristics of the two streams being mixed.

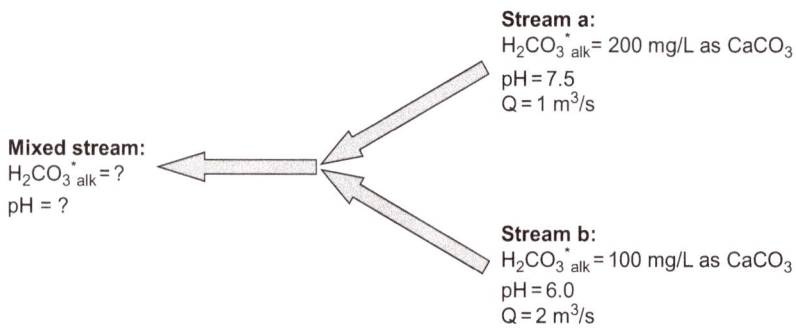

Stream a:
$H_2CO_3^*{}_{alk} = 200$ mg/L as $CaCO_3$
pH = 7.5
Q = 1 m^3/s

Mixed stream:
$H_2CO_3^*{}_{alk} = ?$
pH = ?

Stream b:
$H_2CO_3^*{}_{alk} = 100$ mg/L as $CaCO_3$
pH = 6.0
Q = 2 m^3/s

Fig. 4.1: Schematic of mixing of two natural water sources.

In this example, for the sake of simplicity, calculations on the effects of ionic strength and temperature on the equilibrium constants were ignored. In significant calculations, these effects cannot be ignored since a change in the equilibrium constants can have a significant effect on the result.

Solution This example has practical importance since water from different sources is often mixed in water supply networks (in Israel this is common practice in many water supply systems). The acid-base characteristics of the "mixed" water are mainly important with regard to the waters "chemical stability" which relates to its degree of aggression and corrosion potential – this is addressed in Chapter 9, under Section 9.1 (water stabilization).

In order to solve these types of problems, a minimum number of parameters (that can be considered conservative under the question's terms) should be defined first, on which weighted average between the two water streams can be performed to obtain information about the mixed stream. For the current example, since the system is defined as a closed system (not open to the atmosphere) and since it can be assumed that there is no precipitation or dissolution of solids that contain species comprising alkalinity or acidity, both the alkalinity and acidity values can be considered as conservative parameters. Another conservative parameter that can be used is the total species concentration of the carbonate system – C_T (this parameter is considered conservative under the same assumptions, i.e., that no reactions take place in the non-aqueous phase). However, since it is necessary to know only two independent parameters in order to calculate all the components of the carbonate system, one must decide upon no more than two independent conservative parameters (any two) and stick to them throughout the solution.

Suppose $H_2CO_3^*{}_{alk}$ and $CO_3^{2-}{}_{acd}$ are chosen as the two conservative parameters; the first step of the solution is to calculate the value of $CO_3^{2-}{}_{acd}$ in streams a and b from knowing the pH and the $H_2CO_3^*{}_{alk}$ of each stream using the following relationship (eq. (3.40)):

$$CO_3^{2-}{}_{acd} = \frac{1 + 2 \cdot 10^{-pH}/K'_{C1}}{1 + 2K'_{C2}/10^{-pH}} \cdot \left(H_2CO_3^*{}_{alk} - \frac{K'_w}{10^{-pH}} + 10^{-pH} \right) + 10^{-pH} - \frac{K'_w}{10^{-pH}}$$

Substituting the data into eq. (3.40) yields (make sure to substitute the values of alkalinity and acidity in units of eq/L):

$$CO_3^{2-}{}_{acd} \, (\text{stream a}) = 4.56 \cdot 10^{-3} \, \text{eq/L}$$

$$CO_3^{2-}{}_{acd} \, (\text{stream b}) = 11.2 \cdot 10^{-3} \, \text{eq/L}$$

Since $CO_3^{2-}{}_{acd}$ is a conservative parameter (in a closed system), its weighted average can be taken. Accordingly, the acidity value of the mixed stream will be:

$$CO_3^{2-}{}_{acd} (\text{mixed stream}) = \frac{4.56 \cdot 10^{-3}N \cdot 1m^3/s + 11.2N \cdot 10^{-3} \cdot 2m^3/s}{(1+2)m^3/s} = 8.99 \cdot 10^{-3} \, N \text{ or eq/L}$$

In the same way, it is possible to calculate the alkalinity value of the mixed stream (from knowledge of the alkalinity values of the converging streams):

$$H_2CO_3^*{}_{alk}(\text{mixed stream}) = \frac{4.0 \cdot 10^{-3}N \cdot 1m^3/s + 2.0 \cdot 10^{-3}N \cdot 2m^3/s}{(1+2)m^3/s} = 2.67 \cdot 10^{-3} \, N \text{ or eq/L}$$

Now that the values of two independent parameters of the carbonate system in the mixed stream have been calculated, it is possible to calculate all the other parameters directly from the relationships between the parameters (Chapter 3). There is more than one computational way to do this. For example:

1. The pH value can be calculated using eq. (3.40) after substituting the known Alk_{mix} and Acd_{mix} values:

$$8.99 \cdot 10^{-3} = \frac{1 + 2 \cdot 10^{-pH}/10^{-6.35}}{1 + 2 \cdot 10^{-10.4}/10^{-pH}} \cdot \left(2.67 \cdot 10^{-3} - \frac{10^{-14}}{10^{-pH}} + 10^{-pH} \right) + 10^{-pH} - \frac{10^{-14}}{10^{-pH}} \Rightarrow pH = 6.29$$

2. The C_T value can be calculated using eq. (3.37):

$$C_T = (CO_3^{2-}{}_{acd} + H_2CO_3^*{}_{alk})/2 = 5.83 \cdot 10^{-3} \, M$$

3. It is also possible, of course, to calculate the concentrations of the individual species. For example, in answering the question: "Does the water have a positive or negative potential for $CaCO_{3(s)}$ precipitation?" First, it is possible to calculate the concentration of CO_3^{2-} using the pH and C_T values that have already been calculated (equation listed in Table 2.3). Then, calculate the product of its concentration with the concentration of dissolved calcium, that is, $(Ca^{2+}) \cdot (CO_3^{2-})$, and compare with the K'_{sp} of $CaCO_3$ (For further discussion on solids precipitation, see Chapters 6 and 8):

$$[CO_3^{2-}] = \frac{K'_{C1}K'_{C2}C_T}{K'_{C1}K'_{C2}+K'_{C1}10^{-pH}+10^{-2pH}} = \frac{10^{-6.35}\cdot10^{-10.40}\cdot5.83\cdot10^{-3}}{10^{-6.35}\cdot10^{-10.40}+10^{-6.35}\cdot10^{-6.29}+10^{-(2\cdot6.29)}}$$

$$= 2.11\cdot10^{-7}M$$

Note 1: The above example can be solved using different combinations of two independent conservative parameters, for example, $HCO_3^-{}_{acd}$ and C_T or any other combination. Try for yourself and see that the results work out exactly the same!

Note 2: This question uses slightly different constant values, but no less acceptable than those used in the previous chapters for the carbonate system ($K_{C1}=10^{-6.33}$, $K_{C2}=10^{-10.4}$)

Example 4.2: Mixing two streams of water containing different weak-acid systems in a closed system

This example represents problems involving two (or more) water streams containing several weak-acid systems. The principle of solving such problems is: choose an alkalinity (or acidity) equation that represents all the weak-acid/base systems found in the mixed stream and stick to this definition throughout the calculations.

To demonstrate the basis for solving these types of problems, let us now assume that 1 liter of natural water (that is, water containing only the carbonate system) is mixed with 1 liter of a solution containing the acetic acid system.

Let us mark:

$$CH_3COO^- = Ac^- \text{ and } CH_3COOH = HAc$$

$$(CH_3COOH \Leftrightarrow CH_3COO^- + H^+, \ pK_a = 4.75)$$

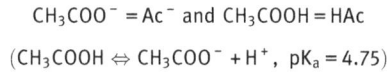

The question presented is: What is the pH of the mixed solution (neglecting the effects of ionic strength and temperature)?

The characteristics of the two solutions:

Natural water: pH = 7.5, $H_2CO_3^*{}_{alk}$ = 200 mg/L as $CaCO_3$

Acetic acid: pH = 4.5, A_T = 120 mg/L as CH_3COOH (A_T represents the sum of the concentration of the two species of the acetic acid system, which is a weak, monoprotic acid)

Solution First the alkalinity equation that represents the mixed solution should be defined. As mentioned, the reference species of this alkalinity equation must represent all the weak-acid/base systems present in the mixed solution (one reference species for each system), in this case, it is the carbonate and acetic acid systems:

$$\text{Alkalinity}_{(H_2CO_3^*,\ HAc)} = 2[CO_3^{2-}] + [HCO_3^-] + [CH_3COO^-] + [OH^-] - [H^+]$$

Of course, an alkalinity equation with different reference species can also be chosen, or alternatively a fitting acidity equation. In both cases, the solution and the end result will be similar. Note that in solving this example, it was decided to define the alkalinity equation representing the mixed stream and not the acidity equation for convenience purposes alone. It is easier to solve with alkalinity because it is already given in this question for natural water.

Next, the conservative quantity which is used for the weighted average calculations should be defined: the first quantity has already been defined in the previous step: Alkalinity$_{(H_2CO_3^*,\ HAc)}$, the additional quantities with which it is most convenient to work are the total concentrations of the two weak acid systems, namely C_T and A_T. As mentioned above, it was possible to choose three different conservative quantities and an identical solution would be obtained.

After defining the conservative quantities, their values should be calculated for each of the solutions:

For the Natural Water (Using the Representative Alkalinity Equation with $A_T = 0$):

Since the concentration of the acetic acid system in the natural water is zero, we can write the total alkalinity equation as follows:

$$\text{Alkalinity}_{(H_2CO_3^*, \; HAc)} = \text{Alkalinity}_{(H_2CO_3^*)} = 2[CO_3^{2-}] + [HCO_3^-] + [OH^-] - [H^+]$$

$$= 200/50,000 = 4 \cdot 10^{-3} \; eq/L$$

To determine the C_T of the water, $CO_3^{2-}{}_{acd}$ should be determined first, which can be calculated directly from the knowledge of $H_2CO_3^*{}_{alk}$ and pH, using eq. (3.40). Now, with the relationship: $H_2CO_3^*{}_{alk} + CO_3^{2-}{}_{acd} = 2C_T$ we can calculate C_T. We obtain: $C_T = 4.27 \cdot 10^{-3}$ M

Finding the pH value obtained in the mixed stream

To calculate the pH value, we must first find the values A_T, C_T and alkalinity$_{(H_2CO_3^*, \; HAc)}$ in the mixed solution by calculating their weighted averages (in the current example the dilution ratio is 1:1):

$$\text{Alkalinity}_{(H_2CO_3^*, \; HAc)} \; (\text{mixed solution}) = (6.88 \cdot 10^{-4} + 4 \cdot 10^{-3})/2 = 2.34 \cdot 10^{-3} \; eq/L$$

$$C_T \; (\text{mixed solution}) = (4.27 \cdot 10^{-3} + 0)/2 = 2.135 \cdot 10^{-3} \; M$$

$$A_T \; (\text{mixed solution}) = (120/60,000 + 0)/2 = 1 \cdot 10^{-3} \; M$$

Finally, the pH value of the mixed stream is calculated using the explicit alkalinity equation of the mixed stream. That is, substituting the appropriate expressions for each of the species as a function of $C_{T(mix)}$ or $A_{T(mix)}$, pH and constants in the following equation (the expressions are given in Table 2.3).

$$\text{Alkalinity}_{(H_2CO_3^*, \; HAc) \; mix} = 2[CO_3^{2-}]_{mix} + [HCO_3^-]_{mix} + [CH_3COO^-]_{mix}$$

$$+ [OH^-]_{mix} - [H^+]_{mix}$$

An equation with one unknown (pH) is obtained:

$$2.34 \cdot 10^{-3} = \frac{2 \cdot K'_{C1}K'_{C2} \cdot 2.135 \cdot 10^{-3} + K'_{C1}10^{-pH} \cdot 2.135 \cdot 10^{-3}}{K'_{C1}K'_{C2} + K'_{C1}10^{-pH} + 10^{-2pH}} + \frac{10^{-4.75} \cdot 1 \cdot 10^{-3}}{10^{-4.75} + 10^{-pH}}$$

$$+ \frac{10^{-14}}{10^{-pH}} - 10^{-pH}$$

Remember to use the values of C_T and A_T that are obtained after mixing of the streams.

If it is not possible to solve for the pH value easily using numerical analysis or special software, it can be calculated using simple iterations. In order to perform as few iterations as possible, it is necessary to come up with an intelligent ballpark figure for the pH value. This can be done by assessing the relative buffer capacities of the two streams and estimating which will be more dominant in determining the pH of the mixed solution. In the current example, for instance, while the concentration of the acetic system is significantly lower than the concentration of the carbonate system, its initial pH is much closer to its pK_a, so its buffer capacity is not negligible. It was therefore possible to predict that the pH of the mixed solution would stabilize more or less in the center between the two pH values of the two solutions.

The numerical solution of the question: pH (mixed solution) = 6.54

For the Acetic Acid Solution (Using the General Equation with $C_T = 0$):

In this case, since $C_T = 0$, the concentrations of the species CO_3^{2-} and HCO_3^- are zero, the total alkalinity equation can be written as follows:

$$\text{Alkalinity}_{(H_2CO_3^*, \; HAc)} = 2[\cancel{CO_3^{2-}}] + [\cancel{HCO_3^-}] + [CH_3COO^-] + [OH^-] - [H^+]$$

And more specifically (using the corresponding expression in Table 2.3 for the concentration of the species CH_3COO^- and the expression for OH^- given in eq. (1.25)):

$$\text{Alkalinity}_{(H_2CO_3{}^*,\,HAc)} = \frac{A_T K_a'}{K_a' + [H^+]} + \frac{K_w'}{[H^+]} - [H^+] = \frac{120/60,000 \cdot 10^{-4.75}}{10^{-4.75} + 10^{-4.5}} + \frac{10^{-14}}{10^{-4.5}} - 10^{-4.5} = 6.88 \cdot 10^{-4}\frac{eq}{L}$$

Example 4.3: Experiment with emission of sulfide gas from water
For a laboratory experiment aimed at measuring the emission rates of $H_2S_{(g)}$ from water, it is necessary to prepare 2 liters of distilled water with a sulfide concentration of 10 mg/L as S and pH7.0. For the preparation of the solution, the chemical $Na_2S \cdot 7H_2O$ was selected.
Neglecting the effects of ionic strength and temperature on the equilibrium constants, calculate:
a. The mass of the chemical to be added to the water to obtain a sulfide concentration of 10 mg/L as S.
b. How many ml of the strong acid HCl at 6N should be added to the solution to reach pH 7 (assume a closed system at this stage)?
c. It is required to add to the solution a phosphate buffer to maintain the pH value in the area of pH 7.0. What is the concentration of phosphate to be added to the solution so that at the end of the H_2S emission from the water the pH should not exceed pH 7.05 (assume that all sulfide is emitted at the end of the experiment)?

Solution
a. The molecular weight of $Na_2S \cdot 7H_2O$ is 204 g/mol. In order to prepare 10 mg/L as S in 2 liters of distilled water:

$$10\frac{mg(S)}{L} \cdot \frac{mol}{32\,g(S)} \cdot \frac{204g(Na_2S \cdot 7H_2O)}{mol} \cdot 2L = 127.5\,mg(Na_2S \cdot 7H_2O)$$

b. The sulfide system is a weak diprotic acid with $pK_1 = 7.00$ and $pK_2 = 12.89$. In terms of the sulfide system, the species added to the water in this example is S^{2-}, that is, the solution will stabilize on the equivalence point of this species (high pH). To transfer the solution to a state where the concentration of the sulfide is divided equally (practically) between the species $H_2S_{(aq)}$ and HS^- (Since pH 7.0 is exactly the first pK value), the S^{2-} should be completely converted to HS^- (and then half of HS^- will be converted to H_2S) by adding a strong acid (HCl). The amount of strong acid to be added to receive a full transformation is equal (in equivalents) to the amount of S^{2-} applied to the solution, that is, 10/32 meq. In addition, in order to continue to convert 50% of HS^- to H_2S, another 50% of this value must be added. A total of 0.46875 meq/L of strong acid should be added to the solution. It follows that for 2 liters, we should add 0.000938 equivalents of strong acid or 0.156 ml of HCl 6N.
    Another way to solve the problem is by using alkalinity and acidity quantities. At the beginning of the experiment, the solution is at the equivalence point of S^{2-} and therefore $Acd(S^{2-}) = 0$. After adding the acid and reaching pH 7, it is known that:

$$[H_2S] = [HS^-] = 0.5\,S_T \text{ (S^{2-} concentration is completely negligible)}$$

Therefore,

$$Acd(S^{2-}) = 2[H_2S] + [HS^-] + [H^+] - [OH^-] = 2 \cdot 0.5\,S_T + 0.5\,S_T = 1.5\,S_T$$

In addition, it is known that the acidity $Acd(S^{2-})$ is equal to the strong acid dose and the same answer is obtained.

c.   One way to solve this problem is to first define an alkalinity type that does not change when the $H_2S_{(g)}$ is emitted from the water. Let us therefore define the following alkalinity equation that does not include the H_2S species:

$$Alk\,(H_3PO_4, H_2S) = 3[PO_4{}^{3-}] + 2\,[HPO_4{}^{2-}] + [H_2PO_4{}^{-}] + 2\,[S^{2-}] + [HS^{-}] + [OH^{-}] - [H^{+}]$$

After defining the alkalinity equation, values can be inserted for the initial state of the solution (that is pH 7.0 and $S_T = 10/32000$ M) and the final state of interest (that is pH 7.05 and $S_T = 0$). Two equations with two unknowns are obtained: P_T and the total alkalinity value (which, as stated, remains constant during the experiment), which can be solved to receive the required P_T. In the current example, the required phosphate concentration is 175 mg/L as P.

4.3 Examples for implementation of the principles of the calculation method for solving problems related to wastewater

Example 4.4: Discharging of wastewater into natural water bodies
A fertilizer plant is requesting permission to discharge two wastewater streams to a river (characterization of the wastewater streams and the river are listed in Table 4.2). Wastewater stream 1 is a strong base that does not include the carbonate system ($C_T = 0$). Wastewater stream 2 contains the ammonia and phosphorous systems alone, without the presence of the carbonate system.

Table 4.2: Characteristics and data of the river stream and wastewater streams 1 and 2.

	Flow rate m^3/s	pH	$H_2CO_3{}^*$ alk mg/L as CaCO$_3$	TDS mg/L	Temp. °C	N_T mg N/L	P_T mg P/L
River	6.5	7.3	150	600	25	0	0
Wastewater stream 1	0.28	13.3	-	7200	25	0	0
Wastewater stream 2	0.28	9.5	-	6000	25	500	40

Determine what are the treatments (acidification? ammonia oxidation?) which the plant must perform before disposing the wastewater into the river, given:

a.   The pH level at the mixing point should not increase more than 0.5 units above the current pH level in the river.

b.   The maximum allowed concentration of $NH_3-N_{(aq)}$ at the mixing point is 0.2 mg/L and the maximum phosphorous concentration allowed in the stream is 2 mg/L as P.

Remarks:
- Assume that the rate of CO_2 exchange with the atmosphere at the mixing point is slow enough for the mixing point to be considered, for calculation purposes, a closed system.
- Perform the calculation for each wastewater stream separately.

Solution:
Mixing the first wastewater stream (strong base) with the river water
This problem is equivalent to dosing a strong base (OH^- ions) into water. When hydroxide ions are dosed into water, the alkalinity ($H_2CO_3{}^*{}_{alk}$) value increases and the acidity ($CO_3{}^{2-}{}_{acd}$) value decreases in direct proportion to the number of equivalents of the hydroxide ions that were added. There is no weak-acid system in the wastewater stream, therefore two independent parameters are sufficient to characterize the mixed stream, which will contain only the carbonate system. The trivial parameter is Alk $H_2CO_3{}^*$ (since it is given), and the additional parameter can be Acd $CO_3{}^{2-}$, since it can be easily calculated. Notice that, since the mixed stream contains only one weak-acid/base system, both the alkalinity and acidity parameters are defined in this case with respect to one reference species (alkalinity with respect to $H_2CO_3{}^*$ and acidity with respect to $CO_3{}^{2-}$).

Characterizing the river water
For the first stage of the solution, we calculate the acidity of the river water. To do so, we must first adjust the equilibrium constants according to the ionic strength of the river.

The activity coefficients for the river water (calculated using the Kemp and Davies equations presented in Chapter 1) are:

$$\gamma_m = 0.882 \quad \gamma_d = 0.602 \quad \gamma_t = 0.325$$

Calculation of the apparent equilibrium constants (note that since the pH electrode measures activity and not concentration, the activity of H^+ is calculated from the pH rather than the concentration, so there is no need to multiply the (activity of) H^+ species found in the equilibrium equation by γ_m):

$$K'_{C1} = \frac{K_{C1}}{\gamma_m} = \frac{10^{-6.35}}{0.882} = 10^{-6.295}$$

$$K'_{C2} = \frac{\gamma_d K_{C2}}{\gamma_d} = \frac{0.882 \cdot 10^{-10.33}}{0.602} = 10^{-10.113}$$

Calculation of $CO_3{}^{2-}{}_{acd}$ for the river water (using eq. (3.40)):

$$CO_{3\,acd}^{2-} = 3.586 \cdot 10^{-3} \text{ eq/L} = 179.3 \text{ mg/L as } CaCO_3$$

Characterization of the first wastewater stream
The calculated activity coefficients for the first wastewater stream are:

$$\gamma_m = 0.735; \quad \gamma_d = 0.292$$

The alkalinity value of the wastewater stream equals to (by good approximation) the concentration of OH^- ions:

$$Alk_{stream1} = [OH^-] = \frac{(OH^-)}{\gamma_m} = \frac{K_W}{\gamma_m (H^+)} = \frac{10^{-14}}{0.735 \cdot 10^{-13.3}} = 0.27 \text{eq/L}$$

Calculation of the alkalinity mass flow rate that is added to the river (or the acidity mass flux being detracted from the river) following the discharge of the untreated wastewater:

$$Alk_{added\ to\ flow} = -Acd_{added\ to\ flow} = 0.27\frac{eq}{L} \cdot 0.28\frac{m^3}{s} \cdot 10^3\frac{L}{m^3} = 75.6\frac{eq}{s}$$

Therefore, the total alkalinity flow rate at the mixing point will be:

$$Alk_{flow\ new} = Alk_{flow\ old} + Alk_{flow\ added} = 0.003\frac{eq}{L} \cdot 6.5\frac{m^3}{s} \cdot 10^3\frac{L}{m^3} + 76\frac{eq}{s} = 95.5\frac{eq}{s}$$

Therefore, the alkalinity concentration at the mixing point is:

$$Alk_{new} = \frac{95.5\frac{eq}{s}}{(6.5+0.28)\frac{m^3}{s}} = 14.09\frac{eq}{m^3} = 14.09\frac{meq}{L} = 704\frac{mg}{L} \text{ as } CaCO_3$$

And the acidity at the mixing point:

$$Acd_{new} = \frac{3.58\frac{meq}{L} \cdot 6.5\frac{m^3}{s} - 76\frac{eq}{s}}{(6.5+0.28)\frac{m^3}{s}} = -7.77\frac{eq}{m^3} = -389\frac{mg}{L} \text{ as } CaCO_3$$

Knowing the new alkalinity and acidity values at the mixing point, using eq. (3.40), we can calculate the pH value at the mixing point. Note that the equilibrium constants must be updated according to the TDS value obtained as a result of the mixing. The pH at the mixing point that would result if the wastewater is not treated is pH 11.81, which of course is too high, so the wastewater stream must be treated by adding a strong acid to lower the pH.

What is the minimum dose of strong acid to be added to the wastewater stream to meet the requirement that the pH does not increase by more than half a unit?

The regulations allow an increase in the pH value at the mixing point in the river to pH 7.8. The calculation presented here assumes that the strong acid is added directly at the mixing point and is therefore defined per liter of river water mixed with wastewater.
    Since the desired maximum pH value, as well as the alkalinity and acidity values at the mixing point prior to dosage of the strong acid are known, and since the addition of acid decreases the alkalinity and increases the acidity equally, the following equation can be used where the unknown x denotes the strong acid dose (in eq/L units):

$$CO_3{}^{2-}_{acd} + x = \frac{1 + \frac{2 \cdot 10^{-7.8}}{K'_{C1}}}{1 + \frac{2K'_{C2}}{10^{-7.8}}}\left((H_2CO_3{}^*{}_{alk} - x) - \frac{K'_w}{10^{-7.8}} + 10^{-7.8}\right)$$
$$- \frac{K'_w}{10^{-7.8}} + 10^{-7.8}$$

Another way to calculate the alkalinity and acidity values at the mixing point prior to the dosage is by using weighted average:

$$H_2CO_3{}^*{}_{alk}(mixture) = \frac{(150/50000) \cdot 6.5 + \{10^{-14}/(10^{-13.3} \cdot 0.735)\} \cdot 0.28}{6.5 + 0.28}$$
$$= 0.0141\ eq/L$$

$$CO_3{}^{2-}{}_{acd}(mixture) = \frac{(179/50000) \cdot 6.5 - \{10^{-14}/(10^{-13.3} \cdot 0.735)\} \cdot 0.28}{6.5 + 0.28}$$
$$= -0.0077\ eq/L$$

In the same way, the TDS value at the mixing point prior to the dosage is calculated:

$$\text{TDS (mixture)} = \frac{600 \cdot 6.5 + 7200 \cdot 0.28}{6.5 + 0.28} = 872.6 \text{ mg/L}$$

From the new TDS value, the new activity coefficients are derived:

$$\gamma_m = 0.864; \quad \gamma_d = 0.557$$

Therefore, $CO_3^{2-}{}_{\text{acd (mix)}}$ and $H_2CO_3^*{}_{\text{alk (mix)}}$ can be placed into eq. (3.40) whose only unknown now is x, and we get the strong acid dose to be added to the stream at the mixing point: 10.96 meq/L. Notice that the solution is normalized to the total flow rate (river + wastewater).

A solution in principle for discharge of the second wastewater stream (containing the ammonia and orthophosphate systems at pH 9.5) to the river
The solution to this "mixing" problem is similar, in principle, to the solution for the first wastewater stream. The only significant difference is the alkalinity and acidity equations, which must now include the species of the additional systems present in water at the mixing point, except for the carbonate system.

This problem can be solved in a number of ways depending on the conservative parameters that we choose to work with. For this example, we choose the following conservative parameters: Alk ($H_2CO_3^*$, H_3PO_4, NH_4^+), C_T, N_T and P_T.

A suitable alkalinity equation at the mixing point is:

$$\text{Alk (H}_3\text{PO}_4\text{, H}_2\text{CO}_3^*\text{, NH}_4^+) = 3[PO_4^{3-}] + 2[HPO_4^{2-}] + [H_2PO_4^-] + 2[CO_3^{2-}]$$
$$+ [HCO_3^-] + [NH_3] + [OH^-] - [H^+]$$

Note that according to the given information, the carbonate system is entirely absent from the wastewater (i.e., $C_T = 0$) and phosphorous and ammonia are absent from the river water (i.e., P_T and N_T are zero). This should not interfere with the definition of alkalinity as it is presented in the equation above, in order to adjust the alkalinity equation to the wastewater, $C_T = 0$ should be placed in it, and in order to adjust it for the river water $N_T = 0$ and $P_T = 0$ should be placed in it, similar to the solution of Example 4.2.

Solution stages:
1. Calculate the total alkalinity (Alk ($H_2CO_3^*$, H_3PO_4, NH_4^+)) and C_T for the river water where it is known that $P_T = 0$ and $N_T = 0$.
2. Use the known P_T and N_T to calculate the total alkalinity (Alk ($H_2CO_3^*$, H_3PO_4, NH_4^+)) of the wastewater where $C_T = 0$.
3. Use weighted averages to find Alk (H_3PO_4, $H_2CO_3^*$, NH_4^+) mix, TDS mix, P_T mix, C_T mix and N_T mix at the mixing point.
4. Rewrite the alkalinity equation and substitute in the values found in stage 3. The only remaining unknown should be pH; find the pH value.
5. Calculate the concentration of $NH_{3(aq)}$ at the mixing point using the equation $NH_{3(aq)} = \frac{K_a' N_T}{K_a' + 10^{-pH}}$

6. Check if $NH_{3(aq)}$mix > 0.2 mgN/L and P_T mix >2 mgP/L. If not – the mixing can be carried out. If P_T mix is too high, determine the P_T concentration to be removed from the wastewater.
7. If $NH_{3(aq)}$ concentration at the mixing point is above the allowed concentration, determine the concentration of acid that must be added to the wastewater in order to lower the concentration of non-ionic ammonia below 0.2 mg/L as N. Calculation method: from knowing the N_T value at the mixing point and using the appropriate equation from Table 2.3, calculate the

maximum pH for which $NH_{3(aq)} < 0.2$ mgN/ L. Use a safety factor of 0.3 pH units when setting the desired pH value at the mixing point. The addition of acid is equivalent to detraction of alkalinity and therefore X in the following equation denotes the amount of acid to be added at the mixing point (for the flow rate of the river + the wastewater). In this equation, the only unknown is X since P_T, N_T and C_T have been calculated and the desired pH is also known:

$$Alk_{(H_3PO_4,H_2CO_3^*,NH_4^+)} - X = \frac{P_T \cdot (3K'_{P1}K'_{P2}K_{P3} + 2K'_{P1}K'_{P2}10^{-pH} + K'_{P1}10^{-2pH})}{K'_{P1}K'_{P2}K'_{P3} + K'_{P1}K'_{P2}10^{-pH} + K'_{P1}10^{-2pH} + 10^{-3pH}}$$

$$+ \frac{C_T \cdot (2K'_{C1}K'_{C2} + K'_{C1}10^{-pH})}{K'_{C1}K'_{C2} + K'_{C1}10^{-pH} + 10^{-2pH}} + \frac{K'_a N_T}{K'_a + 10^{-pH}} + \frac{K'_w}{10^{-pH}} - 10^{-pH}$$

Numeric solution (according to the stages outlined above):
1. Alk $(H_3PO_4, H_2CO_3^*, NH_4^+) = 3 \cdot 10^{-3}$ eq/L; $C_T = 3.293 \cdot 10^{-3}$ M (river water)
2. Alk $(H_3PO_4, H_2CO_3^*, NH_4^+) = 2.52 \cdot 10^{-2}$ eq/L (wastewater stream 2)
3. Alk $_{mix}$ $(H_3PO_4, H_2CO_3^*, NH_4^+) = 3.92 \cdot 10^{-3}$ eq/L; C_T mix $= 3.157 \cdot 10^{-3}$ M; P_T mix $= 5.33 \cdot 10^{-5}$ M; N_T mix $= 1.47 \cdot 10^{-3}$ M; TDS mix $= 823$ mg/L
4. The resulting pH is 8.88

$$[NH_{3(aq)}]_{mix} = \frac{K'_a \cdot 1.47 \cdot 10^{-3}}{K'_a + 10^{-8.88}} = 4.86 \cdot 10^{-4} \, M = 6.8 mgN/L$$

5. The concentration of non-ionic ammonia is significantly higher than the allowed value. $P_T = 1.65$ mg/L, which means that it meets the requirement.
6. The marginal pH value to yield $[NH_{3(aq)}] = 0.2$ mgN/L is pH 7.18. Employing the safety coefficient of 0.3 pH units, we therefore choose that the pH at the mixing point should not exceed pH 6.88. The acid concentration to be added at the mixing point is 1.27 meq/L (calculated for the wastewater and river flow rate).

Example 4.5: Biological treatment of wastewater
In anaerobic biological processes ("anaerobic digesters"), in which the end product is methane gas (CH_4), several bacterial groups (acidogens, acetogens, methanogens) perform successive processes. Figure 4.2 illustrates schematically the decomposition process of organic matter. One of the most

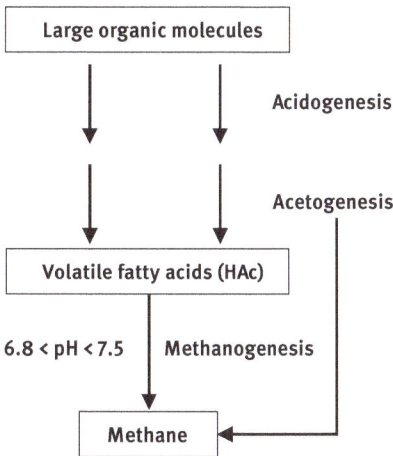

Fig. 4.2: A simplified schematic description for the decomposition processes of organic matter in anaerobic digesters for wastewater treatment.

important intermediate products in the process is a group of weak acids known as volatile fatty acids (VFA). The VFA are composed primarily of acetic acid (CH_3COOH) but also of butyric acid, propionic acid, lactic acid and other monoprotic acids at lower concentrations. For practical purposes, it can be assumed that all the monoprotic weak acids that make up the VFA have an equilibrium constant with a value of $pK_a = 4.75$. The weak acids are an intermediate product in the anaerobic process. Their acidic form is emitted to the water as a result of the decomposition of more complex organic materials by the acidogenic and acetogenic populations. On the other hand, these acids are consumed by the methanogens as substrate and oxidized by them to methane, which is, as stated, the desired final product in the process. The standard operating strategy of anaerobic digesters is to maintain the pH value in the neutral range ($6.8 < pH < 7.5$), which is convenient for the methanogenic bacteria, which are the weakest link in the decomposition chain. Under normal operating conditions, it is generally accepted that the VFA concentration in the reactors should be fairly low – up to about 200 mg/L (maximum). There are two reasons for this: one is that the fatty acids begin to inhibit the activity of methanogenic bacteria at concentrations higher than ~200 mg/L and the second is that an increase in VFA concentration indicates instability between the activity of the various bacterial groups, which is a good indication that the process is on its way to fail. When the pH becomes more acidic than pH 6.8, this usually means that the concentration of fatty acids has increased, indicating that the methanogenic population fed by these acids has reduced its activity. However, the pH itself cannot usually be used as a good indication of the onset of a problem due to the high buffering capacity of the solution in this pH region. In other words, the buffer capacity of the solution is so high that a small change in pH (a change that is sometimes found within the accuracy of the measurement) can indicate an irreversible deterioration of the process and therefore it is not possible to rely on a pH measurement for control purposes. Therefore, it is customary to directly measure VFA concentration on a daily basis in order to monitor the activity of the facility.

After this long introduction, let us formulate the following relevant question:

Assuming that the concentration of the carbonate system in the anaerobic digester is $C_T = 800$ mg/L as $CaCO_3$ and that the concentration of the orthophosphate is $P_T = 100$ mg/L as P, calculate the minimum VFA concentration that will cause the pH to fall below 6.8.

Assume a preliminary condition (before VFA accumulation begins) at which pH = 7.5 and concentration of VFA = 30 mg/L as CH_3COOH.

Solution The relevant alkalinity equation includes the species of the carbonate system, the phosphate system and the volatile fatty acid systems, all represented by the acetate/acetic acid system. The ammonia system, which is also present at a high concentration in the anaerobic digesters, can usually be neglected because its contribution to the alkalinity value in the relevant pH area is very low (remember that the thermodynamic equilibrium constant of the ammonia system is pK = 9.25). A convenient alkalinity equation in this case is the following:

$$Alk\,(H_2CO_3{}^*,\,H_3PO_4,\,HAc) = 3[PO_4{}^{3-}] + 2[HPO_4{}^{2-}] + [H_2PO_4{}^-] + 2[CO_3{}^{2-}] + [HCO_3{}^-]$$
$$+ [Ac^-] + [OH^-] - [H^+]$$

This equation can be written more explicitly, as a function of P_T, C_T and Ac_T (the total concentration of VFA):

$$Alk_{(H_2CO_3{}^*,H_3PO_4,HAc)} = \frac{P_T\left(3K_{p1}K_{p2}K_{p3} + 2K_{p1}K_{p2}\cdot 10^{-pH} + K_{p1}\cdot 10^{-2pH}\right)}{K_{p1}K_{p2}K_{p3} + K_{p1}K_{p2}\cdot 10^{-pH}K_{p1}\cdot 10^{-2pH} + 10^{-3pH}} + \frac{C_T\left(2K_{c1}K_{c2} + K_{c1}\cdot 10^{-pH}\right)}{K_{c1}K_{c2} + K_{c1}\cdot 10^{-pH} + 10^{-2pH}}$$
$$+ \frac{Ac_T\cdot K_a}{K_a + 10^{-pH}} + \frac{K_w}{10^{-pH}} - 10^{-pH}$$

Substituting the given data in the alkalinity equation (neglecting the transformation of equilibrium constants), one obtains that the alkalinity value with the reference species $H_2CO_3^*$, H_3PO_4 and HAc in the reactor solution in the steady state of operation is 0.021 eq/L.

At this point in the solution, it is important to remember that in their metabolism bacteria emit volatile fatty acids in their acidic form, namely CH_3COOH (acetic acid), C_2H_5COOH (propionic acid), C_3H_7COOH (butyric acid), etc. Since these species are not part of the alkalinity equation, their emission to the water does not change the total alkalinity value of the water as defined in the equation above, and it will remain 0.021 eq/L (similar to dissolution of CO_2, which does not affect $H_2CO_3^*{}_{alk}$).

In contrast, the VFA_T value will change as a result of the VFA emission. The VFA_T value (expressed in the equation by Ac_T) will of course increase, while C_T and P_T remain constant.

Accordingly, since the concentration of acidic species increases, the pH value decreases.

In order to determine the concentration of Ac_T in which the pH decreased to 6.8 under the question conditions, the same alkalinity equation should be reintroduced, with the only unknown being Ac_T (pH = 6.8, P_T, C_T and total alkalinity value: unchanged)

The solution to this question is that the concentration of VFA that will cause the pH to drop to 6.8 is 281 mg/L as HAc.

4.4 Using a method based on alkalinity and acidity mass balances to quantify the change in characterization of acid-base properties of water as a result of chemical dosage (deliberate or unintentional)

The alkalinity/acidity mass balance technique allows for *a priori* calculation of the change in water properties as a result of dosing various chemicals to the water; it applies whether the dosage is targeted (i.e., in the framework of any water treatment intended for changing desired properties) or whether it is unintentional. An example of targeted dosing might be reaching a desired pH, while an untargeted dosing might be the change in acid-base properties of the water occurring as a result of unintended oxidation of ammonia (nitrification) in water supply lines in which chloramination was affected, a phenomenon often observed in warm locations during the summer when the supply-water temperature increases. Another possibility is that change in the water characteristics is the result of a process intended for another purpose; for example, chlorination or fluoridation of supply water which are often acidic reactions. In some cases, these processes can significantly affect acid-base properties of the water to the point where the water changes from being stable to being aggressive to the pipeline.

In closed or semi-closed systems, the accuracy of these calculations is very high and can therefore serve as a means for planning treatment processes (in the case of deliberate dosage) or simulations of phenomena that would occur in the system as a result of unintended reactions, allowing advance assessments to solve expected problems.

4.4.1 Solution outline

The basic outline for solving such problems has been already mentioned several times in the text:

1. Characterize the acid/base ratios of the original water using two independent parameters for the carbonate system. If there are other weak-acid systems in the water, it is necessary to know the total analytical concentration of each of them.
2. Define and write the alkalinity/acidity equation that characterizes the water with which you wish to work.
3. Identify and quantify the alkalinity or acidity species that are dosed into the water (whether deliberately or unintentionally). Verify that these species appear in the equations defined in part 2, and determine the appropriate coefficient for each.
4. Calculate the values of the new conservative parameters (e.g., alkalinity/acidity, total concentration of weak-acid system) obtained after the dosage.
5. Calculate the pH value (or any other non-conservative parameter, e.g., the concentration of specific weak-acid species) using the values (obtained in part 4) of the conservative parameters following dosage.

Example 4.6: Changing the characterization of desalinated water in water supply systems
The desalination plant in Ashkelon, Israel produces (as of the end of 2018) desalinated water with the following mean characteristics: TDS = 150 mg/L, $H_2CO_3^*{}_{alk}$ = 48 mg/L as $CaCO_3$, pH 8.15 with dissolved calcium at a concentration of 100–110 mg/L as $CaCO_3$.

In Chapter 6, which deals with aqueous-solid-phase interactions, common indices are developed and described for calculating the degree of "chemical stability" of water as a function of its ability to precipitate $CaCO_{3(s)}$. At this point in the text, before dealing with this topic, it suffices to say that the pH of the water exiting the desalination plant is such that the indices obtained from the water analysis are positive as required, that is, that the values indicate that the water is chemically stable upon exiting the desalination plant.

Immediately after leaving the desalination plant, the water undergoes two treatments: disinfection (by chlorine gas) and fluoridation (by fluorosilicic acid).

Question: What will be the change in the water quality (in terms of acid-base interactions and $CaCO_3$ precipitation potential) following the dosage of fluorosilicic acid to reach a concentration of 1 mg/L of fluoride (as fluoride) in the water?

The solubility constant of $CaCO_3$: $K_{sp}(CaCO_3) = 10^{-8.05}$.

Solution The problem is solved according to the outline detailed above: The acidity in the original water is calculated using eq. (3.40) to obtain $CO_3^{2-}{}_{acd}$ = 48 mg/L as $CaCO_3$. The alkalinity with which one can work is the given alkalinity, that is, $H_2CO_3^*{}_{alk}$.
The following reaction occurs as a result of the fluorosilicic acid dosage:

$$H_2SiF_6 + 2H_2O \rightarrow 6H^+ + 6F^- + SiO_2 \tag{4.3}$$

From eq. (4.1), it is obtained that for each mole of fluoride (as F) dosed, 1 mole of H^+ is released, that is, 1 eq of alkalinity is consumed from the water while 1 eq of acidity is added to it. In the current example, 1/19 mM of fluoride ($M_W(F)$ = 19 g/mol) is dosed to the water, therefore 1/19 mM of H^+ will be released which is $5.26 \cdot 10^{-5}$ eq/L of H^+, or 2.6 mg/L as $CaCO_3$. The alkalinity and acidity values following the dosage will therefore be:

$$Alk_{new} = 48 - 2.6 = 45.4 \text{ mg/L as } CaCO_3$$

$$Acd_{new} = 48 + 2.6 = 50.6 \text{ mg/L as } CaCO_3$$

The pH of the water following dosage of the acid can now be calculated using eq. (3.40). It is also possible to find C_T using the relationship: $Alk + Acd = 2C_T$. Next, the product of the carbonate and calcium activities can be calculated, i.e. $(Ca^{2+}) \cdot (CO_3^{2-})$ which is $[Ca^{2+}] \cdot [CO_3^{2-}] \cdot \gamma_d^2$, since the calcium concentration is given and the carbonate concentration can be calculated using C_T and pH which were calculated using the relationship given in Table 2.3. This product is important for the precipitation and dissolution potentials of the mineral $CaCO_{3(s)}$, as explained in detail in Chapter 9. At this point, it suffices to say that when the product is greater than the solubility constant ($K_{sp}(CaCO_3)$) the water has a tendency to precipitate the mineral (i.e., the water is chemically stable, as required). Whereas if the product is smaller than the constant, the water tends to dissolve the mineral – and this state is not desirable in water supply lines.

Table 4.3 presents the water quality conditions after its release from the desalination plant and the water quality after dosing the fluorosilicic acid (the results of the calculations described above). Calcium concentration of 100 mg/L as $CaCO_3$ was assumed. A significant change in pH between the two stages can be seen, which results in a significant change in the value of the concentrations product. While before the acid dose, the product is greater than the solubility constant, after the dose the product is smaller than the constant. The reason for the sharp change in the precipitation potential value as a result of a relatively small change in the alkalinity and acidity values is due to the very low buffer capacity of the water released from the Ashkelon desalination plant. This issue is dealt with extensively in Chapter 6.

Table 4.3: Water quality data before and after fluoridation using fluorosilicic acid.

	$H_2CO_3^{*}{}_{alk}$ mg/L as $CaCO_3$	$CO_3^{2-}{}_{acd}$ mg/L as $CaCO_3$	C_T M	pH –	$(Ca^{2+})(CO_3^{2-})$	$(Ca^{2+})(CO_3^{2-})$ $- K_{sp}$
Prior to fluorosilicic acid dosage	48	48.5	$9.65 \cdot 10^{-4}$	8.15	$10^{-7.75}$	>0
Following fluorosilicic acid dosage	45.4	51.1	$9.65 \cdot 10^{-4}$	7.50	$10^{-8.42}$	<0

In the same way, it is possible to calculate the change in water quality as a result of other processes occurring in the water. For example, eq. (4.4) describes a chlorination reaction (intended dose of chlorine to water for disinfection), and eq. (4.5) describes an unintentional biological oxidation reaction of ammonia to nitrite (the first stage of the nitrification process), which are sometimes observed in water supply lines to which chloramines are added to prevent re-growth of bacteria in the line:

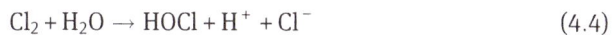

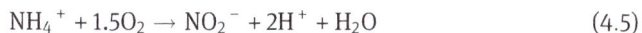

$$Cl_2 + H_2O \rightarrow HOCl + H^+ + Cl^- \tag{4.4}$$

$$NH_4^+ + 1.5O_2 \rightarrow NO_2^- + 2H^+ + H_2O \tag{4.5}$$

From eqs. (4.4) and (4.5), one can derive the expected change in the water quality
in terms of acid-base equilibrium as a result of the reactions. Note that for each of
the above cases, a weak-acid/base system is added to the water. Therefore, in principle, an alkalinity/acidity equation should be defined with respect to the new reference species. For example, in terms of chlorination (described by eq. (4.4)), it
should be recalled that chlorine gas undergoes hydrolysis in water to hypochlorous
acid (HOCl), which itself is a weak acid that is found in equilibrium with the ion
OCl^- (hypochlorite), with $pK_a = 7.5$ and therefore the appropriate total alkalinity
equation will be:

$$Alk_{(H_2CO_3^*,\ HOCl)} = 2[CO_3^{2-}] + [HCO_3^-] + [OCl^-] + [OH^-] - [H^+]$$

Note that the total alkalinity changes as a result of H^+ released into the water but
does not change as a result of the addition of the species HOCl to the water. On the
other hand, Cl_T (the total concentration of the hypochlorite system species) will
change (its concentration increases). Similarly, in terms of ammonia oxidation (described in eq. (4.5)) recall that the ammonium ion comprises the acidity equation,
whose reference species are CO_3^{2-} and NH_3. The decision whether to neglect these
reactions depends primarily on the conditions outlined in the question. For example, if the chlorination process (the first reaction) takes place in water around pH 6,
practically all chlorine will break down to form HOCl and the concentration of OCl^-
will be negligible. As a rule, if you are not sure what can be neglected, it is best not
to neglect anything. In any case, most calculations today are performed by computer rather than by hand, therefore the error that can be caused from erroneously
neglecting a species is usually greater than the complication involved in solving
without neglecting that species.

Example 4.7: Preventing spontaneous precipitation during groundwater extraction
Drawing groundwater in pressurized lines sometimes involves spontaneous deposition of $CaCO_{3(s)}$
inside the line, causing a reduction in the effective diameter of the pipe, an increase in energy
losses during operation and a drop in the nominal flow rate of the line.
    Calculate the acid dose required to lower the pH of the groundwater characterized below to
pH 7.5, the highest pH value for which $CaCO_3$ should not precipitate under these water conditions
(Chapter 6 describes how this pH value is calculated).
    In order to reduce the pH, two chemicals are available, it is necessary to calculate the dosage
required from each of them to reach the required pH.
1.   Concentrated solution of strong acid, H_2SO_4.
2. $CO_{2(g)}$

Groundwater parameters prior to dosage: pH 8.3, $H_2CO_3^*{}_{alk}$ = 280 mg/L as $CaCO_3$, TDS = 1100 mg/L.

Solution Note: As a preliminary step, the equilibrium constants must be converted based on the
ionic strength calculated from the given TDS concentration.
    Solving the question for the H_2SO_4 dose is performed in two stages, in which the equation
linking $H_2CO_3^*{}_{alk}$, $CO_3^{2-}{}_{acd}$ and pH eq. (3.40) is used.

The first step is to substitute the known $H_2CO_3{}^*{}_{alk}$ and pH into eq. (3.40) to find $CO_3{}^{2-}{}_{acd}$ before the acid dose (to obtain $CO_3{}^{2-}{}_{acd} = 5.6$ meq/L). In the second step, eq. (3.40) is used again, with the knowledge that the addition of strong acid increases the acidity and reduces the alkalinity value to the same extent (as in Example (4.4), for example) and to substitute pH 7.5 into the equation.

One equation with a single unknown is obtained:

$$5.54 \cdot 10^{-3} + x = \frac{1 + \frac{2 \cdot 10^{-7.5}}{K'_{C1}}}{1 + \frac{2K'_{C2}}{10^{-7.5}}} \left\langle \left(5.6 \cdot 10^{-3} - x\right) - \frac{K'_w}{10^{-7.5}} + 10^{-7.5} \right\rangle - \frac{K'_w}{10^{-7.5}} + 10^{-7.5}$$

Solution: $x = 3.35 \cdot 10^{-4}$ eq/L or 16.4 mg/L as H_2SO_4.

To solve the problem for the $CO_{2(g)}$ dosing, we must first understand what is changing in the conservative parameters of the water as a result of the CO_2 dosage.

When $CO_{2(g)}$ is dosed to the water, it can be assumed that $H_2CO_3{}^*$ is dosed (since these two parameters are nearly identical). We will mark the addition of $H_2CO_3{}^*$ as x, therefore:

$$\Delta C_T = x$$

$$\Delta CO_3{}^{2-}{}_{acd} = 2x$$

$$\Delta H_2CO_3{}^*{}_{alk} = 0$$

And therefore:

$H_2CO_3{}^*{}_{alk}$ (initial) $= H_2CO_3{}^*{}_{alk}$ (after dosage) $= 5.56 \times 10^{-3}$ eq/L

$$5.54 \cdot 10^{-3} + 2x = \frac{1 + \frac{2 \cdot 10^{-7.5}}{K'_{C1}}}{1 + \frac{2K'_{C2}}{10^{-7.5}}} \left\langle \left(5.6 \cdot 10^{-3}\right) - \frac{K'_w}{10^{-7.5}} + 10^{-7.5} \right\rangle - \frac{K'_w}{10^{-7.5}} + 10^{-7.5}$$

Solution: The $CO_{2(g)}$ dose required to lower the pH to 7.5 is $x = 3.55 \cdot 10^{-4}$ M, which is 17.73 mg/L as $CaCO_3$.

Notably, when $CO_{2(g)}$ is used, more equivalents are required than when using H_2SO_4 to reach the same pH. The reason for this lies in the fact that the addition of CO_2 increases not only the acidity but also the C_T, which in turn increases the buffer capacity. Another way to look at it is that dosing the acid has two effects on the water: increasing the acidity and lowering the alkalinity, while CO_2 only increases the acidity.

Example 4.8: Removal of nitrogen compounds from wastewater
As part of the wastewater treatment process, it is often necessary to remove almost completely the dissolved nitrogen compounds (ammonia, nitrite, nitrate) from the wastewater before the effluents are discharged to the receiving water body. The combined biological process, termed nitrification–denitrification, is based on autotropic aerobic bacteria (i.e., bacteria that utilize inorganic carbon for the buildup of their cells) that oxidize ammonia to nitrite and nitrate ("nitrification") and heterotrophic bacteria (that utilize organic carbon as an electron donor and for the buildup of their cells) which operate under anoxic conditions for the purpose of nitrate reduction ("denitrification"). The process is presented schematically in Fig. 4.3. eqs. (4.6) and (4.7) are used to describe the processes of nitrification and denitrification in the process of wastewater treatment:

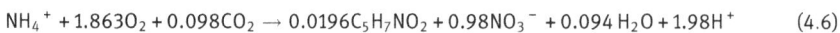

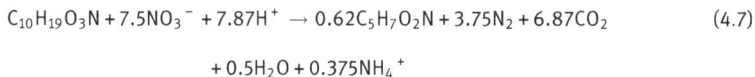

$$NH_4{}^+ + 1.8630_2 + 0.098CO_2 \rightarrow 0.0196C_5H_7NO_2 + 0.98NO_3{}^- + 0.094H_2O + 1.98H^+ \qquad (4.6)$$

$$C_{10}H_{19}O_3N + 7.5NO_3{}^- + 7.87H^+ \rightarrow 0.62C_5H_7O_2N + 3.75N_2 + 6.87CO_2 \qquad (4.7)$$

$$+ 0.5H_2O + 0.375NH_4{}^+$$

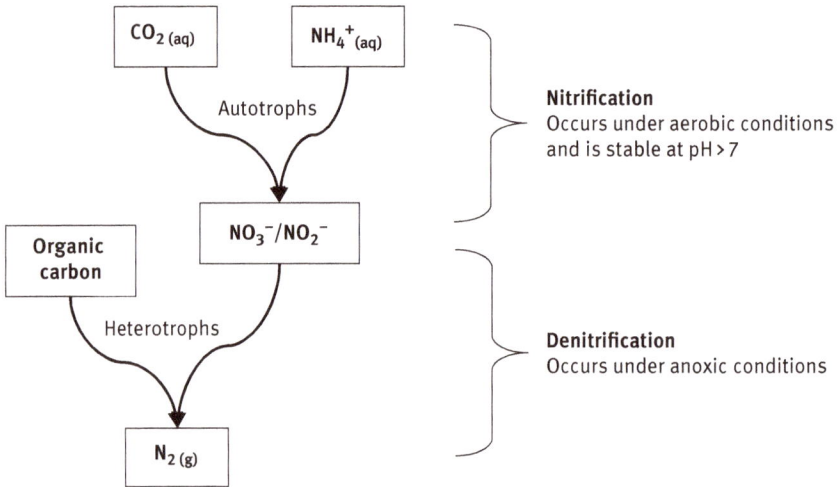

Fig. 4.3: Schematic description of nitrification and denitrification processes.

Equation (4.8) can be used in a practical way to describe an approximation of the overall reaction (nitrification–denitrification):

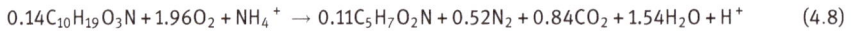

$$0.14C_{10}H_{19}O_3N + 1.96O_2 + NH_4^+ \rightarrow 0.11C_5H_7O_2N + 0.52N_2 + 0.84CO_2 + 1.54H_2O + H^+ \quad (4.8)$$

It can be seen from eq. (4.8) that as a result of the oxidation of the organic matter, CO_2 is emitted into the water. What cannot be seen from the equation is that CO_2 is also released from the water into the atmosphere in the aerobic part of the treatment facility (in which oxygen is introduced using air, into the water, which causes a strong stripping of CO_2 into the atmosphere). Under the (reasonable) assumption that the concentration of $CO_{2(aq)}$ in the wastewater during the process is 10 mg/L as CO_2 (which results in parallel from the significant absorption of CO_2 in the water as a result of the decomposition of organic matter and from the CO_2 being released to the atmosphere due to the massive aeration), calculate the minimum alkalinity required in the raw wastewater so that the pH value of the wastewater in the treatment process will not fall below pH 7.0 under the assumption that the dissolved nitrogen compounds are completely removed.

This question has a significant practical meaning since (a) the nitrifying bacteria slow their activity when the pH becomes acidic and (b) due to the entry into operation of many desalination plants the alkalinity value of the raw wastewater may in the future reach a low value to the point of concern about the stability of the biological treatment, and the need to add basic chemicals as part of the treatment process, which are expensive. In the raw sewage, the pH value is given as 7.8 and the overall ammonia concentration ($[NH_3]+[NH_4^+]$) is 50 mg/L as N.

Solution Note: In the solution shown here, the effects of ionic strength and temperature on the equilibrium constants were neglected.

From eq. (4.8), it can be seen that in the overall process for each mole of ammonium released eventually to the atmosphere as nitrogen gas, one mole of H^+ is emitted into the water, that is, the process (within the wastewater treatment facility) is acidic.

In terms of mass balances for alkalinity and acidity, it is clear that for each mole of ammonium removed, the alkalinity decreases by 1 eq and the acidity increases by 1 eq due to the emission of protons (independent of the alkalinity/acidity type which was selected). In contrast to the alkalinity value, the acidity value is also affected by CO_2 emissions that are part of the nitrification–denitrification process. In addition, if the acidity is defined against the reference species NH_3, then the consumption of NH_4^+ decreases the acidity.

In the treatment process, $50/14000 = 0.003571$ molar of ammonia are removed. It follows that a similar number of equivalents per liter of H^+ are emitted into the water (see eq. (4.8)).

At the end of the treatment process $N_T = 0$, so the overall alkalinity at this stage can only be defined against a reference species of the carbonate system, namely $H_2CO_3^*{}_{alk}$. Therefore, from the (empirical) knowledge of the CO_2 concentration and the minimum pH value of interest, the stabilizing alkalinity value during the treatment process can be calculated by explicitly assigning an alkalinity equation in which the CO_3^{2-} and HCO_3^- species are expressed using the known $H_2CO_3^*$ species (11.4 mg/L as $CaCO_3$). Expressing the values of the bicarbonate and carbonate species using $H_2CO_3^*$ is done by manipulating the thermodynamic equilibrium equations between the three carbonate system species. The following equation is obtained:

$$Alk_{H_2CO_3^*, NH_4^+} = Alk_{H_2CO_3^*} = 2[CO_3^{2-}] + [HCO_3^-] + [OH^-] - [H^+]$$

$$= \frac{2K_1K_2[H_2CO_3^*]}{[H^+]^2} + \frac{K_1[H_2CO_3^*]}{[H^+]} + \frac{K_w}{[H^+]} - [H^+]$$

Substituting $[H^+] = 10^{-7}$ M and $[H_2CO_3^*] = 11.4$ mg/L, the value of Alk $H_2CO_3^*$ is obtained as $1.02 \cdot 10^{-3}$ eq/L which is 50.8 mg/L as $CaCO_3$.

Therefore, in order to calculate the minimum initial alkalinity required for the raw wastewater, the alkalinity remaining at the end of the process must be summed up as well as the alkalinity destroyed in the process, that is:

$$Alk_{H_2CO_3^*, NH_4^+ \text{(raw sewage)}} = Alk_{H_2CO_3^*, NH_4^+ \text{(remaining at steady state)}} + Alk_{H_2CO_3^*, NH_4^+ \text{(destroyed in process)}}$$
$$= 1.016\,meq/L + 3.571\,meq/L = 229.4\,mg/L \text{ as } CaCO_3$$

Note that this value includes the carbonate and ammoniacal alkalinity (which is essentially the concentration of $NH_{3(aq)}$ in the raw wastewater) at the beginning of the process. It is therefore possible to calculate the contribution of each system to the total alkalinity:

$$Alk_{NH_4^+ \text{(raw Sewage)}} = NH_{3(aq)} = \frac{K_a \cdot N_T}{K_a + (H^+)} = \frac{10^{-9.25} \cdot (50/14000)}{10^{-9.25} + 10^{-7.8}} = 1.22 \cdot 10^{-4}\,eq/L = 6.1\,mg/L \text{ as } CaCO_3$$

$$Alk_{H_2CO_3^* \text{(raw sewage)}} = Alk_{Total \text{(raw sewage)}} - NH_{3(aq) \text{(raw seawage)}} = 229.4 - 6.1 = 223.3\,mg/L \text{ as } CaCO_3$$

Note: In eq. (4.6), which describes the ammonium oxidation, it was assumed that the pH of the wastewater is lower than 8.3 and therefore all the forms of the ammoniacal weak acid system are in the ammonium ion form (remember that $pK(NH_4^+/NH_3) = 9.25$). However, this assumption does not change the final answer to the given question. In other words, even if the pH was higher and the species in the system were mostly ammonia (NH_3), the destruction of alkalinity in the process would be the same. Explanation: record eq. (4.8) assuming that ammonia is the dominant form in the ammoniacal system:

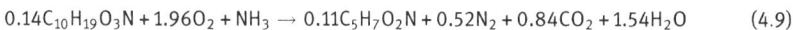

$$0.14C_{10}H_{19}O_3N + 1.96O_2 + NH_3 \rightarrow 0.11C_5H_7O_2N + 0.52N_2 + 0.84CO_2 + 1.54H_2O \quad (4.9)$$

In order for the equation to be balanced, protons are not emitted into the water when a mole of ammonia is oxidized. However, consumption of one mole of ammonia reduces the alkalinity by one equivalent. Therefore, the overall change in alkalinity resulting from the process remains the same, regardless of pH.

5 Equilibrium between the aqueous and gas phases and implications for water treatment processes

5.1 Introduction

A significant portion of the engineered systems used for water and wastewater treatment is exposed to the atmosphere and to transfer of gases from the aqueous phase to the gas phase and vice versa. The supply of atmospheric oxygen as the final electron acceptor in aerobic biological systems for wastewater treatment or intensive ponds for fish farming, stripping of carbon dioxide from aquaculture ponds or the production and collection of methane in anaerobic digesters are a number of prominent examples of such systems, and there are many others. There are instances in which the gas transfer is not deliberate, such as when carbon dioxide ($CO_{2(g)}$) volatilizes from groundwater when exposed to the atmosphere, or when hydrogen sulfide ($H_2S_{(g)}$) is released to the gas phase in wastewater pumping stations or gravitational sewage lines. In other cases, gas transfer is built into the process such as in aeration to supply oxygen or volatilize carbon dioxide for the purpose of pH control or preventing accumulation of toxic concentrations of $CO_{2(aq)}$. Either way, whether the gas transfer is deliberate or occurs naturally, understanding the mechanisms that govern it is very important for any theoretical calculation of the chemical reactions expected in the aqueous phase and for optimal design of the treatment process.

The present chapter initially focuses on basic definitions of concentrations of components in the gas phase. Later, Henry's law is defined, and then several examples are presented to elucidate the importance of gas compounds in water treatment processes. It should be noted that this chapter deals with the equilibrium relationship between the aqueous phase and the gas phase, without explicit reference to the rate of transfer, that is, the kinetics. However, it is important to understand that unlike reactions occurring in the aqueous phase, where kinetics are very fast and the equilibrium equations can be treated as accurate representations of reality, in reactions involving gases and liquids, the rate of approaching equilibrium is often rather slow, thus equilibrium equations do not necessarily represent what actually happens. For example, only in rare occasions it can be assumed that the dissolved carbon dioxide concentration in natural waters or in water treatment systems is in equilibrium with the carbon dioxide concentration in the atmosphere. In most natural systems, carbon dioxide is either supersaturated due to the respiration of microorganisms emitting it into the water and its subsequent relatively slow emission into the atmosphere or undersaturated due to intensive photosynthesis during the day. Therefore, to calculate the exact concentration of a specific gas component in

https://doi.org/10.1515/9783110603958-005

the aqueous phase, in addition to the equilibrium equation, one must also know the rates of the processes involved which are sometimes difficult to measure, especially in complex systems. Equilibrium equations, however, can certainly provide a good benchmark for the direction in which biological-chemical processes advance and hence their great importance.

5.2 Expressions describing concentrations of components in the gas phase

Concentrations in the gas phase are usually expressed in units of mass or mole per unit of volume, as is customary in aqueous solutions. The most common units are mg/L, microgram per cubic meter and mole per liter. It is often common and convenient to also use units of partial pressure to describe the concentration of a component in the gas phase, as explained below.

Under standard environmental conditions (low pressure and ambient temperature), it is possible with acceptable accuracy, to treat all gases which obey the equation of state for ideal gas, as ideal gases:

$$C_{g,i} = \frac{n_{g,i}}{V_g} = \frac{P_i}{RT} \qquad (5.1)$$

where,

$C_{g,i}$ = The concentration of component i in the gas phase in units of number of moles in the gas phase ($n_{g,i}$) per volume (V_g),

P_i = The partial pressure of component i = the pressure exerted by component i,

R = The universal gas constant $0.082 \frac{atm \cdot L}{mol \cdot K} = 8.3143 \frac{J}{mol \cdot K} = 8.3143 \frac{Pa \cdot m^3}{mol \cdot K}$

T = The absolute temperature (K).

Eq. (5.1) shows that at constant temperature, the partial pressure of a certain component P_i depends only on its concentration in the gas phase:

$$P_i = R T C_{g,i} \qquad (5.2)$$

The total pressure of a system containing a number of gases is the sum of the partial pressures of all gases.

$$P_{tot} = \sum_{j=1}^{N} P_j = RT \sum_{j=1}^{N} C_{g,j} \qquad (5.3)$$

Another way to express concentrations, which is common mainly in literature related to chemical engineering, is by mole fractions:

$$y_i \equiv \frac{P_i}{P_{tot}} = \frac{n_{g,i}}{\sum\limits_{j=1}^{N} n_{g,j}} = \frac{C_{g,i}}{\sum\limits_{j=1}^{N} C_{g,j}} \tag{5.4}$$

where y_i is the mole fraction of component i. Thus:

$$P_i = y_i P_{tot} = R T C_{g,i} \tag{5.5}$$

The use of mole fractions requires multiplication by the total pressure to obtain partial pressures. For open systems, this pressure is approximately 1 atm at sea level (or equivalently, 101.3 kilopascals or 1,013 millibars). Warning: Do not assume out of habit that the system is open, since in systems that are not open to the atmosphere the total pressure can of course be different than 1 atm.

5.3 Henry's law

The transition of molecules from the aqueous phase to the gas phase is termed volatilization or stripping. The reverse process is termed absorption or dissolution. The tendency of a substance to transition from the dissolved phase to the gas phase is defined as its volatility. Henry's law is named after William Henry, an English chemist who formulated the law in the year 1803.

According to Henry's law, one can describe the transition of a specific component from the aqueous phase to the gas phase using a normal chemical equilibrium equation as follows:

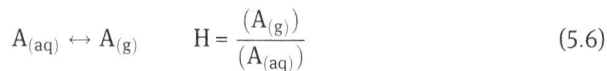

$$A_{(aq)} \leftrightarrow A_{(g)} \qquad H = \frac{(A_{(g)})}{(A_{(aq)})} \tag{5.6}$$

where
$(A_{(aq)})$ describes the activity of the dissolved molecule of component A and
$(A_{(g)})$ describes the activity of A in the gas phase.

H is defined as the equilibrium constant of the reaction under standard conditions and is called "Henry's law constant". The higher the value of H, the more component A tends to be in the gas phase, or in other words it is more volatile.

One must pay close attention to the units of Henry's law constant when using values from literature due to the existence of another convention for describing the following equilibrium, which is the inverse of the previous convention:

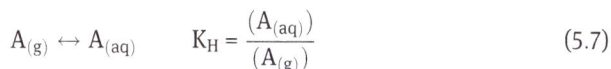

$$A_{(g)} \leftrightarrow A_{(aq)} \qquad K_H = \frac{(A_{(aq)})}{(A_{(g)})} \tag{5.7}$$

Henry's law constant in this case is specified by K_H and its relationship to H is of course:

$$H = 1/K_H \qquad (5.8)$$

Since only non-charged species can exist in the gas phase at significant concentrations, the activity coefficients of these species when dissolved (i.e., $\gamma_{A(aq)}$) are usually very close to 1.0. Therefore, the equilibrium equation of eq. (5.6) can also be expressed using the species concentrations:

$$H = \frac{\left(A_{(g)}\right)}{\left(A_{(aq)}\right)} = \frac{\left[A_{(g)}\right]}{\left[A_{(aq)}\right]} \qquad (5.9)$$

Henry's law constant, H, appears in the literature in many forms, using different units for concentration both in the gas phase and in the aqueous phase. Table 5.1 presents several examples of this, found in the literature:

Table 5.1: Various types of units used for Henry's law constant (H).

Units for concentration of component i under standard conditions in the aqueous phase (aq)	Units for concentration of component i under standard conditions in the gas phase (g)	Henry's law constant units, H
Microgram per liter solution $\left(\frac{\mu g}{L_l}\right)$	Microgram per liter gas $\left(\frac{\mu g}{L_g}\right)$	$\frac{L_l}{L_g}$
Milligram per liter solution $\left(\frac{mg}{L_l}\right)$	Partial pressure (bar_i)	$\frac{bar_i \cdot L_l}{mg}$
Mole per liter solution $\left(\frac{mol_{(aq)\,i}}{L_l}\right)$	Partial pressure (bar_i)	$\frac{bar_i \cdot L_l}{mol_{(aq)\,i}}$
Mole per liter solution $\left(\frac{mol_{(aq)\,i}}{L_l}\right)$	Mole fraction (moles of gas i divided by the total moles in the gas phase) $\left(\frac{mol_{(g)\,i}}{mol_{(g)\,total}}\right)$	$\frac{L_l}{mol_{(g)\,total}}$
Mole fraction (moles of component i divided by the total moles in the solution) $\left(\frac{mol_{(aq)\,i}}{mol_{(aq)\,total}}\right)$	Mole fraction (moles of gas i divided by the total moles in the gas phase) $\left(\frac{mol_{(g)\,i}}{mol_{(g)\,total}}\right)$	$\frac{mol_{(aq)\,total}}{mol_{(g)\,total}}$

As shown in Table 5.1, Henry's law constant can be expressed without any dimensions, when the concentrations in the gas phase and the aqueous phase are expressed using the same units. This may in some cases lead to ambiguity in the use of constants, and therefore it should always be noted which units refer to the gas and which refer to the liquid.

Example 5.1: Converting Units

Henry's law constant for methane is $776 \frac{bar(CH_4)}{mol(CH_4)/L_l}$. Express Henry's law constant using the concentration units shown in Table 5.2.

Table 5.2: Concentrations in the gas phase and dissolved in the water.

	Concentration in the gas phase	Concentration dissolved in the water
a	$bar(CH_4)$	$mol(CH_4)/m^3_l$
b	$bar(CH_4)$	$mg(CH_4)/L_l$
c	$mg(CH_4)/L_g$	$mg(CH_4)/L_l$
d*	Mole fraction	Mole fraction

* Assume a molar density in the solution of 55.6 M and a total pressure of 1 bar.

Solution

a. $\dfrac{776 \frac{bar(CH_4)}{mol(CH_4)/L_l}}{1000 \frac{L_l}{m^3_l}} = 0.776 \dfrac{bar(CH_4)}{mol(CH_4)/m^3_l}$

b. $\dfrac{776 \frac{bar(CH_4)}{mol(CH_4)/L_l}}{16000 \frac{mg(CH_4)}{mol(CH_4)}} = 0.0485 \dfrac{bar(CH_4)}{mg(CH_4)/L_l}$

c. Partial pressure can be converted to concentration in mg/L by dividing by RT and multiplying by the molecular weight (see eq. (5.1)). Using the value obtained in part b, yields:

$$H\left(\frac{bar}{mg/L}\right) \frac{M_W\left(\frac{mg}{mol}\right)}{R\left(\frac{bar \cdot L}{mol \cdot K}\right) \cdot T(K)} = H\left(\frac{L}{L}\right)$$

Substituting the appropriate values yields:

$$\frac{\left(0.0485 \frac{bar(CH_4)}{mg(CH_4)/L_l}\right) \cdot \left(16000 \frac{mg(CH_4)}{mol(CH_4)}\right)}{\left(0.082 \frac{bar(CH_4) \cdot L_g}{mol(CH_4) \cdot K}\right) \cdot (293 \text{ K})} = 32.3 \frac{mg(CH_4)/L_g}{mg(CH_4)/L_l} = 32.3 \frac{L_l}{L_g}$$

d. Henry's law constant for methane, $776 \frac{bar(CH_4)}{mol(CH_4)/L_l}$, means that methane gas at a partial pressure of 1 bar, will be in equilibrium with $\frac{1}{776} mol(CH_4)/L_l$ (In other words, methane at a pressure of 776 bar in the gas phase will be in equilibrium with 1 mol of methane per liter in the solution). The mole fraction of methane in the solution is:

$$\frac{\frac{1}{776} mol(CH_4)/L_l}{55.6 \text{ total moles in solution}/L_l} = 2.3 \times 10^{-5} \frac{mol(CH_4)}{mol \text{ solution}}$$

Henry's law constant in the desired units would therefore be:

$$\frac{1 mol(CH_4)/mol \text{ gas}}{2.3 \times 10^{-5} \frac{mol(CH_4)}{mol \text{ solution}}} = 43,145 \frac{mol \text{ solution}}{mol \text{ gas}}$$

Note: The mole fraction of methane in the gas phase is 1.0 (mole methane per mole of gas) since the partial pressure of methane is given as 1 bar, and the total pressure is also 1 bar.

5.4 Factors affecting henry's law constant

Polarity of the gas. As described in Chapter 1, the solubility of substances in water is due to the polarity properties of the water and the electrostatic bonds between the solute and the water molecules. Bonds between charged molecules (ions) and water are therefore very strong and prevent the passage of ions to the gas phase. As a result, Henry's law constant for ions is close to zero. Of the same considerations, neutral molecules with a polar structure, such as NH_3, H_2S, CO_2, will be more soluble (lower Henry's law constant) than non-polar neutral molecules (CH_4, N_2, H_2, O_2), which dissolve in water due to dipole-London dispersion interactions.

Henry's constant values for several gases of environmental importance are shown in Table 5.3.

Table 5.3: Henry's constant value for several environmentally important gases at 20 °C. From Stumm and Morgan [4].

Compound	$H\left(\frac{bar \cdot L}{mol}\right)$	Compound	$H\left(\frac{bar \cdot L}{mol}\right)$
Nitrogen	1560	Hydrogen sulfide H_2S	9.8
Hydrogen	1260	Chloroform	4
Carbon monoxide	1050	Sulfur dioxide	0.81
Oxygen	730	Bromoform	0.7
Methane	776	Benzene	0.22
Ozone	107	Cyanide (HCN)	0.04
Carbon dioxide	30	Ammonia	0.017
Tetrachloroethylene	20	Acetic acid	0.0013
Trichloroethylene	11		

Temperature. Raising the temperature significantly increases the volatility and Henry's constant (i.e., decreases its solubility in the solution) of the gas. Adjusting Henry's law constant to a non-standard temperature (standard temperature being 25 °C) is carried out using the Van't Hoff equation, also mentioned in Chapter 1 (eq. 1.45).

The following website is a good and regularly updated source for Henry's law constants: http://www.henrys-law.org.

Example 5.2: Dissolution of O_2 in Water
a. Estimate the oxygen concentration in a water reservoir at 20 °C assuming that the aqueous phase is in equilibrium with the atmosphere.
b. What would be the oxygen concentration at a temperature of 10 °C provided that the standard enthalpy of volatilization ($\Delta H^o_{aq \to g}$) for oxygen is 14.9 kJ/mol?

Solution
a. The mole fraction of oxygen in the atmosphere is approximately 0.21. Assuming the atmospheric pressure at the water surface is exactly 1 bar, then the partial pressure of oxygen is

0.21 bar. The oxygen concentration in the water can be calculated directly from Henry's law (Henry's law constant is given in Table 5.3):

$$\left(O_{2(aq)}\right) = \frac{P_{O_2(g)}}{H_{O_2}} \cdot M_{W(O_2)} = \frac{0.21 bar}{730\left(\frac{bar \cdot L_l}{mol}\right)} \cdot 32,000 \frac{mg}{mol} = 9.2 \frac{mg}{L}$$

b.   Substituting the appropriate values into the Van't Hoff equation to obtain Henry's law constant for Oxygen at 10 °C:

$$\ln \frac{K_{eq(T_2)}}{K_{eq(T_1)}} = \frac{\Delta H^o_{aq \to g}}{R} \cdot \left(\frac{1}{T_1} - \frac{1}{T_2}\right)$$

$$\ln K_{eq(283)} = \ln(730) + \frac{14,900}{8.314} \cdot \left(\frac{1}{293} - \frac{1}{283}\right) = 6.377$$

$$K_{eq(283)} = 588 \left(\frac{bar \cdot L_l}{mol}\right)$$

Substituting to the equilibrium equation (Henry's law):

$$\left(O_{2(aq)}\right) = \frac{P_{O_2(g)}}{H_{O_2}} \cdot M_W(O_2) = \frac{0.21 bar}{588\left(\frac{bar \cdot L_l}{mol}\right)} \cdot 32,000 \frac{mg}{mol} = 11.4 \frac{mg}{L}$$

As expected, a decrease in temperature increases the oxygen solubility.

5.4.1 The effect of the solution Ionic strength on Henry's constant

As a rule, Henry's law constant (H) increases (solubility of the gas component in the water decreases and the gas becomes more volatile) as the salt concentration in the water increases. This can be seen in Fig. 5.1, which describes the oxygen

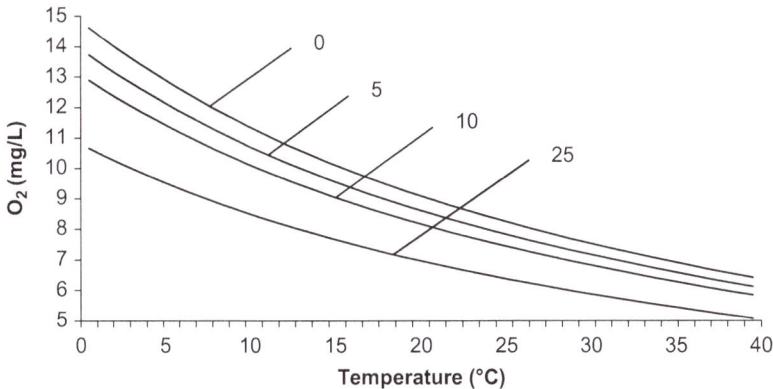

Fig. 5.1: Solubility of oxygen in water as a function of temperature and salinity. The values labeling the curves describe the chloride concentration in units of grams per kg (e.g., concentration in sea-water = 19.3 g/kg).

concentration in the water depending on salinity and temperature. This phenomenon is explained by the fact that the greater the concentration of salts, the fewer available water molecules to dissolve the gas.

5.5 Systems that are closed to the atmosphere

In Example 5.2, it was assumed that the oxygen concentration in the gas phase does not change as a result of reaching equilibrium with the water. In other words, it was assumed that the gas phase is infinite and therefore its composition does not change as a result of adding or removing a specific gas from a finite volume of solution. This is a reasonable assumption when the gas phase is the entire atmosphere of the earth.

When the volume of the gas phase is limited, or when high concentrations of specific gases are found in a limited area in the atmosphere (e.g., a plant emitting sulfur dioxide), we can estimate the change of concentrations in the gas phase using mass balances.

Example 5.3 One hundred ml of air are in contact with 1 L of a solution at 10 °C, in a closed container. The initial oxygen concentration in the water is zero. What will be the dissolved oxygen concentration in the water and its partial pressure in the gas phase inside the container at equilibrium?

Solution Since the system is closed, the partial pressure of oxygen in the air at equilibrium is not necessarily 0.21 atm. For this problem, we can perform a mass balance on the oxygen in both phases and solve using equilibrium equations. We begin by calculating the number of moles of oxygen in the system at time $t = 0$. This value remains constant throughout the process, due to conservation of mass and the assumption of a closed system.

$$n_{tot,\ O_2} = n_{(g),\ O_2,\ initial} + n_{(aq),\ O_2,\ initial} = n_{(g),\ O_2,\ initial} = \frac{P_{O_2(g)} V}{RT}$$

Substituting the appropriate values:

$$n_{tot,\ O2} = n_{gas,\ O_2,\ initial} = \frac{P_{O_2(g)} V}{RT} = \frac{0.21\,bar \cdot 0.1\,L}{[0.082\,(bar \cdot L)/(mol \cdot K)]283\,K} = 9.05 \cdot 10^{-4}\,mol O_2$$

At equilibrium, this quantity is divided between the two phases according to Henry's law.

$$n_{tot,\ O_2} = n_{(g),\ O_2\ eqilibrium} + n_{(aq),\ O_2\ eqilibrium} = \frac{P_{O_2(g)} V_{(g)}}{RT} + \frac{P_{O_2(g)}}{H_{O_2}} V_{solution}$$

$$9.05 \cdot 10^{-4}\,mol = \frac{P_{O_2(g)} \cdot (0.1\,L)}{\{0.082\,(bar \cdot L)/(mol \cdot K)\} \cdot (283\,K)} + \frac{P_{O_2(g)}}{730\,(bar \cdot L)/mol} \cdot 1.0\,L$$

$$\Rightarrow P_{O_2(g)} = 0.16\,bar \quad \Rightarrow \quad \{O_2(aq)\} = \frac{P_{O_2}}{H_{O_2}} \cdot M_{w\,(O_2)} = 7.0\,\frac{mg}{L}$$

5.6 The carbonate system in the context of gas-liquid phase equilibrium equations

The carbonate system discussed extensively in the previous chapters has of course a representative of great significance in the gas phase – carbon dioxide ($CO_{2(g)}$), which is in dynamic equilibrium with dissolved carbon dioxide ($CO_{2(aq)}$), which is the main component (99.7%) of the fictitious dissolved species $H_2CO_3^*$.

The following provides, as an example, the various effects on the carbonate system as a result of gas (carbon dioxide) transfer from the gas phase to the aqueous phase and vice versa. Similar reactions may occur with any gas that dissociates into ionized species in water, and the choice of CO_2 and the carbonate system for the example stems from their tremendous importance to the existence of life processes, environmental processes in general and processes in chemical/environmental engineering in particular.

The partial pressure of atmospheric $CO_{2(g)}$ is (in 2018) approximately 4×10^{-4} bar and can be assumed to be constant (in the current example and throughout this chapter the upward creep of carbon dioxide concentration in the atmosphere resulting from anthropogenic activities is ignored). Therefore, according to Henry's law, the $CO_{2(aq)}$ concentration in the water will also be constant if equilibrium exists between the phases. Henry's law constant for CO_2 is 30.2 bar·L/mol; it follows that the carbon dioxide concentration in water that is in equilibrium with the atmosphere at a temperature of 25 °C is:

$$\left[CO_{2(aq)}\right] = \frac{P_{CO_{2(g)}}}{H_{CO_2}} \cdot MW_{CO_2} = \frac{4.0 \cdot 10^{-4} \text{bar}}{30.2 \left(\frac{\text{bar·L}}{\text{mol}}\right)} \cdot 44,000 \frac{\text{mg}}{\text{mol}} = 0.583 \frac{\text{mg}}{\text{L}}$$

Therefore, water exposed to the atmosphere in which the carbon dioxide concentration is greater than 0.583 mg/L is defined as supersaturated with respect to $CO_{2(aq)}$, and CO_2 will tend to transfer to the gas phase in accordance with Henry's law. Similarly, water with a concentration of carbon dioxide lower than 0.583 mg/L that is exposed to the atmosphere will be undersaturated, and the CO_2 will tend to transfer to the aqueous phase from the gas phase.

As noted, these states of super and under saturation are possible because of the relatively slow kinetics of the reactions between the phases.

Since in water that is in equilibrium with the atmosphere the concentration of the carbon dioxide is as stated constant, in determining the dissolved species concentrations of the carbonate system, one degree of freedom is removed and we can fully characterize the carbonate system by means of one variable – for example, pH, alkalinity concentration (with respect to a specific reference species), acidity, etc. As mentioned in Chapter 2, the concentration of dissolved carbon dioxide ($CO_{2(aq)}$) is three orders of magnitude greater than the carbonic acid concentration ($H_2CO_{3(aq)}$) dissolved in the water, and it is customary to treat these two species as one with the symbol $H_2CO_3^*$, which is mostly composed of $CO_{2(aq)}$.

Therefore, henceforth we will refer only to the concentration of $H_2CO_3^*$ and assume that its concentration in equilibrium with the atmosphere is the same as the molar concentration of $CO_{2(aq)}$ in equilibrium, that is, $1.32 \cdot 10^{-5}$ M according to:

$$[CO_{2(aq)}] = 0.583 \frac{mg}{L} / 44,000 \frac{mg}{mol} = 1.32 \cdot 10^{-5}\,M = {\sim}10^{-4.88}\,M$$

If we plot $\log(H_2CO_3^*)$ on a graph of log(species) as a function of pH (where the word "species" represents the concentration of the various forms), we obtain a straight line with a value of −4.88. As mentioned above, the concentration of carbon dioxide depends only on Henry's law constant and the concentration of carbon dioxide in the atmosphere. Since these two factors are not dependent on pH, it follows that the concentration of carbon dioxide is not dependent on pH and its value will be $10^{-4.88}$ M for any pH value. Any deviation from this value will cause the water system to break out of equilibrium with the atmosphere, and stripping or dissolution of carbon dioxide will occur to restore the carbon dioxide concentration in the water to its previous state.

At the same time, the equilibrium between the forms of the carbonate system must exist simultaneously. Thus, in aqueous solutions exposed to the atmosphere, the concentrations of bicarbonate and carbonate can be calculated as a function of pH, based on the equilibrium equations of the carbonate system, Henry's law and carbon dioxide concentration in the atmosphere, as follows:

The first equilibrium reaction and its corresponding equation:

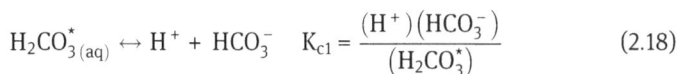

$$H_2CO_{3\,(aq)}^* \leftrightarrow H^+ + HCO_3^- \quad K_{c1} = \frac{(H^+)(HCO_3^-)}{(H_2CO_3^*)} \tag{2.18}$$

Solving this equation for the bicarbonate concentration and expressing the carbon dioxide concentration using Henry's law:

$$(HCO_3^-) = \frac{K_{c1}(H_2CO_3^*)}{(H^+)} = \frac{K_{c1}\frac{P_{CO_2(g)}}{H_{CO_2}}}{(H^+)} \tag{5.10}$$

Substituting in the constant values yields:

$$(HCO_3^-) = \frac{10^{-6.35} \cdot 10^{-4.88}}{(H^+)} = \frac{10^{-11.23}}{(H^+)}$$

$$\Rightarrow \log(HCO_3^-) = -11.23 - \log(H^+) = -11.23 + pH$$

Similarly, the second equilibrium reaction and its corresponding equation:

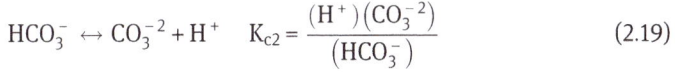

$$HCO_3^- \leftrightarrow CO_3^{-2} + H^+ \quad K_{c2} = \frac{(H^+)(CO_3^{-2})}{(HCO_3^-)} \tag{2.19}$$

$$(CO_3^{2-}) = \frac{K_{c2}(HCO_3^-)}{(H^+)} = \frac{K_{c1}K_{c2}(H_2CO_3^*)}{(H^+)^2} = \frac{K_{c1}K_{c2}\frac{P_{CO_2(g)}}{H_{CO_2}}}{(H^+)^2} \tag{5.11}$$

$$(CO_3^{2-}) = \frac{10^{-6.35} \cdot 10^{-10.33} \cdot 10^{-4.88}}{(H^+)^2} = \frac{10^{-21.56}}{(H^+)^2}$$

$$\Rightarrow \log(CO_3^{2-}) = -21.56 - 2\log(H^+) = -21.56 + 2pH$$

Figure 5.2 shows a graphic description of the distribution of species on a log (species) vs pH diagram for the case where the aqueous solution is in equilibrium with the atmosphere.

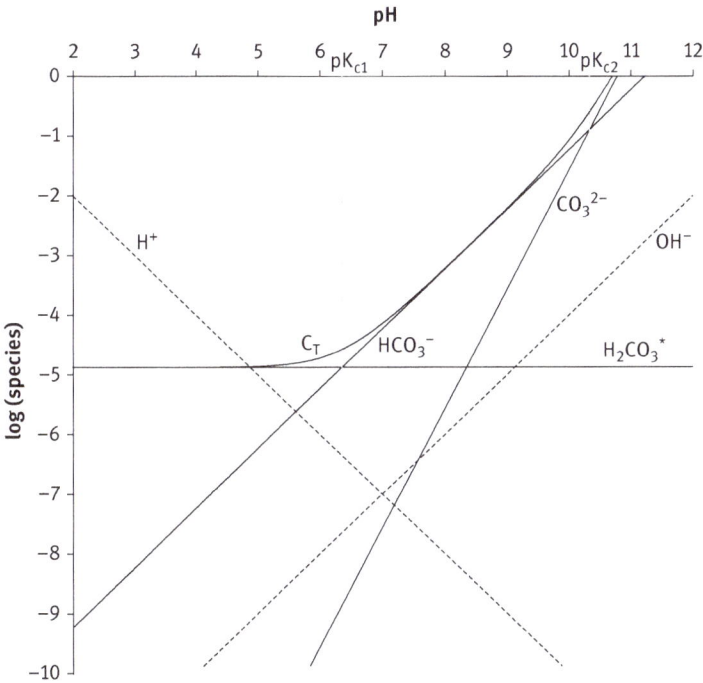

Fig. 5.2: log (species) vs. pH diagram for aqueous solution in equilibrium with atmospheric $CO_{2(g)}$. The carbonate system species are marked with solid lines, water species in dashed lines, and C_T with a thick line.

One of the conclusions from Fig. 5.2 is that when a system is in equilibrium with the atmosphere and at a pH value above pH5.5, the pH value has a great effect on the total carbonate system species concentration, C_T. This is in contrast to the systems discussed in the previous chapters, in which C_T was more or less constant.

The use of the alkalinity value ($H_2CO_3{}^*_{alk}$) is particularly convenient in solving problems of gas-liquid equilibrium. This is because alkalinity with respect to said reference species remains unchanged as a result of dissolution or stripping of $CO_{2(g)}$ as explained in Chapters 3 and 4.

As noted, all properties of the carbonate system can be defined by one variable, such as the pH, and thus the alkalinity ($H_2CO_3{}^*_{alk}$) of water in equilibrium with the atmosphere can be defined using the expressions derived in eqs. (5.10) and (5.11):

$$alk_{(H_2CO_3{}^*)} = 2\,[CO_3^{2-}] + [HCO_3^-] + [OH^-] - [H^+]$$

$$= \frac{2K'_1K'_2\frac{P_{CO_2(g)}}{H_{CO_2}}}{[H^+]^2} + \frac{K'_1\frac{P_{CO_2(g)}}{H_{CO_2}}}{[H^+]} + \frac{K'_w}{[H^+]} - [H^+] \qquad (5.12)$$

Use of this relationship is demonstrated in the following example.

Example 5.4 From a water sample analysis, we obtain:
Alkalinity ($H_2CO_3{}^*$) = 80 mg/L as $CaCO_3$, pH = 6.3

The sample was brought into equilibrium with the air. What will be the resulting pH? How much CO_2 was transferred and in which direction?

H = 30.2 bar/(mol/L), assume a negligible ionic strength and a temperature of 25 °C.

Solution Substitute the appropriate figures from the question into eq. (5.12) and solve for the partial pressure at which the solution is in equilibrium with the gas phase:

$$Alk = \frac{2K_1K_2\frac{P_{CO_2(g)}}{H_{CO_2}}}{[H^+]^2} + \frac{K_1\frac{P_{CO_2(g)}}{H_{CO_2}}}{[H^+]} + \frac{K_w}{[H^+]} - [H^+]$$

$$\frac{80}{50000} = \frac{2 \cdot 10^{-6.35} \cdot 10^{-10.33}\frac{P_{CO_2(g)}}{30.2}}{10^{-2 \cdot 6.3}} + \frac{10^{-6.35}\frac{P_{CO_2(g)}}{30.2}}{10^{-6.3}} + \frac{10^{-14}}{10^{-6.3}} - 10^{-6.3}$$

$$\Rightarrow P_{CO_2(g)} = 5.42 \cdot 10^{-2}\,bar$$

The partial pressure of carbon dioxide in the atmosphere is approximately $4 \cdot 10^{-4}$ bar, therefore the water is supersaturated with respect to carbon dioxide. In order to calculate the concentration of carbon dioxide emitted into the atmosphere, one must calculate the change in a conservative parameter as a result of the emission of carbon dioxide. The parameter chosen here to use is acidity. It is also possible to calculate the change in C_T, but this is slightly more complicated. Note that as explained in Chapter 4, emission of CO_2 has no affect on the alkalinity so this parameter cannot be used to calculate the change in the dissolved carbon dioxide concentration.

Using eq. (3.40), one can find that the acidity value in the original solution is Acd = $5.19 \cdot 10^{-3}$ eq/L. In order to calculate the acidity following the emission of CO_2, the pH when the solution is at equilibrium with the atmosphere must be first calculated.

The alkalinity did not change as a result of the CO_2 emission. One can therefore calculate the new pH using the partial pressure of carbon dioxide in the atmosphere, as follows:

$$Alk = \frac{2K_1K_2\frac{P_{CO_2(g)}}{H_{CO_2}}}{[H^+]^2} + \frac{K_1\frac{P_{CO_2(g)}}{H_{CO_2}}}{[H^+]} + \frac{K_w}{[H^+]} - [H^+]$$

$$\frac{80}{50000} = \frac{2\cdot10^{-6.35}\cdot10^{-10.33}\frac{0.0004}{30.2}}{[H^+]^2} + \frac{10^{-6.35}\frac{0.0004}{30.2}}{[H^+]} + \frac{10^{-14}}{[H^+]} - [H^+]$$

$$\Rightarrow pH = 8.39$$

Now, based on this pH and on the given alkalinity, we can find the acidity in equilibrium to be:

$$Acd\ CO_3{}^{2-} = 1.58\cdot10^{-3}\ eq/L.$$

Therefore:

$$\Delta Acd = 5.19 - 1.58 = 3.61\frac{meq}{L}$$

And the concentration of CO_2 volatized is equal (in equivalence units) to the change in acidity:

$$\Delta(CO_2) = \Delta Acd = 3.61\frac{meq}{L}\cdot22\frac{mg}{meq} = 79.42\frac{mg}{L}$$

Note, it is a common mistake to calculate the volatized CO_2 concentration based on the difference in CO_2 concentrations between the initial and final state (equilibrium state). This calculation is incorrect because the concentration of a single species is not a conserved parameter, as defined in Chapter 4. In other words, as CO_2 is emitted, the pH changes, and accordingly, the distribution of the carbonate system species is changed. As a result, it is incorrect to say that $H_2CO_3{}^*{}_{init} - H_2CO_3{}^*{}_{equilibrium} = H_2CO_3{}^*{}_{removed}$.

5.7 Distribution of species as a function of pH for systems that are in equilibrium with the gas phase

Working with acid-base systems which one of its species can transfer to the gas phase is very similar to working with regular acid-base systems. The difference lies in the fact that the total concentration of the acid-base system does not remain constant, and is in fact dependent on the concentration of the species in the gas phase and the pH. Henry's equation provides a solution to the problem.

Example 5.5 Find the pH and species distribution of the carbonate system in rain water (assume that rain water contains no compounds apart from CO_2 and H_2O). Assume equilibrium with atmospheric carbon dioxide.

Solution This example can be solved by writing a proton balance equation (see Chapter 2):

$$[H^+] = 2[CO_3{}^{2-}] + [HCO_3{}^-] + [OH^-]$$

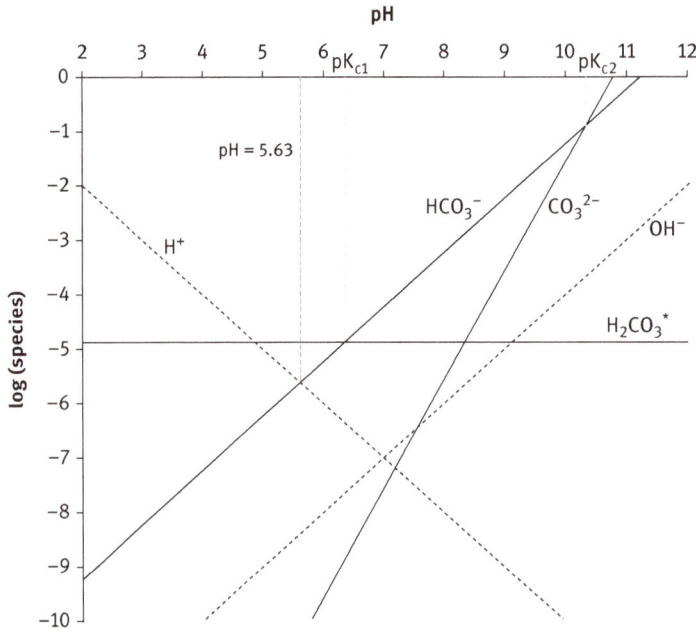

Fig. 5.3: pH-log(species) diagram for aqueous solution in equilibrium with atmospheric $CO_{2(g)}$. The thick dashed line indicates the approximate solution for the proton balance equation (pH5.63).

The solution of this equation on a pH-log(species) diagram in equilibrium with the atmosphere (Fig. 5.3) is found at the intersection point between the line representing the left-hand side of the equation and the line representing the right-hand side. The pH at this point is 5.63. The species concentrations of the carbonate system at this point are $10^{-4.88}$ M, $10^{-5.59}$ M, and $10^{-10.30}$ M for $H_2CO_3^*$, HCO_3^- and CO_3^{2-}, respectively. The concentrations of the species can be found using eqs. (5.10) and (5.11).

Another way to solve the problem is by considering the alkalinity value. Distilled water in equilibrium with carbon dioxide means dissolution of $H_2CO_3^*$ in distilled water, therefore, by definition, Alk = alk($H_2CO_3^*$) = 0. In addition, since we can assume an acidic pH, the concentrations of OH^- and CO_3^{2-} in the alkalinity equation are negligible and one obtains: Alk = $[HCO_3^-]$ – $[H^+]$ = 0. So, we have reached the same solution in which we found the intersection point between $[H^+]$ and $[HCO_3^-]$, or, in other words the equivalence point of $H_2CO_3^*$.

Finally, the problem can be solved analytically, by substituting Henry's law in the equation obtained (i.e., in $[HCO_3^-] = [H^+]$). This yields an equation with a single unknown, pH:

$$\frac{K_{c1}P_{CO_2(g)}}{(H^+)H_{CO_2}} = (H^+)$$

Example 5.6: Emission of $H_2S_{(g)}$ to the Gas Phase of Sewer Pipes
Background: Gravity wastewater pipes are designed to flow most of the time in an incomplete flow section, allowing the emission of hydrogen sulfide gas ($H_2S_{(g)}$) from the aqueous phase to the gas phase within the pipe. Dissolved hydrogen sulfide is often discharged into wastewater because of industrial activity, but in most cases it is formed (as $H_2S_{(aq)}$) in gravity and pressure pipes because of reduction of sulfate (SO_4^{2-}) under anaerobic conditions by heterotrophic bacteria (oxidants of organic compounds). The sulfide generated in the reaction (beyond its distribution to various forms of the weak-acid system according to the pH value) can undergo several reactions in the wastewater: (1) precipitation with metal ions such as divalent zinc ions to form ZnS or Cd^{2+} to form CdS; (2) Oxidation in the body of the wastewater by a metal ion (e.g., trivalent iron, oxidizing hydrogen sulfide to elementary sulfur) or by oxygen (normally in this case the end product is sulfate); and (3) Emission of the acidic species $H_2S_{(aq)}$ to the gas phase, where the emission rate is mainly a function of the flow conditions (the more turbulent the flow is, the greater the surface area of the water in contact with the gas phase and the emission is faster) and pH, which affects the distribution of species and therefore the concentration of the dissolved species ($H_2S_{(aq)}$) before volatilization, that is, the distance from aqueous-gas equilibrium.

$H_2S_{(g)}$ emitted from wastewater is a nuisance for three reasons: One is its unpleasant smell (rotten eggs smell), the second is that this gas is toxic to mammals even at very low concentrations of ppm in the gas phase ($H_2S_{(g)}$ has a high affinity to hemoglobin in the blood and can, at very low concentrations, cause suffocation to mammals in a very short period) and the third reason, which this question focuses on, is the damage caused to concrete lines as a result of the acidic conditions formed during the oxidation of sulfide by bacteria in the presence of dissolved oxygen, as explained below.

Figure 5.4 shows a schematic outline of a gravity sewage pipe. In the three areas marked on the diagram (the volume of the pipe at the meeting points between the water level and the gas phase and the crown of the pipe fed by condensate water plus debris), an aerobic bacterial population, composed also of sulfide oxidants, develops. These areas are exposed to the destruction of concrete or corrosion of iron as a result of acidic conditions resulting from oxidation of the sulfide. In these regions, there are bacterial populations that absorb the sulfide gas (which volatizes from the wastewater according to Henry's law and the kinetics that correspond to the flow conditions in the line). The bacteria oxidize the sulfide as described simplistically in the following equation:

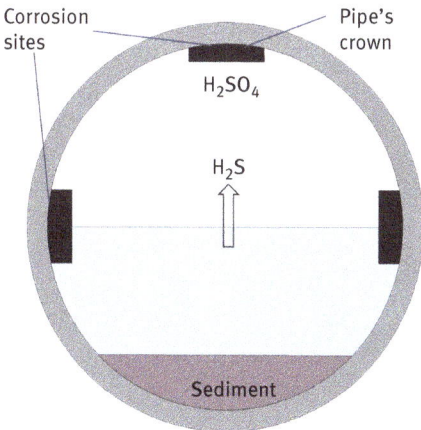

$$H_2S_{(aq)} + 2O_2 \xrightarrow{\text{Bacteria}} SO_4^{2-} + 2H^+$$

Corrosion sites

Pipe's crown

H_2SO_4

H_2S

Sediment

Fig. 5.4: Description of the areas in gravitational sewage pipes in which acidic conditions lead to corrosion (in iron pipes) or dissolution (in concrete pipes).

In the process of oxidation, protons are released into the water. In the raw wastewater, there is a relatively high buffer capacity and therefore the concern from formation of acidic conditions is minimal. In contrast, in the areas marked in Fig. 5.4, the buffer capacity is low because the contact with the wastewater is intermittent (the pipe walls) or nonexistent (the crown). The walls of the pipe are in contact with the wastewater as function on the flowrate of the sewage. During the day, there is usually a supply of wastewater and therefore a buffer capacity, but at night, when the sewage-level decreases, there is no supply of alkalinity, while the supply of sulfide and the process of its oxidation continues. In the pipe crown (the upper point in Fig. 5.4)), the situation is more problematic, since the supply of alkalinity is solely the result of a sewage spray. For this example, it is possible to assume that 0.4 thousandth (1/2500) of the flow rate of wastewater is sprayed to the pipe crown (i.e., 0.4 L for every cubic meter of wastewater flowing in the pipe).

In the following example, the pH value that can occur in the pipeline crown under routine operating conditions is calculated, as a result of the effects described above.

For the purposes of the example, let us assume that the wastewater flow is 1 cubic meter per second, the total concentration of the sulfide system species dissolved in the water (S_T) is 2 mg/L and the pH value of the wastewater is 7.5. In addition, it is given that due to limitations of kinetics, the sulfide mass emitted to the gas phase is only one-third of the maximum mass that can be ejected. In other words, equilibrium between the phases is not reached as described by Henry's law but only 33% of it undergoes the reaction $H_2S_{(aq)} \rightarrow H_2S_{(g)}$. We can also assume that 90% of the sulfide emitted to the gas phase is absorbed by the bacteria and oxidized to sulfate, and that for each cubic meter of wastewater only 0.4 L (as stated) are absorbed in the biofilm that covers the pipe crown.

Under these conditions, calculate the approximate pH value that will develop within the biofilm if the alkalinity of the wastewater is 400 mg/L as $CaCO_3$ (for simplicity, assume that all alkalinity is $H_2CO_3^*{}_{alk}$). Note that the alkalinity value is necessary in two ways: First, in order to assume that the buffer capacity of the wastewater is high enough thus there to be no significant change in pH as a result of the $H_2S_{(aq)}$ stripping. Second, it is the alkalinity provided to the pipe crown (half a liter per second) and it slightly reduces the pH drop in this area.

Neglect the effects of ionic strength and temperature.

Solution The stages of the solution are:

a. Calculate the concentration of the acidic (non-ionic) form of the sulfide system in the wastewater.
b. Calculate the sulfide flux into the air, assuming aqueous-gas equilibrium.
c. Calculate the sulfide flux into the air, taking into account the kinetics and calculation of the sulfide flux to the biofilm.
d. Calculate the alkalinity value in the pipe's crown.
e. Calculate C_T in the pipe's crown.
f. Find the pH in the crown.

First calculate the concentration of the acidic (non-ionic) species of the sulfide system in the wastewater. For this purpose, let us substitute S_T ($S_T = 2/34{,}000 = 5.88 \cdot 10^{-5}$ M) and the pH (pH = 7.5) given for the wastewater, in the relevant equation from Table 2.3.

This calculation yields: $[H_2S] = 1.41 \cdot 10^{-5}$ M.

In the next stage, the following equations are used to calculate the sulfide gas flux that would have been obtained assuming that there was equilibrium between the phases:

$$H_2S_{(g)\,equilibrium} = [H_2S]_{(aq)} \cdot H_{H_2S}$$

$$H_2S_{(g)\,equilibrium\,flux} = flux_{sewage} \cdot H_2S_{(g)\,equilibrium}$$

To calculate the actual gas flow (according to the question's data), this flux should be multiplied by one-third. In addition, unit conversion should be performed by dividing by RT (see Example 5.1).

$$H_2S_{(g)\,actual\,flux} = 1\frac{m^3}{s} \cdot 1.41 \cdot 10^{-2}\frac{mol}{m^3} \cdot 9.8\frac{bar}{mol} \cdot \frac{1}{0.082\frac{bar\cdot L}{mol\cdot K} \cdot 293K} \cdot \frac{1}{3} = 1.92\frac{mmol}{s}$$

The flow of the sulfide to the biofilm is 90% of this flow, namely:

$$H_2S_{biofilm\,flux} = 1.92\frac{mmol}{s} \cdot 0.9 = 1.73\frac{mmol}{s}$$

According to the oxidation equation of sulfide to sulfate, it can be seen that for each mole of sulfide oxidized, two equivalents of protons are emitted. Therefore, the proton flow to the wastewater in the pipe crown is 3.46 meq/s and this is also the flow in which the alkalinity value decreases. The spray flux to the pipe crown is 5.0 L per second. This spray provides alkalinity at a concentration of 400/50000 = 8 meq/L. Based on these data, we can calculate the alkalinity obtained in the water adjacent to the biofilm in the vicinity of the pipe crown:

$$Alk_{new} = Alk_{spray} - \frac{[H^+]}{flux_{spray}} = 8\frac{meq}{L} - \frac{3.46meq/s}{0.4L/s} = -0.65\frac{meq}{L}$$

The pH at the pipe crown can be extracted from an explicit equation of alkalinity defined by H_2S and $H_2CO_3^*$. To do this, we must first calculate the C_T and S_T values in the pipe crown, which are equal to their value in the wastewater itself. C_T is calculated from the alkalinity, pH and S_T in the wastewater, from the alkalinity equation:

$$Alk\,(H_2S,\,H_2CO_3^*) = 2[S^{2-}] + [HS^-] + 2[CO_3^{2-}] + [HCO_3^-] + [OH^-] - [H^+]$$

Now in place of the species concentrations let us place expressions as a function of pH, S_T and C_T (from Table 2.3). We obtain an equation in which C_T is the only unknown. From its solution, we obtain: $C_T = 8.5$ mM.

Now let us write a similar equation for the wastewater at the crown, in which we place: Alk = −0.65 meq/L, C_T = 8.5 mM, S_T = 5.88·10^{-5} M and solve for pH. We obtain: pH = 2.84, which is, of course, a pH that causes rapid dissolution of the cement, which eventually leads to the destruction of the pipe.

Example 5.7: Partial Pressure and Carbon Dioxide Concentration in Carbonated Beverages
For the preparation of certain carbonated beverages, it is common to dose distilled water with a mixture of CO_2 at a high concentration and phosphoric acid (H_3PO_4) to create a strong acidic environment (pH<3.0). Note – in this question, neglect effects of ionic strength.
a. Assuming the partial pressure of $CO_{2(g)}$ in the gas phase in the bottle is 2.5 bar (without phosphoric acid), what is the pH value in the solution (assuming a temperature of 25 °C)?
b. How much phosphoric acid (in meq/L units in the drink solution) should be added to the solution in which the concentration of the CO_2 is as calculated in section a, in order to stabilize the pH at pH2.8?
c. If we wanted to reduce the pH to pH2.8 with $CO_{2(g)}$ alone, what is the required partial pressure? Explain in terms of buffer capacity why this approach does not make sense.

d. If it is necessary to reduce the pH to pH2.8 using hydrochloric acid (HCl) instead of phosphoric acid, what is the concentration of HCl required (in meq/L units in the beverage solution)? Why is it different from the H_3PO_4 concentration that was calculated in section b?

e. When a bottle of a carbonated beverage (pH2.8) is opened, CO_2 is released into the atmosphere. What will be the pH value of the drink after the CO_2 has reached equilibrium with the atmosphere if pH has been lowered to 2.8 by a strong acid (HCl) and a weak acid (H_3PO_4)? It can be assumed that H_3PO_4 is not released from the beverage (assume a temperature of 25 °C).

Solution

a. Since the $CO_{2(aq)}$ is dosed to distilled water, the alkalinity concentration ($H_2CO_3{}^*$alk) in the water is by definition zero. The equation that relates alkalinity to pH can therefore be used from the knowledge of the partial pressure of $CO_{2(g)}$.

$$Alk = 0 = \frac{2K_1K_2\text{H}_2\text{CO}_2{}^{\frac{2.5}{30.2}}}{[\text{H}^+]^2} + \frac{K_1\text{H}_2\text{CO}_2{}^{\frac{2.5}{30.2}}}{[\text{H}^+]} + \frac{K_w}{[\text{H}^+]} - [\text{H}^+] = \frac{2 \cdot 10^{-6.35} \cdot 10^{-10.40} \frac{2.5}{30.2}}{10^{-2 \cdot pH}} + \frac{10^{-6.35} \frac{2.5}{30.2}}{10^{-pH}}$$
$$+ \frac{10^{-14}}{10^{-pH}} - 10^{-pH}$$

⇓

pH = 3.72

b. This section can be solved by assigning an alkalinity equation with H2CO3 and H3PO4 as reference species:

$$Alk\,(H_3PO_4, H_2CO_3{}^*) = 3[PO_4{}^{3-}] + 2[HPO_4{}^{2-}] + [H_2PO_4{}^-] + 2[CO_3{}^{2-}] + [HCO_3{}^-] + [OH^-] - [H^+]$$

In this equation, an explicit expression can be substituted for each of the forms of the phosphate system as a function of pH, constants and P_T. For the carbonate system, the species concentrations can be substituted as a function of the partial pressure of $CO_{2(g)}$ as done in section a. Since the addition of H_3PO_4 does not change the alkalinity value (which remains therefore 0), from the solution of the equation for pH2.8, P_T can be extracted, which is the concentration of phosphoric acid required in the water.
Solution of the section: $P_T = 0.026$ M

c. Using the alkalinity equation in Section a where the pH is known (pH2.8) and the partial pressure is the only unknown, it is possible to calculate that an extremely high partial pressure of 170 atmospheres of $CO_{2(g)}$ is required. The reason for the high (and impractical) $CO_{2(g)}$ pressure required is that at low pH values the distance from the pK becomes significant and therefore the concentration of the basic species is very low. In other words, at pH<5, the dissociation of $H_2CO_3{}^*$ to $HCO_3{}^-$ is negligible and therefore raising the dissolved CO_2 concentration in the water will result in the release of a very low concentration of protons, and the pH will hardly decrease.

d. When HCl (a strong acid) is dosed to water the alkalinity decreases and the acidity increases by the value of the acid which was dosed (in meq/L units). It is therefore possible to solve for the new alkalinity value (for example) obtained after lowering the pH to 2.8 by the strong acid (negative alkalinity value). Since the alkalinity value before adding the acid was zero, the difference would be the concentration of the strong acid to be added. Computationally, the following equation can be used:

$$Alk_{H_2CO_3^*} = \frac{2K_1K_2\bar{H}_{CO_2}^{2.5}}{[10^{-2.8}]^2} + \frac{K_1\bar{H}_{CO_2}^{2.5}}{[10^{-2.8}]} + \frac{K_w}{[10^{-2.8}]} - [10^{-2.8}] =$$

$$\frac{2\cdot10^{-6.35}\cdot10^{-10.40}\frac{2.5}{30.2}}{10^{-5.6}} + \frac{10^{-6.35}\frac{2.5}{30.2}}{10^{-2.8}} + \frac{10^{-14}}{10^{-2.8}} - 10^{-2.8} = -0.00156 \text{ eq/L}$$

Therefore, the concentration of strong acid to be added to reduce the pH to pH2.8 is $1.56\cdot10^{-3}$ eq/L.

The result shows that to obtain the same pH, one should dose less HCl than H_3PO_4 (see section b). This difference stem from the fact that HCl is a strong acid that is fully ionized in water while H_3PO_4 is a weak acid.

e. Since the alkalinity value (with $H_2CO_3^*$ as a reference species) does not change when CO_2 is emitted from the water, the alkalinity values obtained in section b (alkalinity = 0) and d (alkalinity = −0.0156 meq/L) can be used. From knowing the partial pressure of CO_2 in the atmosphere (0.00037 atm was assumed in this question), the pH in equilibrium with the atmosphere can be calculated in both cases from the appropriate alkalinity equations for each case.

Because at the given pH, the dominant component of the carbonate system is $H_2CO_3^*$, CO_2 volatilization will not be expressed in the absorption of protons in the solution and therefore its effect on the pH will be negligible (as long as the pH does not exceed ~4.5). In terms of buffer capacity, in the given pH, the buffer capacity of the water is relatively high and in the case H_3PO_4 is used there is also a high buffer capacity of the phosphoric system (pK$_1$ = 2.16). Therefore, from this aspect too one should not expect to observe a change in the pH of the beverage. Therefore, the result is not surprising, in which after the CO_2 has reached equilibrium with the atmosphere, the pH of the solution will rise only to 2.805 and 2.807 for the case where the pH is lowered to 2.80 by a weak acid (H_3PO_4) and a strong acid (HCl), respectively.

Test yourself: How much will the answers change if the temperature at which the beverage is prepared is 4 °C?

6 Principles of equilibrium between the aqueous and solid phases with emphasis on precipitation and dissolution of $CaCO_{3(s)}$

6.1 Introduction

$CaCO_3$ has special significance in the context of water treatment processes, especially with regard to hard water softening processes on the one hand and the stabilization of soft (and corrosive) water on the other. $CaCO_{3(s)}$ is also in many instances the main chemical fouling membranes and heat exchange surfaces in a variety of industrial applications. The present chapter deals with the precipitation and dissolution of $CaCO_{3(s)}$ as part of these processes, but to maintain the generality of the discussion, general equilibrium equations between the aqueous phase and the solid phase and the main factors influencing it are first defined. While the discussion later in the chapter focuses on $CaCO_{3(s)}$, the equilibrium concepts associated with the precipitation/dissolution of other solids are similar in principle.

To evaluate the state of the solution relative to a particular solid, i.e. to assess whether a solid has the potential to precipitate or dissolve, one must compare the quotient of the activity of the ions that comprise the solid, with its solubility constant. The ion activity product comprising a slightly soluble solid exceeds the value of the solubility constant of the solid (K solubility product is marked K_{sp}), in principle the solid will precipitate. The time needed to reach equilibrium between the aqueous phase and the solid phase is (often) very long, in contrast to the reactions in the aqueous phase where the time to reach equilibrium is very short. Thus, the use of K_{sp} allows only for a qualitative evaluation of the ability of the solution to dissolve or precipitate a given solid. This potential will be realized or not realized, among other things, depending on the time given for the reaction to occur. In addition, if dissolution potential exists, its realization also depends on the presence of the solid and the contact between it and the solution with the dissolving potential. Of course, if the solution with the positive dissolution potential does not come in contact with the solid, dissolution will not take place. In summary, for practical purposes, K_{sp} enables a qualitative assessment of the state of the solution relative to the solid (i.e. the answer to the question of whether there is precipitation or dissolution potential), and depending on the additional parameters presented further in this chapter, it also allows for quantitative calculation of the precipitation potential (or dissolution potential) of a solid in a given solution.

https://doi.org/10.1515/9783110603958-006

Example: Given is the equilibrium equation of gypsum (according to the convention, the solid is always present on the left-hand side of the equation and so the solubility constant K_{sp} is defined):

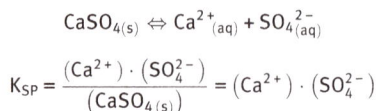

$$CaSO_{4(s)} \Leftrightarrow Ca^{2+}_{(aq)} + SO_{4(aq)}^{2-}$$

$$K_{SP} = \frac{(Ca^{2+}) \cdot (SO_4^{2-})}{(CaSO_{4\,(s)})} = (Ca^{2+}) \cdot (SO_4^{2-})$$

The activity of a solid is defined as unity. Therefore, the solubility constant is equal to the product of the solute activities. Therefore, in order for precipitation to occur, the ion activity product should be larger than the equilibrium constant or alternatively, the ion concentration product should be larger than the apparent equilibrium constant (K'_{sp}). Such a situation is defined as "supersaturation" of the solution relative to the solid. The opposite situation (the activity product is lower than K_{sp}) is defined as "undersaturation" of the solution relative to the solid, as summarized in the following table (Table 6.1).

Table 6.1: Defining the saturation state by comparing the solute product to the solubility constant.

Based on a thermodynamic constant	Based on an apparent constant	Saturation state
Product of solute activities > K_{sp}	Product of solute activities > K'_{sp}	Supersaturation
Product of solute activities < K_{sp}	Product of solute activities < K'_{sp}	Undersaturation
Product of solute activities = K_{sp}	Product of solute activities = K'_{sp}	Saturation

General representation of the solubility product constant (K_{sp}):

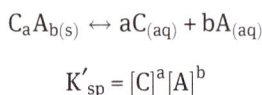

$$C_aA_{b(s)} \leftrightarrow aC_{(aq)} + bA_{(aq)}$$

$$K'_{sp} = [C]^a[A]^b$$

Note: $C_aA_{b(s)}$ added to water when the solution is saturated or supersaturated with respect to the same solid will not change the solute concentration.

It is common to assume that for most solids, precipitation in clear solutions is not immediate, but consists of two stages:

a. Nucleation (formation of colloidal sediment particles).
b. Expansion of the particles as a result of the kinetic preference for further precipitation over the already formed solids.

In addition, for solutions that are at low supersaturation, that is, solutions that are close to equilibrium with the solid phase, the initial rate of solid formation can be quite slow and can even last for several weeks (e.g., for precipitation of $CaCO_{3(s)}$).

However, it should be understood that the precipitation rate is unique to each solid and the conditions of the solution in which it precipitates (e.g., presence of nucleation seeds which accelerate precipitation, presence of precipitation inhibitors or catalysts, etc.) and any generalization in this context is problematic. In general, it is known that the rate of precipitation depends both on the distance from equilibrium and on physical and chemical conditions, such as the degree of mixing and the rate of nucleation.

6.2 The effect of ionic strength on the solubility constants

Realistic activity coefficients (i.e. smaller than unity) increase the "solubility" of the solid, i.e. the product of the active ion concentrations required for precipitation to occur.

$$CA_{(s)} \leftrightarrow C_{(aq)} + A_{(aq)}$$

$$K_{SP} = (C)(A) = \gamma[C]\gamma[A] = \gamma^2 K'_{SP}$$

$$K'_{sp} = K_{sp}/\gamma^2$$

Since γ^2 is smaller than unity, K'_{sp} is larger than K_{sp}, which means that the solid is more soluble.

Example 6.1

a. Calculate the solubility constant of gypsum in water with salinity (TDS) of 400 and 3500 mg/L.

b. What will be the calcium and sulfate concentration in this water in equilibrium with gypsum (assuming the concentrations of the two ions at the beginning of the dissolution are zero and that the temperature is 25 °C).

Solution

$$CaSO_{4(s)} \leftrightarrow Ca^{2+}_{(aq)} + SO_4^{2-}_{(aq)} \qquad K_{sp}(25\,°C, I \rightarrow 0) = 10^{-4.85}$$

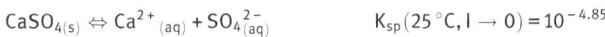

a. Calculation of ionic strength in both solutions according to eq. (1.19):

$$I = 2.5 \cdot 10^{-5}(TDS - 20) = 2.5 \cdot 10^{-5}(400 - 20) = 0.0095\ M$$

$$I = 2.5 \cdot 10^{-5}(TDS - 20) = 2.5 \cdot 10^{-5}(3500 - 20) = 0.087\ M$$

Finding the activity coefficients (diprotic) in each of the solutions from eq. (1.21):

$$\log \gamma_d(TDS = 400mg/L) = -1.82 \cdot 10^6 (78.3 \cdot 298)^{-3/2}(2)^2 \left(\frac{\sqrt{0.0095}}{1+\sqrt{0.0095}} - 0.2 \cdot 0.0095\right) \Rightarrow \gamma_d = 0.66$$

$$\log \gamma_d(TDS = 3500mg/L) = -1.82 \cdot 10^6 (78.3 \cdot 298)^{-3/2}(2)^2 \left(\frac{\sqrt{0.087}}{1+\sqrt{0.087}} - 0.2 \cdot 0.087\right) \Rightarrow \gamma_d = 0.37$$

$K'_{sp}(TDS = 400mg/L) = K_{sp}/(\gamma_d)^2 = 10^{-4.85}/(0.66^2) = 10^{-4.49}$ i.e., 2.29 times higher than K_{sp}

$K'_{sp}(TDS = 3500mg/L) = K_{sp}/(\gamma_d)^2 = 10^{-4.85}/(0.37^2) = 10^{-3.39}$ i.e., 7.28 times higher than K_{sp}

b. Solution with TDS = 400 mg/L

The atomic weight of calcium is 40.08 g/mol and the molecular weight of SO_4^{2-} is 96.06 g/mol.

$$K'_{sp} = 3.34 \cdot 10^{-5} = [Ca^{2+}] \cdot [SO_4^{2-}] = x^2$$

$$\rightarrow x = 0.00577\,M = 231\,mg\ Ca^{2+}/L \text{ and } 520\,mg\ SO_4^{2-}/L$$

Solution with TDS = 3500 mg/L

$$K'_{sp} = 1.37 \cdot 10^{-4} = [Ca^{2+}] \cdot [SO_4^{2-}] = x^2$$

$$\rightarrow x = 0.0117\,M = 469\,mg\ Ca^{2+}/L \text{ and } 1123\,mg\ SO_4^{2-}/L$$

It can be seen that the change in ionic strength has a very significant effect on the solubility constant of gypsum and therefore also on equilibrium concentrations. More generally, since most of the solids encountered by environmental and chemical engineers are composed of two and three-valence ions, the values of the activity coefficients of these ions (i.e., γ_d and γ_t) are most sensitive to ionic strength (note that in this case the valence values are raised to the power of two in the Davies equation (eq. (1.21))).

6.3 Effect of temperature on the solubility constant

The effect of the temperature on solid solubility constants is identical to its effect on "regular" equilibrium constants in the aqueous phase. To calculate the effect, the integrative solution of the Van't Hoff equation described in Chapter 1 can be used (eq. (1.46)). Note that the solubility of certain solids (sugars, for example) increases with temperature and the solubility of others decreases with temperature (e.g., gypsum, $CaSO_4$). For calculating the change in the solubility constant of gypsum with temperature, see Example (1.11).

6.4 Effect of the addition of one of the solid components on the concentration of the other component in equilibrium (common ion effect)

Example 6.2 Calculate the concentration of calcium ions in distilled water that were brought to equilibrium with $CaCO_{3(s)}$ and compare this value with the calcium concentration after adding 1 g/L of Na_2CO_3 (strong electrolyte). Temperature = 25 °C, neglect effects of ionic strength.

Solution

Step 1: The concentration of calcium ions before adding Na_2CO_3:

$$CaCO_{3(s)} \leftrightarrow Ca^{2+}_{(aq)} + CO_{3\,(aq)}^{2-}$$

$$K_{sp} = [Ca^{2+}][CO_3^{2-}] = 8.9 \cdot 10^{-9} = X^2$$

$$X = [Ca^{2+}] = 0.94 \cdot 10^{-4}\,M$$

$$0.94 \cdot 10^{-4}\,mol/_L \cdot 40,080\,mg/_{mol} = 3.78\,mg/_L\,Ca^{2+}$$

Step 2: After adding sodium carbonate:
For sake of simplicity, in this example, the distribution of the carbonate system species will be neglected. In other words, the effect of the reactions in which HCO_3^- and $H_2CO_3^*$ are formed from the additional CO_3^{2-} concentration, are neglected.
  The charge balance equation is thus as follows:

$$[Na^+] + 2[Ca^{2+}] = 2[CO_3^{2-}]$$

Molar concentration of Na_2CO_3 dosed to the solution:

$$\frac{1gNa_2CO_3/_L}{106gNa_2CO_3/_{mol}} = 9.43 \cdot 10^{-3}mol/_L$$

The concentration of sodium ions after Na_2CO_3 has completely dissolved is twice the salt concentration:

$$[Na^+] = 2(9.43 \cdot 10^{-3}) = 0.0188 \text{ M}$$

Now the charge balance equation is multiplied by $[Ca^{2+}]$ on both sides and divided by two:

$$[CO_3^{2-}][Ca^{2+}] = \frac{[Ca^{2+}][Na^+] + 2[Ca^{2+}]^2}{2}$$

In equilibrium, the left-hand side of the equation obtained is equal to K'_{sp} of $CaCO_{3(s)}$ which is $10^{-8.48}$ (for calcite) and since the sodium concentration is known it can be directly solved for $[Ca^{2+}]$ (neglecting ionic strength effects).
  The concentration of dissolved calcium in equilibrium under these conditions is 0.014 mg/L.

Example 6.3 The solubility constant of PbI_2 is $K_{sp} = 7.9 \cdot 10^{-9}$. What will be the solubility of lead in distilled water and what will be its solubility after the addition of 0.08 molar of the strong electrolyte NaI?
  Equations:

$$PbI_{2(s)} \Leftrightarrow Pb^{2+}_{(aq)} + 2I^-_{(aq)}$$

$$NaI_{(s)} \rightarrow Na^+_{(aq)} + I^-_{(aq)}$$

Solution Denote the concentration of lead in equilibrium by x and write for distilled water alone:

$$K_{sp} = 7.9 \cdot 10^{-9} = x \cdot (2x)^2 \Rightarrow x = [Pb^{2+}] = 0.00125 \text{ M} = 260.0 mg/L$$

Now mark the concentration of lead in equilibrium after adding NaI at a concentration of 0.08 M by the letter y.

$$K_{sp} = 7.9.10^{-9} = y \cdot [0.08 + 2y]^2$$

The concentration of lead dissolved in water under these conditions is $1.23 \cdot 10^{-6}$ M, that is, only 0.256 mg/L (The atomic weight of lead is 207.2 g/mol).
  The phenomenon described in Examples 6.2 and 6.3 is called "common ion effect". In general, the phenomenon presented in these examples describes a situation in which a certain component of the solid is added to the water. As a result, the solubility of the solid decreases. In contrast, the common ion effect also describes the situation in which an ion comprising a solid participates in the water in reactions with other constituents, resulting in its concentration decreasing, and the solubility of the solid increases. The phenomenon is further described in Section 6.5.

6.5 Effect of side reactions on solubility of solids

Any reaction involving the product/s of a solid precipitation equation reduces the activity of the ion/s that appear/s in the solute product and therefore increases solid solubility.

The following are examples of such reactions:

1. Solubility of solids where one of their ions (or more) are species of any weak acid is affected by pH. For example, reactions involving carbonates (CO_3^{2-}), phosphates (PO_4^{3-}) or sulfides (S^{2-}).

   For example:

   $$CaCO_{3(s)} \Leftrightarrow Ca^{2+}_{(aq)} + CO^{2-}_{3\,(aq)}$$

   $$+$$

   $$H^{+}_{(aq)}$$

   $$\Updownarrow$$

   $$HCO^{-}_{3\,(aq)}$$

   $$+$$

   $$H^{+}_{(aq)}$$

   $$\Updownarrow$$

   $$H_2CO^{*}_{3\,(aq)}$$

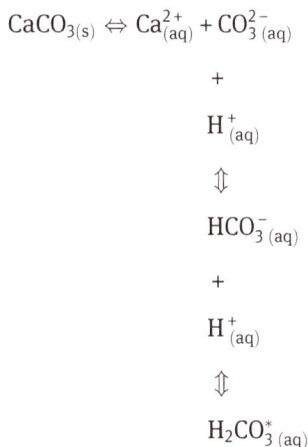

2. Reactions involving ion pairing (ion complexation).

   Complexes are soluble species composed of a metal with a positive charge and a ligand with a negative charge. Metals appear in the aqueous phase as hydrated ions (i.e., the ion complex + water molecules alone) or as a complex with another ligand.

   To be able to predict the concentration of the metal cation participating in the precipitation reaction, the speciation of the metal must be performed, and the free ion concentration and the concentration of its different complexes must be determined. In water with high concentrations of ions (such as seawater) or those with high concentrations of organic matter capable of being used as ligands for the formation of complexes with metallic cations, complexation has great significance in the context of solid solubility, which cannot be neglected.

   Quantification of complex equations in the context of solid precipitation is not covered by the current text.

6.6 Precipitation/dissolution of CaCO₃: qualitative and quantitative assessment of the saturation state

6.6.1 Langelier saturation index (LSI)

A qualitative assessment of the saturation state comes to answer the question "Is the precipitation or dissolution of a given solid in a given solution expected?"

The simplest method still commonly used to determine qualitatively the saturation state relative to $CaCO_3$ was formulated by Langelier in 1936 [7] and is known as the Langelier saturation index (LSI). Since its formulation, additional techniques have been defined on its basis in attempt to improve it [8–11] and parallel methods have been proposed, the main ones being the Driving force index [12], the Momentary excess [10] and the Aggressiveness Index [13]. But the basic method is still the most accepted. The Langelier method is accepted as one of the parameters for determining the quality of water at the exit of desalination plants, where it is often formulated quantitatively, which is theoretically not a viable approach (see below).

Langelier defined a fictitious parameter termed pH_L. The goal of the mathematical development is to calculate pH_L and then to subtract it from the true pH measured in water. Defining the saturation state is determined by:

$$LSI = pH - pH_L$$

where If LSI < 0 water is undersaturated with respect to $CaCO_{3(s)}$, If LSI = 0 water is saturated with respect to $CaCO_{3(s)}$ and If LSI > 0 water is supersaturated with respect to $CaCO_{3(s)}$.

6.6.1.1 Mathematical development of the formula for calculating pH_L

Langelier limited the method to be used for "natural" water, that is, drinking water from groundwater or surface water in which he assumed that the pH range is 5.0 < pH < 9.5.

In this pH range, it can be assumed that Alkalinity($H_2CO_3^*$) ≅ $[HCO_3^-]$.

Assuming that natural water is controlled by the carbonate system and that the concentration of bicarbonate is replaced by the concentration of alkalinity:

$$K'_{C2} = \frac{(H^+)[CO_3^{2-}]}{[HCO_3^-]} = \frac{(H^+)[CO_3^{2-}]}{Alk}$$

Since in the saturated state: $[Ca^{2+}] [CO_3^{2-}] = K'_{sp}$ the carbonate ion concentration can be replaced with $K'_{sp}/[Ca^{2+}]$ and it is obtained:

$$K'_{C2} = \frac{(H^+)_L}{Alk_{measured}} \cdot \frac{K_{sp}}{[Ca^{2+}]_{measured}}$$

where (H_L) represents the activity of the protons in equilibrium.

Operating a log on both sides, the equation for simple calculation of pH_L is obtained:

$$\log(\acute{K}_{C2}) = -pH_L - p\acute{K}_{sp} - \log(Alk) - \log([Ca^{2+}])$$

$$\Downarrow$$

$$\mathbf{pH_L = p\acute{K}_{C2} - p\acute{K}_{sp} - \log(Alk) - \log([Ca^{2+}])}$$

Therefore, it is only necessary to determine the concentration of alkalinity and calcium (in addition to the temperature and EC required to convert the constants) to calculate the pH_L and from it to determine whether precipitation or dissolution of $CaCO_3$ is expected.

Note pH_L has the following significance: pH_L is the pH value of a solution with alkalinity and a given $[Ca^{2+}]$, when this solution is in solid-liquid equilibrium relative to $CaCO_3$. Therefore, if the pH measured in the solution is equal to pH_L, then the solution is exactly saturated with respect to $CaCO_3$. Given that the dissolution of $CaCO_3$ in natural water causes an increase in pH, it can be concluded that if the pH measured in the given solution is lower than pH_L, the solution is undersaturated with $CaCO_{3(s)}$. Similarly, if $pH_L < pH$, it means that the solution is supersaturated with respect to $CaCO_{3(s)}$.

Before the computer age, the Langelier method was a very convenient procedure for assessing the saturation state. It should be noted that in the manner in which the Langelier method is formulated, there is no significance to the exact result of subtracting the pH_L value from the measured pH, except for the sign obtained from this subtraction (i.e., a negative result means undersaturation and vice versa). The only quantitative significance is if it is obtained LSI = 0 where the system is actually saturated (this is also the only case where the saturation assumption underlying the development of the pH_L equation is correct).

6.6.1.2 The inherent problems in the Langelier method

- The method does not provide information about the distance from saturation but only the direction (there is no quantitative information on the amount of material that would precipitate or dissolve). It follows that this method should not be used to obtain a quantitative indication. That is, it makes no sense to define quality standards based on LSI as a quantitative parameter but only as a response to the question of whether precipitation or dissolution of $CaCO_{3(s)}$ is expected. In addition, as explained at the beginning of the chapter, the distance from equilibrium has an effect on the rate of precipitation/dissolution. Therefore, this distance is of practical importance to determine whether precipitation/dissolution will occur. This is another reason why using LSI as a parameter for water quality is problematic. The perception that it is not possible to use the parameter of pH differences as a quantitative parameter, also stems from the material studied in previous chapters, in which it was emphasized many times that in order to characterize natural water (controlled by the carbonate system) two independent parameters must be known (i.e., alkalinity and pH or any other pair), and in any case, it is clear that water quality cannot be characterized by only one parameter (pH in this case).

– The computation of the Langelier method is based on the alkalinity and calcium values present in the water prior to precipitation/dissolution of $CaCO_3$. On the other hand, the equation was developed under the assumption that the water reaches saturation relative to $CaCO_{3(s)}$. When the water reaches saturation, the concentrations of calcium and alkalinity will necessarily be different from those measured, which results in a built-in error in the procedure of calculating pH_L, except for the special case where the water is exactly saturated with $CaCO_{3(s)}$.

– The assumption that $[HCO_3^-]$ = Alkalinity is not true in every situation. At relatively high pH values above pH9, this assumption is not accurate.

– In many cases (in practice), the equilibrium constants are not converted according to the ionic strength and temperature, which can lead to errors in water with high ionic strength, even reversing the actual saturation state.

Example 6.4 Calculate the saturation state using the Langelier index (neglecting ionic strength and temperature) in water characterized by:

$$[Ca^{2+}] = 130 \text{ mg/L as } CaCO_3, \text{Alk} = 130 \text{ mg/L as } CaCO_3, \text{pH} = 8.2$$

Solution

$$pH_L = pK_{C2} - pK_{sp} - \log \text{Alk} - \log [Ca^{2+}]$$

$$pH_L = 10.33 - 8.05 - (-2.59) - (-2.89) = 7.75$$

$$\text{LSI} = 8.2 - 7.75 > 0 \Rightarrow \text{Supersaturation}$$

6.6.2 Precise quantification of CaCO₃ precipitation/dissolution potential (CCPP method)

The potential for precipitation (or dissolution) is defined as the exact concentration of $CaCO_3$ that can potentially precipitate (or dissolve) from a given solution (or to it) until equilibrium is reached.

When calculating the $CaCO_3$ precipitation potential of a given solution, a numerical calculation is required. The initials of the relevant index are CCPP (calcium carbonate precipitation potential) and can be calculated by any aquatic chemistry computer program. The value of CCPP can be positive or negative. When it is negative, it can also be defined as positive dissolution potential.

6.6.3 Determination of the precipitation potential (numerical method)

The technique is based on determining the saturation state and then numerical crawling from the current state to saturation (assuming precipitation or dissolution of infinitesimal concentration of $CaCO_3$), while calculating the change in water properties as a result of any crawling iteration performed by the computer on the way to equilibrium. The calculation of the water characteristics is done based on the knowledge of how different parameters change in water as a result of the precipitation/dissolution assumption. The principles of the method are general and can therefore be used to calculate the precipitation potential of any solid (not necessarily $CaCO_3$).

Possible Procedure for Calculating the Precipitation Potential of $CaCO_{3(s)}$

Given: $[Ca^{2+}]$, $H_2CO_3{}^*$ alk, pH, EC and temperature.

The procedure:

1. Calculation of ionic strength, I; The activity coefficients, γ_m and γ_d; and the substituted constants.
2. Calculate the product $[Ca^{2+}] \cdot [CO_3{}^{2-}]$ and compare it to K'_{sp}.
3. Determination of the saturation state: undersaturation, supersaturation or saturation.
4. In the case of supersaturation (for example), the computer "assumes" precipitation of an infinitesimal concentration of $CaCO_3$ from the solution and, knowing the concentration that precipitated, the ionic strength is recalculated (and from it the new activity coefficients), alkalinity, dissolved calcium concentration and C_T. For example, the new alkalinity concentration after precipitation is equal to the result of subtracting two times the molar concentration of the $CaCO_3$ that "precipitated" (since the precipitation of $CaCO_3$ means the carbonate ions exit the aqueous phase, multiplied in the alkalinity equation with a factor of 2 eq/mol) from the previous alkalinity concentration (i.e., the one that was calculated in the previous iteration). Similarly, CT and Ca^{2+} concentrations will be reduced in direct proportion to the concentration that precipitated. For the pH, an internal loop of iterations is performed to calculate it from the alkalinity equation and knowledge of C_T. From the new pH calculated, the species distribution in the carbonate system is recalculated to accurately calculate the new value of ionic strength, I.
5. The new $CO_3{}^{2-}$ concentration is calculated from the new pH and the new C_T.
6. Repeat sections (2)–(5) until in section (3) it is obtained that the product of the two ions is equal to K'_{sp} (with accuracy determined by the programmer). The concentration of $CaCO_3$ that precipitated to this point represents the precipitation potential (PP) or CCPP.

Remarks:

1. The same calculation can be performed in other ways. For example, it is known that when CaCO$_3$ precipitates or dissolves, the acidity (CO$_3^{2-}$ acd) does not change. It is therefore possible to determine the initial acidity from the knowledge of pH and alkalinity, and its value will remain constant throughout the calculation. Therefore, in order to determine the change of the new pH between iterations, it will be necessary to calculate the change of only one parameter (alkalinity or C$_T$).
2. Note that from the first step of the computational procedure (comparison of the product of the ions to K'$_{sp}$), the same information that Langelier index provides is obtained. In modern times, there is no logic to using the index or its derivatives.
3. The precipitation potential is a thermodynamic parameter based on equilibrium equations, as its name implies – it gives an indication of potential that will not necessarily materialize under real conditions. Moreover, as noted, when water is defined as having a negative precipitation potential, no reaction should occur unless the water comes in contact with solid CaCO$_3$.
4. The index does not give any kinetic information.
5. Since CCPP is a quantitative parameter with a precise mathematical definition, it can serve as an unambiguous index in water quality regulations. Indeed, the Israeli Ministry of Health regulations regarding the quality of water exiting desalination plants specify that the range of the CCPP index would be between 3 and 10 mg/L as CaCO$_3$. This issue is dealt with extensively in Chapter 9.

6.6.4 Comparison of LSI values and CCPP in a given solution

In the section that discusses the Langelier index, it is stated explicitly that only the sign of the index is important, that is, it being negative or positive, and that the value of the index itself has no quantitative significance. To illustrate this, Table 6.2 shows five different water combinations for which their LSI is +0.47. On the other hand, it can be seen in Table 6.2 that the CCPP values of the five solutions are completely different. The reason for the large difference stems mainly from the

Table 6.2: Comparison of LSI and CCPP values calculated from five different solutions (TDS = 1000 mg/L Temperature 25 °C).

pH	Alkalinity mg/L as CaCO$_3$	[Ca^{2+}] mg/L as CaCO$_3$	LSI	CCPP mg/L as CaCO$_3$
7.3	289	420	0.47	49.2
7.9	188	162	0.47	14.9
8.3	120	101	0.47	6.9
9.0	65	37	0.47	6.2
9.3	40	30	0.47	5.7

buffer capacity, which is significantly different between the solutions. The higher the buffer capacity, the pH change in pH resulting from the precipitation (or dissolution) of $CaCO_3$ is smaller, allowing for more $CaCO_3$ to precipitate to reach equilibrium and it is expressed with higher precipitation potential.

6.6.5 Determining the precipitation potential (CCPP) graphically

6.6.5.1 Modified Caldwell–Lawrence (MCL) diagrams

Prior to the computer age and to facilitate the tedious calculations involved in the manual determination of CCPP, a graphical method was developed to calculate the precipitation potential of $CaCO_3$. The method, registered in the name of the scientists Caldwell and Lawrence [14], is based on a multi-phase diagram that includes information on the aqueous phase (the carbonic system), the solid phase ($CaCO_{3(s)}$) and the gas phase ($CO_{2(g)}$).

Although in the modern era this method is not widely used, for teaching and clarification purposes, it is still very useful. From a teaching perspective, this method is important because it allows for the practice and understanding of chemical processes occurring in the aqueous phase as well as the reciprocal effects of processes occurring in the three phases. The MCL method is especially useful for understanding the factors affecting precipitation potential and for explaining the changes of parameters in the aqueous phase (alkalinity, acidity, CCPP, pH, etc.) as a result of the addition (or subtraction) of chemicals to the solution. It is important to note that these principles are also applied in the understanding of other phenomena and processes related to the chemistry of water and not only in the calculation of CCPP. In this chapter, the method will be displayed as well as how it can be used to determine instantly both the saturation state and CCPP. Chapter 8 (water softening) explains how to use this method to plan the dose of chemicals required to reach the desired water quality in the "Lime-Soda Ash" softening process.

6.6.5.2 MCL graph development

The developers of the method understood that in order to quantitatively characterize water that is undersaturated relative to $CaCO_{3(s)}$, two parameters have to be defined whose values do not change when $CaCO_3$ precipitates or dissolves in water and to place them as two axes in the solution graph. The two parameters chosen were the acidity (CO_3^{2-} acd) and another parameter they called AMC, which is simply the difference between the alkalinity concentration and the calcium concentration dissolved in water, both in mg/L as $CaCO_3$ units:

1. CO_3^{2-} acidity → the acidity value, which does not contain $[CO_3^{2-}]$ as a species in the equation, does not change with the precipitation or dissolution of $CaCO_3$.
2. AMC = Alkalinity – $[Ca^{2+}]$ → when $CaCO_3$ precipitates or dissolves, the concentrations of dissolved calcium and alkalinity vary equally (in units of eq/L or mg/L as $CaCO_3$), so AMC (the difference between them) does not change.

Numerous Caldwell–Lawrence diagrams appear in the literature, each of which has been developed for constant temperature and ionic strength conditions. To cover a wide range of values, entire booklets appeared, including hundreds of such graphs.

The MCL graph in Fig. 6.1 is a solution of the following system of seven equations. The Caldwell–Lawrence diagram is a graphic description of the solution to the equations for the pH, alkalinity and $[Ca^{2+}]$ parameters, where the X-axis is Alk – $[Ca^{2+}]$ and the Y axis is the acidity. Therefore, in order to produce the curves on the graph, the equations were arranged so that these three parameters ($H_2CO_3{}^*$ alk, $[Ca^{2+}]$ and pH) were isolated as a function of the two parameters, which are not constant when $CaCO_3$ precipitates or dissolves, namely Acd and AMC (all this for a constant temperature and TDS). The solution of the equations for the four parameters appears on the graph as nonlinear curves.

The seven equations are:

1. $K'_{C1} = \dfrac{(H^+)[HCO_3{}^-]}{[H_2CO_3{}^*]}$
2. $K'_{C2} = \dfrac{(H^+)[CO_3{}^{2-}]}{[HCO_3{}^-]}$
3. $(H^+)[OH^-] = K'_w$
4. $[Ca^{2+}][CO_3{}^{2-}] = K'_{sp}$
5. $H_2CO_3{}^*alk = [CO_3{}^{2-}] + [HCO_3{}^-] + [OH^-] - [H^+]$
6. $CO_3{}^{2-} acd = [H_2CO_3{}^*] + [HCO_3{}^-] + [H^+] - [OH^-]$
7. $AMC = H_2CO_3{}^*Alk - [Ca^{2+}]$

Schematically, the equations can be described as follows (where F_1, F_2 and F_3 represent nonlinear functions):

$$pH = F_1(AMC, Acd)$$

$$Alk = F_2(AMC, Acd)$$

$$[Ca^{2+}] = F_3(AMC, Acd)$$

In total, seven equations and ten unknowns can be formulated (three species of the carbonate system, two of the water system, alkalinity, acidity, dissolved calcium and AMC). Therefore, to solve, knowledge of at least three parameters is required. In practice, it is necessary to know the following parameters: **alkalinity $H_2CO_3{}^*$ alk, pH and $[Ca^{2+}]$**.

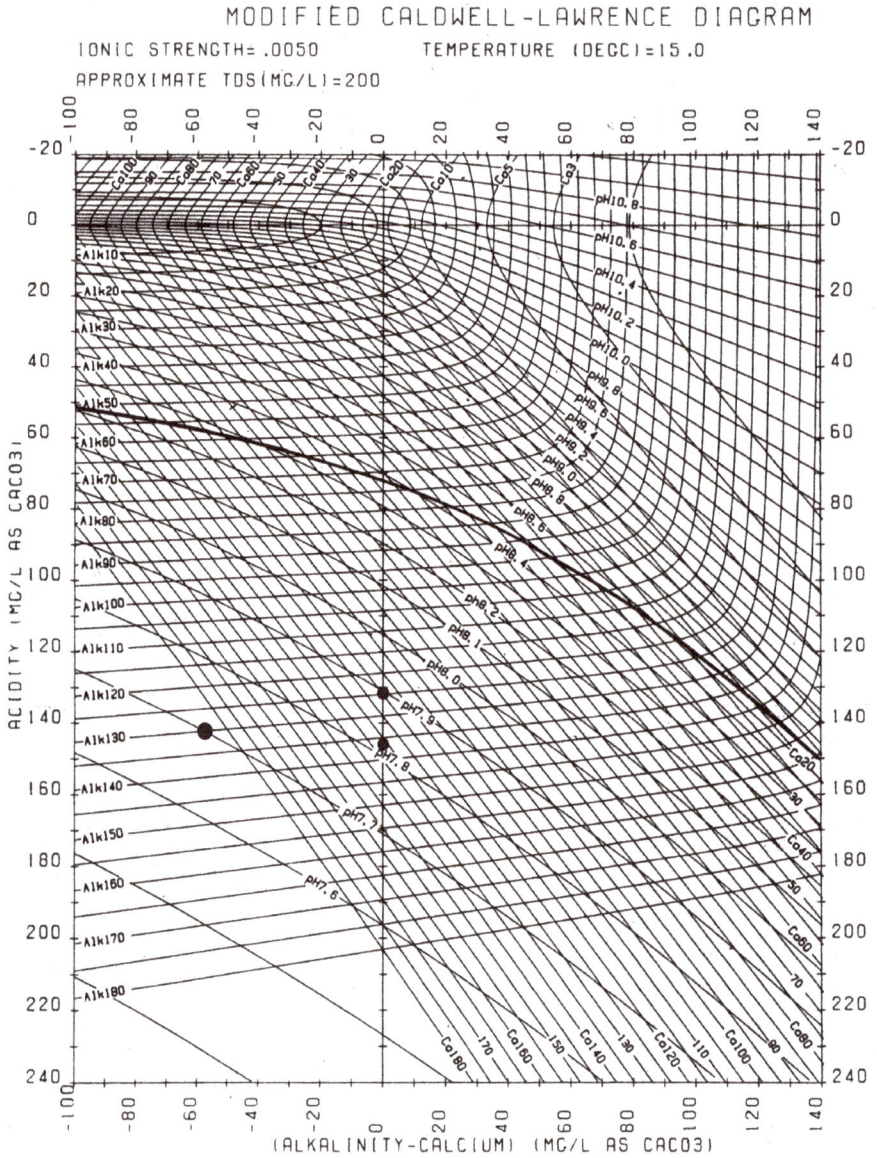

Fig. 6.1: Example of Caldwell–Lawrence diagram. The black line in the center of the graph represents equilibrium of CO_2 with the atmosphere (i.e. equilibrium with the gas phase, assuming an open system). The source of all Caldwell–Lawrence diagrams appearing in this book is from Loewenthal et al. [14].

With the help of the alkalinity and pH data, the acidity can be derived (using the graph).

1. As explained in Chapter 3, two independent parameters (e.g., alkalinity and pH or acidity and pH) define the chemical equilibrium state in the liquid phase only. Since the kinetics of reactions in the aqueous phase is almost instantaneous, this state can always be considered correct.
2. In a state of equilibrium between the liquid phase and the solid phase, the curves representing the pH, (Ca^{2+}) and the alkalinity of the given water meet at the same point on the graph. If they do not meet, the conclusion is that there is no equilibrium between $CaCO_3$ and the solution and there is either undersaturation or supersaturation.
3. The relationship among alkalinity, acidity and pH is always true, regardless of the equilibrium state of the solution-solid phases.

First Step for Using the MCL Diagram: Determination of the Saturation State
- Mark the lines that represent the concentrations of alkalinity, calcium and pH.
- Define the equilibrium state in the aqueous phase, that is, the junction point between the alkalinity line and the pH line. Find the acidity from the graph (by stretching a horizontal line to the Y axis representing the acidity).
- Determine the saturation state by comparing the measured calcium concentration (given) with the calcium concentration line passing through the equilibrium point of the aqueous phase.
- If the measured calcium concentration is higher than the equilibrium concentration, it is supersaturated. If the measured concentration is lower than the equilibrium concentration, then the solution is undersaturated relative to $CaCO_3$.

Example 6.5
Given:

$$Alk = 130 \text{ mg/L as } CaCO_3$$

$$[Ca^{2+}] = 130 \text{ mg/L as } CaCO_3$$

$$pH = 8.2$$

$$TDS = 200 \text{ mg/L}$$

$$T = 15 \,°C$$

a. What is the saturation state?
b. What would the saturation state be if the measured pH was 7.7 (the calcium and alkalinity concentrations remain unchanged)?

Solution
a. The solution appears on Fig. 6.2. Note that the appropriate graph is selected in terms of temperature and TDS conditions. The intersection point of the alkalinity line and the pH line defines the acidity (pull a horizontal line to the Y axis) = 130 mg/L as $CaCO_3$. See point 1 on the graph.
   The calcium curve that passes through this point = 60 mg/L as $CaCO_3$, meaning the calcium concentration at saturation is 60 mg/L as $CaCO_3$ compared with the current measured calcium concentration of 130 mg/L, meaning that the solution is defined as supersaturated in relation to $CaCO_3$.

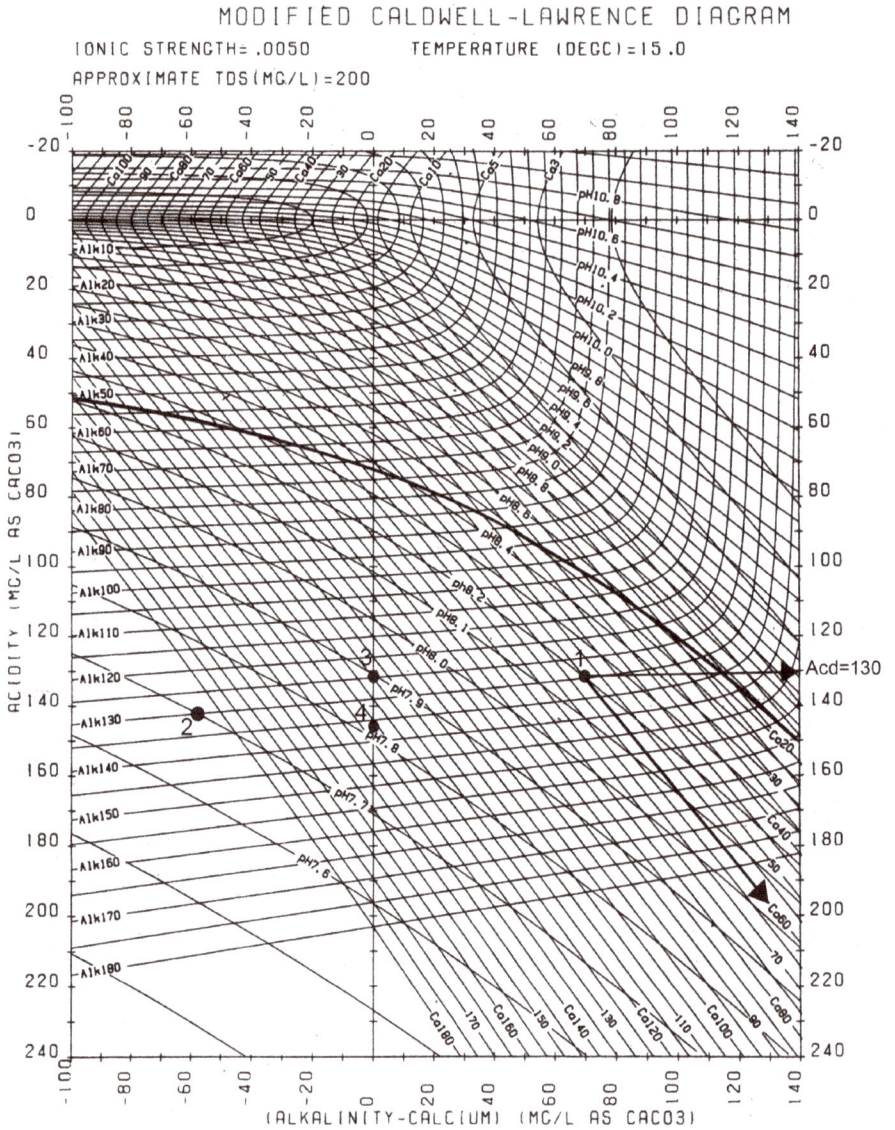

Fig. 6.2: MCL diagram on which are outlined solutions for Examples 6.5, 6.6 and 6.7.

b. See point 2 on the graph. Since the solution can contain a calcium concentration greater than 180 mg/L as CaCO₃, in this situation, the solution is defined as undersaturated with CaCO₃.

Quantitative question solution: How much CaCO₃ can precipitate or dissolve up to reaching equilibrium (or in other words: What is the CCPP)?

As mentioned, when $CaCO_3$ precipitates or dissolves, two parameters remain constant: Acd and AMC.

Meaning, the intersection point of the acidity and the AMC lines unequivocally define the state of saturation. The Ca^{2+} line passing through this point is the concentration of calcium at saturation.

Example 6.6 Given the data of the previous example: pH = 8.2, Ca^{2+} = 130, Alk = 130. What is the CCPP?

Solution The steps of the solution:
- Set the initial state in the dissolved phase (point 1 in Fig. 6.2).
- Calculate AMC: AMC = Alk – $[Ca^{2+}]$ = 130 – 130 = 0.
- Find the junction point of the lines of Acd and AMC (point 3 in Fig. 6.2). This point represents the state in the liquid phase (i.e. the characterization of the water) at a solid–liquid equilibrium.
- From point 3, it is obtained: Alk_s = 125, Ca^{2+}_s = 125, Acd_s = 130, pH_s = 7.9 (the index 's' represents a saturation state relative to $CaCO_3$).

Conclusion: The solution began with 130 mg/L Ca^{2+} and finished with 125 (in equilibrium), i.e. up to reaching equilibrium 5 mg/L as $CaCO_3$ precipitated. Meaning the initial state will be defined as having a positive precipitation potential of 5 mg/L $CaCO_3$.

Example 6.7 Given: pH = 7.7, $[Ca^{2+}]$ = 130, Alk = 130 (all concentrations in mg/L as $CaCO_3$).
In Example 6.5, it was shown that the above solution is undersaturated, how many mg/L of calcium can dissolve in it?

Solution
- Draw the lines representing alkalinity, calcium and pH.
- Find the acidity value from the junction point of alkalinity and pH on the graph (it is obtained Acd = 144 mg/L as $CaCO_3$).
- Calculate the AMC of the given water: AMC = 130 – 130 = 0
- Intersect the lines Acd = 144 and AMC = 0 to obtain the point representing the solid–liquid equilibrium (point 4 on Fig. 6.2).
- Through the above point in the diagram, it is obtained: pH = 7.83, $[Ca^{2+}]$ = 136 and Alk = 134.
- Conclusion: The precipitation potential (PP) is equal to 130 – 136 = –6 mg/L as $CaCO_3$ (or a positive dissolution potential of 6 mg/L).

---

7 Computer software for calculations in the field of aquatic chemistry and water treatment processes, with an emphasis on the Stasoft4.0 program

7.1 Introduction

There are several computer programs in the market that are designed to perform calculations in the water-chemistry field and water treatment processes, at high speed and with low effort to the user. These programs are designed to facilitate the engineer and save him the need to perform the many calculations required to obtain a picture of the aqueous solution status due to a dosage of a given chemical, mixing several aqueous streams, gas emissions (e.g., CO_2) or when there is an iterative calculation (determining pH or CCPP, for example). Like any designated computer software, the garbage in garbage out (GIGO) principle holds true also for software designed for planning and simulating water treatment processes. In other words, these programs can only be used when the theory on which the calculations are based is completely clear to the user.

There are several programs in the market, but the two main ones are (1) RTW model, version 4, a commercial software based on an Excel spreadsheet developed by AWWA (American Water Works Association) and sold for ~$500 per unit. (2) Stasoft4.0 software. A software originally developed in the University of Cape Town, South Africa, by Prof. Richard Loewenthal and Mr. Ian Morrison and sold for a nominal fee of US $50 (an online order can be made at http://www.wrc.org.za/). A more recent version, Stasoft 5.0, was further developed by the South African Water Research Commission is available online, free of charge, (see https://www.youtube.com/watch?v= Xhq5dlULLGw). Both programs (RTW and Stasoft) have similar capabilities. This chapter describes only the Stasoft4 program, but the principles for working with the other programs are almost identical.

7.2 Principles of calculation and limitations

The principles of computation in Stasoft4 are based on the material covered by this book. An earlier version of the software is described by Loewenthal et al., [15]. The article describing the software can be downloaded freely from the Internet

The software is sold with a helpful manual, which can be used when operational and professional questions arise. The software uses thermodynamic equilibrium constants from various sources. As the constants are updated and change a bit overtime, there may be a small difference between the results of Stasoft4 and other

https://doi.org/10.1515/9783110603958-007

software (older or newer), but this difference is usually negligible in terms of engineering design. The principle of calculation of the software is based on mass balances performed on conservative parameters, and calculation of the new state created in the aqueous phase using the binding equations between the conservative elements. For example, when the designer wishes to add x mg/L of HCl, the software considers the corresponding changes in the following conservative parameters: the increase in the acidity concentration (in its various forms), decrease in alkalinity and increase in TDS and chloride concentrations (which appear as a parameter in the software). According to the change in TDS, the activity coefficients are recalculated, and the thermodynamic constants are converted according to the new ionic strength in the water. The software distinguishes between processes with fast time constants, such as changing characteristics in the aqueous phase, whose calculations are performed immediately, without the need for a specific command, and processes with long time constants, such as the arrival of the aqueous phase to equilibrium with the solid or gas phase. For characterizing the quality of water as a result of reaching equilibrium with the solid phase and/ or the gas phase, an explicit order from the user is required, whereas the characteristic change in the aquatic phase after the addition of chemicals is done automatically without requiring the user's approval.

Stasoft4 has a number of limitations: (1) The highest TDS concentration it receives as an input is 20,000 mg/L, meaning that it is not possible to perform calculations on sea water, for example. The reasons for this are that at very high concentrations of TDS: (a) equilibrium constants cannot be corrected using the Davies equation (see Chapter 1) and therefore other equations that link salinity to the values of equilibrium constants should be used. (b) The effects of forming complexes in the aqueous phase become highly significant at very high TDS concentrations and the software does not contain sufficient information on most of the complexes that can potentially form. The program does consider in its calculations the creation of a limited number of simple complexes defined as ion pairs: $CaHCO_3^+$, $CaCO_3^0$, $CaOH^+$, $MgHCO_3^+$, $MgCO_3^0$, $MgOH^+$. The program does not consider the pairs formed between calcium and magnesium to sulfate nor the pairs formed between sodium ions to bicarbonate and sulfate. However, the last pairs have little meaning when it comes to drinking water and the software's developers preferred to give them up and risk a very small loss of precision and in turn they significantly reduced the complexity of the solution algorithm; (2) The second limitation is also related to ionic strength. The program calculates ionic strength from empirical equations that link the TDS or the EC to the ionic strength (eqs. (1.19) and (1.20), respectively) rather than calculating the ionic strength from the given ion concentrations, since the concentrations of all ions are often unknown, but the TDS or the EC are known. Similarly, the TDS itself is not directly calculated from the given ion concentrations; (3) The third limitation is that the software database includes only two weak-acid systems: the carbonate system and the hypochlorite system ($HOCl/OCl^-$). The program does not include the phosphoric, sulfide and boric systems or weak organic acid systems, which may be present in drinking water, although not often and usually at very low concentrations. (4) Another limitation relates to the fact that the program does not have the possibility to calculate the chemistry associated with dissolved concentrations of sulfide, iron and divalent manganese that sometimes appear in reduced groundwater. (5) The last limitation is inherent to programs based on equilibrium equations: the time-dependent change in water quality is not taken into account, even if the relevant process has slow kinetics. For example, it is possible to calculate the pH obtained when groundwater reaches equilibrium with the atmosphere using the program, but it is not possible to obtain an accurate picture of the water quality as a function of time before reaching equilibrium.

7.3 How to use the software

The start page of the program is shown in Fig. 7.1. The left most column on the screen lists the parameters determined in the water. Each of the other columns represents the water quality after undergoing a process (the process appears on the top row of the column). The default of the program shown on the start page (Fig. 7.1) is that no process is performed on the water and therefore only one column appears (besides the parameters), representing the initial water quality. Each row on the screen represents a water quality parameter. From the default state, seven water quality characteristics appear on the screen: temperature ($^{\circ}$C), electrical conductivity (mS/m), dissolved calcium concentration (mg/L), pH, Alkalinity in mg/L as $CaCO_3$ (meaning $H_2CO_3{}^*$ alk), total concentration of the carbonate system – C_T (mg/L as CO_2), denoted carbonic species and the precipitation potential of $CaCO_3$ – CCPP (mg/L as $CaCO_3$).

A left click on the mouse when the arrow is located on one of the gray cells of the water quality characteristics opens a window from which one can select another seven quality parameters from a list of 19 parameters in total. Figure 7.2 shows the home screen to which seven additional parameters were added, including Acidity ($CO_3{}^{2-}$ acd), total concentration of dissolved chlorine species and several ion concentrations. Note that the screen shown in Fig. 7.2 shows four parameters of the carbonate system: pH, Alk, Acd, C_T. Since two independent parameters are sufficient to define the quality of natural water, the designer is required to define the parameters that he wants to enter to the program (defined as independent) and from these the dependent parameters (which cannot be modified) are calculated by the program. To do so, one should click on the mouse and set the dependent parameters to be "not editable", while the independent parameters should be defined as "editable".

The program allows entering the dissolved concentrations in two ways: TDS and EC. These two quantities are related to each other by a constant product (TDS $_{mg/L}$ = 6.7 EC $_{mS/m}$) and one automatically changes when the other is entered by the user.

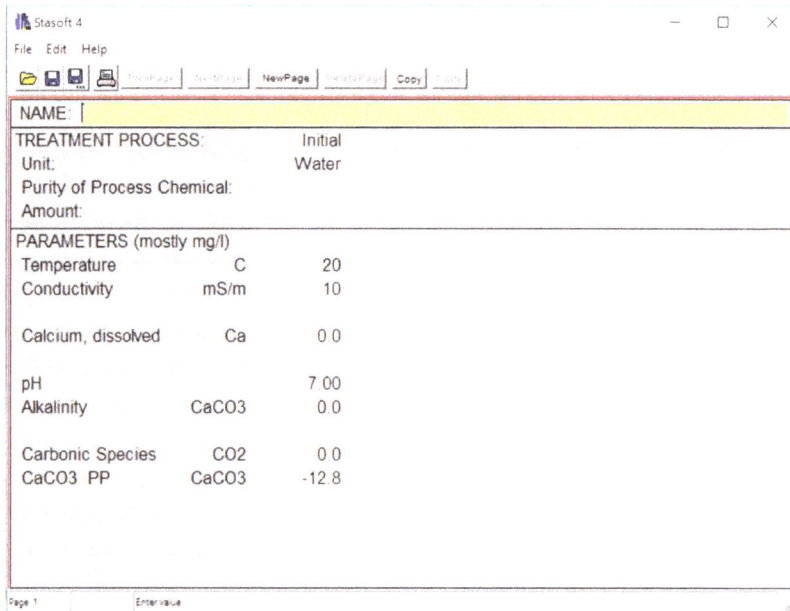

Fig. 7.1: The default window obtained when opening Stasoft4 program.

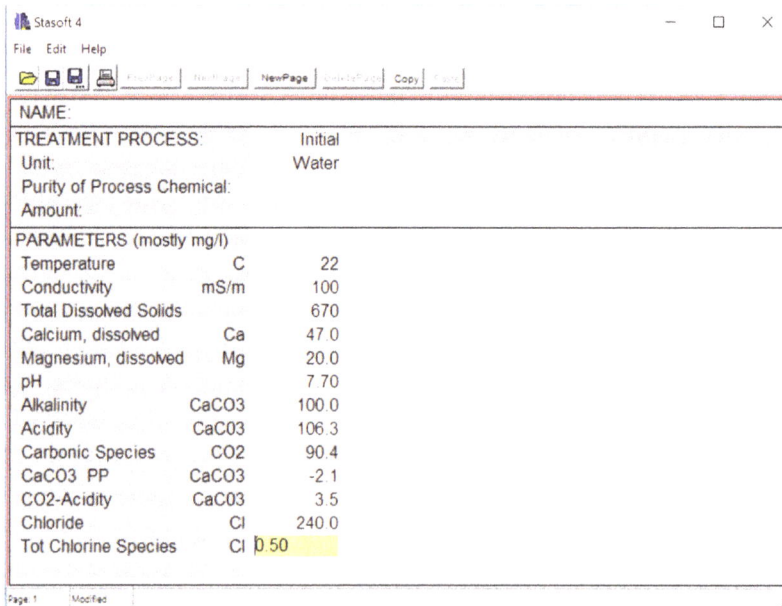

Fig. 7.2: Entering the data to determine the initial water quality.

To demonstrate the initial use of the program, a model water quality is now used as an example representing the (partial) water quality distributed in the National Water Carrier of Israel:

Alk = 100 mg/L as $CaCO_3$; pH7.7; $[Ca^{2+}]$ = 47 mg/L; $[Mg^{2+}]$ = 20 mg/L; EC = 100 mS/m; Temperature = 22 °C, $[Cl^-]$ = 240 mg/L; $[Cl_2]_T$ = 0.5 mg/L.

The data entered appears in Fig. 7.2. Note that the water quality data contains parameters entered by the user and parameters calculated automatically by the program such as the concentration of TDS, CCPP, Acd and the value of CO_2 acidity (as it is termed in the program) that is equivalent to the value of HCO_3^- acidity, as denoted by the terminology in this book. It can be seen that the water of the national carrier, in the data as shown in the current example, is not stabilized (negative CCPP value). This quality was characteristic of this water up to 2009, when Mekorot (national water company of Israel) began to stabilize the water by increasing the pH to pH8.0 by dosage of NaOH. As a result, today the CCPP value with which the water is supplied to the national system is positive.

So far, it was demonstrated how the user defines the initial water quality. Now the operations/processes which can be performed on the water using the program are defined. There are 23 such options including the addition of 14 different chemicals (lime, soda, caustic soda, strong acids and bases, sodium bicarbonate, chlorine compounds, etc.) and a variety of processes such as reaching equilibrium with the atmosphere, reaching equilibrium with solid $CaCO_{3(s)}$ and $Mg(OH)_{2(s)}$, filtration, evaporation and mixing solutions.

For example, to examine how the program calculates the effect of an addition of NaOH at a given concentration. First, an addition of 1 mg/L to the water quality shown in Fig. 7.2 is considered. It can be seen that the pH value increased to pH7.83 and the CCPP value became slightly positive (0.1 mg/L as $CaCO_3$). The addition of another 1 mg/L of NaOH increases the pH value to pH7.98 and the CCPP value to 2.4 mg/L as $CaCO_3$. This simulation is shown in Fig. 7.3. Note also that, as expected, the alkalinity value increased, and the acidity value decreased as a result of the dose. Logically, the new values of TDS and sodium concentrations (not defined in the current set of parameters) are calculated by the program following the addition of the base. As a result of the change in TDS, the activity coefficient values are recalculated, and the equilibrium constants used for calculation purposes are transformed.

Technically, the program allows up to five processes in a row. In practice, this limitation can be circumvented by transferring the water (e.g., by diluting with 100% of the previous water) to a new page (by clicking on the New page tab) and performing four additional processes, and so on. The dilution option is illustrated later in this chapter.

Now a set of examples that will demonstrate some of the capabilities of the program in the context of the material presented in the previous chapters will be demonstrated. All the examples in this chapter rely on examples that have been

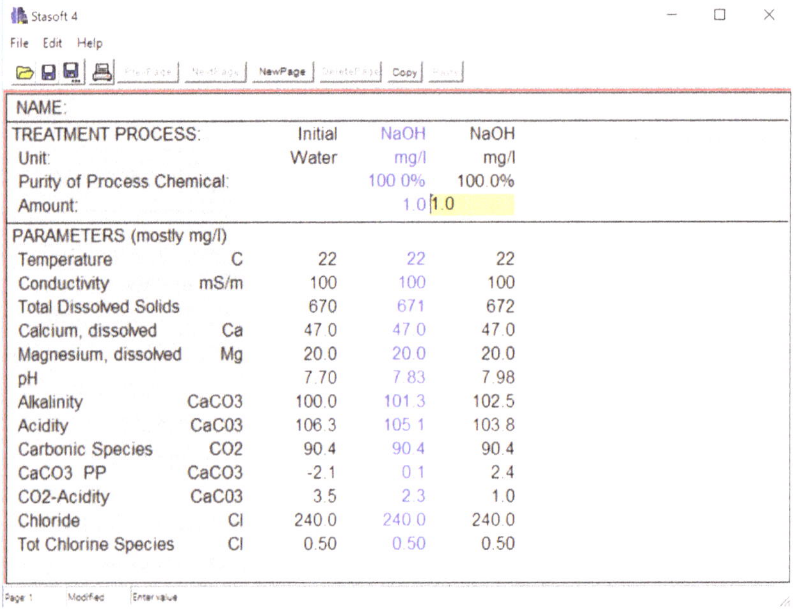

NAME:				
TREATMENT PROCESS:		Initial	NaOH	NaOH
Unit:		Water	mg/l	mg/l
Purity of Process Chemical:			100.0%	100.0%
Amount:			1.0	1.0
PARAMETERS (mostly mg/l)				
Temperature	C	22	22	22
Conductivity	mS/m	100	100	100
Total Dissolved Solids		670	671	672
Calcium, dissolved	Ca	47.0	47.0	47.0
Magnesium, dissolved	Mg	20.0	20.0	20.0
pH		7.70	7.83	7.98
Alkalinity	CaCO3	100.0	101.3	102.5
Acidity	CaCO3	106.3	105.1	103.8
Carbonic Species	CO2	90.4	90.4	90.4
CaCO3 PP	CaCO3	-2.1	0.1	2.4
CO2-Acidity	CaC03	3.5	2.3	1.0
Chloride	Cl	240.0	240.0	240.0
Tot Chlorine Species	Cl	0.50	0.50	0.50

Fig. 7.3: Obtained water quality after dosing 2 mg/L NaOH (100%).

solved in previous chapters in analytical methods. Note that Stasoft4 solutions may differ slightly from the solutions provided in the original examples because (1) the computer solution does not neglect anything, (2) slightly different constants may be used and (3) because the program considers the formation of certain ion pairs, which is not done with the analytic examples appearing in the book. However, as mentioned, the differences in the solutions should be quite small.

Example 7.1: Illustration of the use of a program to solve Example 4.1

Figure 4.1 shows two water supply lines that converge into a single pipe. Calculate the acid-base characteristics of the mixed water stream (pH, Alkalinity, Acidity and C_T).

Solution Using Stasoft4 program

a. Assume a temperature of 25 °C and a low electrical conductivity of 100 and 50 mS/m, respectively, for each water source. Enter the data, each on a separate page in the program (to go to the second page, press NewPage on the top bar).
b. On the second page, select from the Actions column the "Blend – add Page 1 water" option (third option before the end of the list). Since the flow from water source b is twice that of water source a, set the program to 33% dilution with water source a (assuming water quality a is entered on the first page).

The calculation result is shown in Fig. 7.4. For the mixed solution, pH6.28 was obtained while in the original manual calculation, the result was pH6.29. The alkalinity concentration, of course, is identical, regardless of the calculation method (since it is only a weighted average calculation). The result of the acidity calculation is slightly different: 414.4 in the program compared to 449.5 in the manual calculation. These differences result from conversion of thermodynamic equilibrium constants to apparent ones, according to salinity (EC), calculations that were neglected in the manual solution.

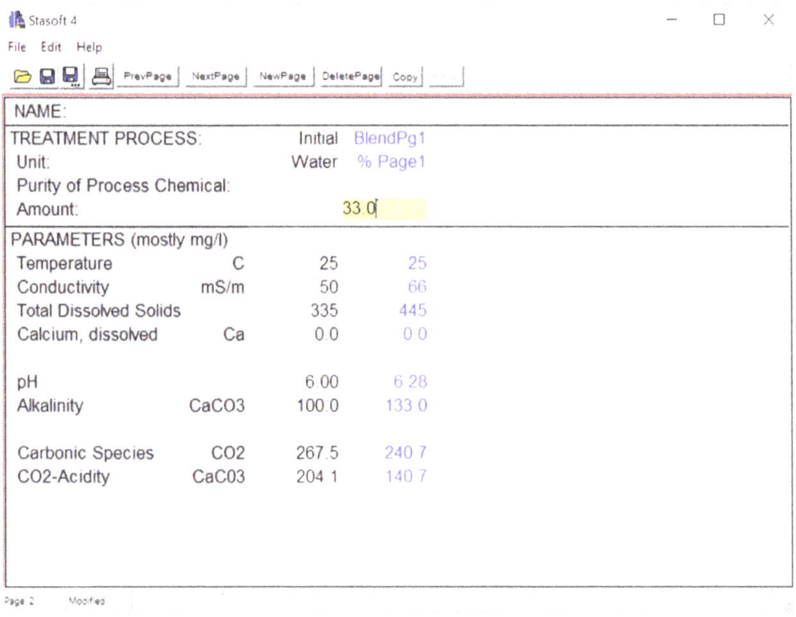

```
Stasoft 4                                                          —  □  ✕
File  Edit  Help
  PrevPage  NextPage  NewPage  DeletePage  Copy

 NAME:
 TREATMENT PROCESS:              Initial   BlendPg1
 Unit:                           Water     %  Page1
 Purity of Process Chemical:
 Amount:                              33.0
 PARAMETERS (mostly mg/l)
 Temperature          C          25          25
 Conductivity         mS/m       50          66
 Total Dissolved Solids          335         445
 Calcium, dissolved   Ca         0.0         0.0

 pH                              6.00        6.28
 Alkalinity           CaCO3      100.0       133.0

 Carbonic Species     CO2        267.5       240.7
 CO2-Acidity          CaC03      204.1       140.7


Page 2      Modified
```

Fig. 7.4: Results of the computer solution for example 4.1.

Example 7.2: Illustration of the use of the program for solving Example 4.4 Section a:

A fertilizer plant is requesting permission to discharge a stream of wastewater to the river. The characteristics of the two water streams are given in Table 7.1. The wastewater stream is a strong base that does not contain the carbonate system ($C_T = 0$). Determine:
(a) It is given that the pH at the mixing point is not to exceed 0.5 pH units above the current pH in the river. Should the plant acidify the wastewater before discharging it to the river?
(b) What concentration of HCl will be required to acidify the given wastewater so that after mixing the pH value will not exceed pH7.8? Note that it is necessary to dose the acid into the waste-water itself rather than to the mixed stream.

Table 7.1: Characteristics of the two water streams.

	Flowrate m^3/s	pH	$H_2CO_3^*$ alk mg/L as $CaCO_3$	TDS mg/L	Temp. °C
River	6.5	7.3	150	600	25
Wastewater	0.28	13.3	-	7200	20

Solution This question is solved as a mixing problem, similar to Example 7.1. The only difference is that beyond the solution of the question with the existing data, it is necessary to determine the concentration of the strong acid (HCl) to be added in order to meet the pH requirements at the mixing point.

The stages of the solution:

a. Enter the river and wastewater data to pages 1 and 2, respectively.
b. Perform mixing where the mixing ratio is 95.87% river water. The pH value at the mixing point is higher than pH7.8 (i.e., more than 0.5 units above the current pH in the river). Therefore, it is not possible to discharge the wastewater to the stream without first adding acid.
c. Add acid treatment using HCl before calculating the results of mixing, and increase HCl doses until pH7.8 is reached in the mixed stream, as required in section b.

The simulation results are shown in Fig. 7.5. Note that to simulate the wastewater, the program is required to perform the addition of a strong base NaOH to reach pH13.3, since it is not possible to directly determine the pH of the water without knowing their alkalinity. The pH value obtained after mixing is pH11.56, which is much higher than the maximum permissible value.

The concentration of acid to be added to the wastewater is 6385 mg/L, in order to neutralize the wastewater up to pH11.73. After mixing, the pH value stabilizes at exactly 7.8. The exact amount of acid to be added to the water is found by trial and error, as follows: Increase the acid concentration in stages and monitor the pH value in the column of the blending with the first page stream (i.e., river water) until pH7.80 is reached, as required. Similarly, the concentration of the base required to obtain a pH value of 13.3 in the wastewater was obtained.

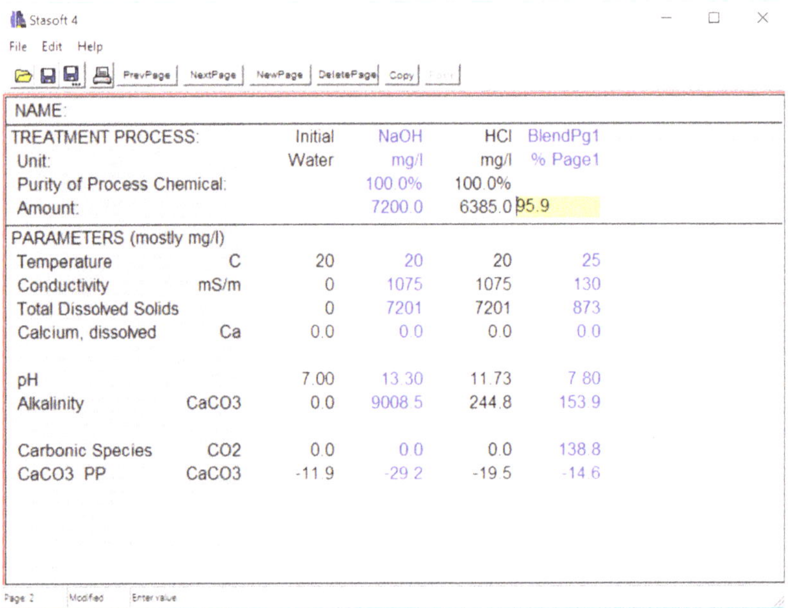

TREATMENT PROCESS:		Initial	NaOH	HCl	BlendPg1
Unit:		Water	mg/l	mg/l	% Page1
Purity of Process Chemical:			100.0%	100.0%	
Amount:			7200.0	6385.0	95.9
PARAMETERS (mostly mg/l)					
Temperature	C	20	20	20	25
Conductivity	mS/m	0	1075	1075	130
Total Dissolved Solids		0	7201	7201	873
Calcium, dissolved	Ca	0.0	0.0	0.0	0.0
pH		7.00	13.30	11.73	7.80
Alkalinity	CaCO3	0.0	9008.5	244.8	153.9
Carbonic Species	CO2	0.0	0.0	0.0	138.8
CaCO3 PP	CaCO3	-11.9	-29.2	-19.5	-14.6

Fig. 7.5: Results of the computer solution of Example 7.2 (computer solution of Example 4.4 section a).

Example 7.3: Illustration of the use of software to solve Example 4.6

The desalination plant in Ashkelon, Israel produces (as of the end of 2018) desalinated water with following mean characteristics: TDS = 150 mg/L, pH8.15, $H_2CO_3^*{}_{alk}$ = 48 mg/L as $CaCO_3$ and dissolved calcium concentration of 110 mg/L as $CaCO_3$.

What will be the change in the water quality (in terms of acid base interactions and $CaCO_3$ precipitation potential) following the dosage of fluorosilicic acid to reach a concentration of 1 mg/L of fluoride (as fluoride) in the water?

Solution Stasoft4 does not support the addition of fluorosilicic acid to water. However, from the knowledge of the stoichiometric equation for the dissociation of fluorosilicic acid (a strong acid) in water it is possible to deduce the concentration of protons emitted to the water as a result of its addition. It is therefore possible to perform an approximate simulation of the addition of the protons by adding an equivalent amount of a strong acid (e.g., HCl).
The dissociation equation of H_2SiF_6 is:

$$H_2SiF_6 + 2H_2O \rightarrow 6H^+ + 6F^- + SiO_2 \tag{7.1}$$

Therefore, for each mole of fluoride (F$^-$, molecular weight: 19 g/mol) dosed to the solution, 1 mol (or one equivalent) of H$^+$ is released. It follows that an addition of 1 mg/L of fluoride to the water is equivalent to an addition of 1/19 millimolar of H$^+$ which is $5.26 \cdot 10^{-5}$ equivalents of H$^+$, or 2.6 mg/L as $CaCO_3$ of acidity. It is therefore possible to calculate using the program, the change in the pH of the water as a result of a dosage of 1.92 mg/L of HCl (molecular weight 35.5 g/mol). The results are shown in Fig. 7.6. It can be seen that the results of the manual calculations are almost identical (as expected) to the computer solution.

🏛 Stasoft 4	— □ ✕	
File Edit Help		
📂 🖫 🖫 🖨 NewPage Copy		
NAME:		
TREATMENT PROCESS:	Initial	HCl
Unit:	Water	mg/l
Purity of Process Chemical:		100.0%
Amount:	1.9	
PARAMETERS (mostly mg/l)		
Temperature C	25	25
Conductivity mS/m	22	22
Total Dissolved Solids	150	151
Calcium, dissolved Ca	44.0	44.0
pH	8.15	7.54
Alkalinity CaCO3	48.0	45.4
Acidity CaCO3	47.7	50.3
Carbonic Species CO2	42.1	42.1
CaCO3 PP CaCO3	1.0	-4.0
Page 1 Modified Enter value		

Fig. 7.6: Solution of Example 7.3 (original Example 4.6).

Example 7.4 Partial pressure and carbon dioxide concentration in carbonated beverages (original Example 5.7 a, c, e)

For the preparation of certain carbonated beverages, it is common to dose distilled water with a mixture of CO_2 at a high concentration and phosphoric acid (H_3PO_4) to create a strong acidic environment (pH<3.0). Note – in this question, neglect effects of ionic strength.

a.   Assuming the partial pressure of $CO_{2(g)}$ in the gas phase in the bottle is 2.5 bar (without phosphoric acid), what is the pH value in the solution (assuming a temperature of 25 °C)?

c.   If it was required to reduce the pH to pH2.8 with $CO_{2(g)}$ alone, what is the required partial pressure?

e.   When a bottle of a carbonated beverage (pH2.8, CO_2 partial pressure of 2.5 bar, and HCl dosage) is opened, CO_2 is released into the atmosphere. What will be the pH value of the drink after the CO_2 has reached equilibrium with the atmosphere? (assume a temperature of 25 °C).

Solution

a.   Figure 7.7 shows the Stasoft solution to this problem. The process "Equilibrate with Carbon Dioxide" was chosen, and the relevant partial pressure (in atmosphere units) was inserted. The computed pH value is practically the same as the one calculated manually.

c.   For this case too, the only process chosen in the simulation is Equilibrate with Carbon Dioxide. The required CO_2 partial pressure is found by trial and error as follows: from section a,

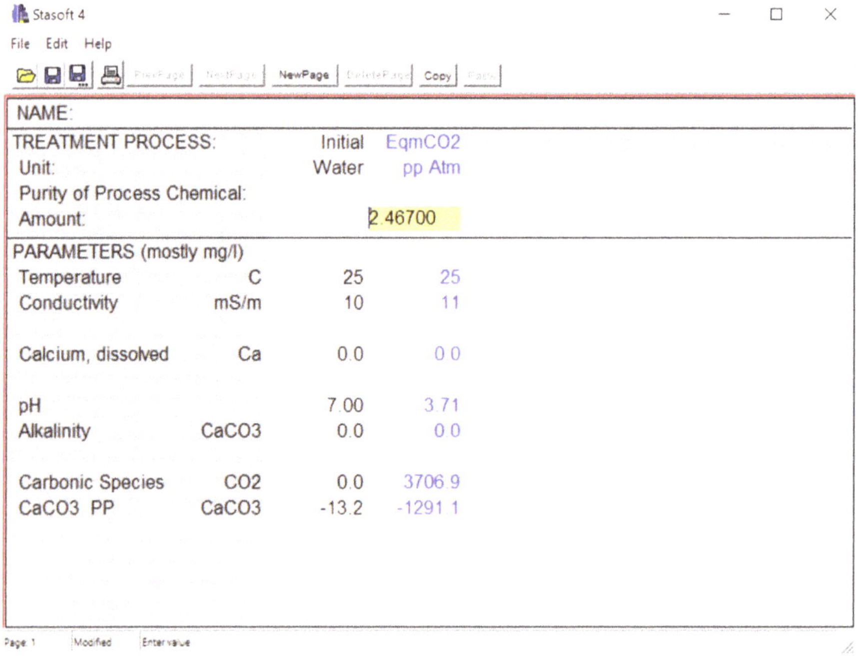

Stasoft 4   — □ ×

File Edit Help

[toolbar icons] PrevPage NextPage NewPage DeletePage Copy Paste

NAME:			
TREATMENT PROCESS:		Initial	EqmCO2
Unit:		Water	pp Atm
Purity of Process Chemical:			
Amount:			2.46700
PARAMETERS (mostly mg/l)			
Temperature	C	25	25
Conductivity	mS/m	10	11
Calcium, dissolved	Ca	0.0	0.0
pH		7.00	3.71
Alkalinity	CaCO3	0.0	0.0
Carbonic Species	CO2	0.0	3706.9
CaCO3 PP	CaCO3	-13.2	-1291.1

Page 1   Modified   Enter value

Fig. 7.7: Solution of Example 7.4 (original Example 5.7a).

it is known that a 2.5 atm CO_2 partial pressure results in pH of 3.71. Thus, a much higher partial pressure is needed for attaining pH2.8. The user should elevate CO_2 partial pressure values in steps and follow the pH obtained. While doing so, the nature of weak acid is displayed: since the carbonic acid pKa is 6.3, significant dosages are required for every small reduction of pH at this low pH range, i.e., below pH3.5. Finally, at equilibrium with CO_2 partial pressure of 163 atm, the required pH is attained (Fig. 7.8).

e. Let us first enter the three required process steps to the simulation: equilibrium with $CO_{2(g)}$, HCl dosage, and equilibrium with atmospheric CO_2 (Fig. 7.9). Note, the order of the first two process steps is insignificant. That is, one may first simulate the equilibrium of the water with 2.5 atm of CO_2 and then find the correct dosage of acid. Alternatively, one can insert the acid dosage step first, and then equilibrium with CO_2. In both cases, the required pH should be attained after the two processes are complete, i.e., in the second process column. In addition, in both cases, the HCl dosage for attaining pH2.8 is found by trial and error (second process column in Fig. 7.9). The last process step is "equilibrate with air". However, to compute the final water quality, the air partial pressure of CO_2 given by the program should be updated, as it is 0.0004 nowadays. Naturally, the solution C_T is reduced considerably by this process step (Fig. 7.9). However, as mentioned in Example 5.7, the exposure to air in such conditions hardly affects the solution pH. In order to notice the small difference in the pH value due to CO_2 vaporization, one should click on the pH in the PARAMETERS column, and elevate the default decimals from 2 to at least 3.

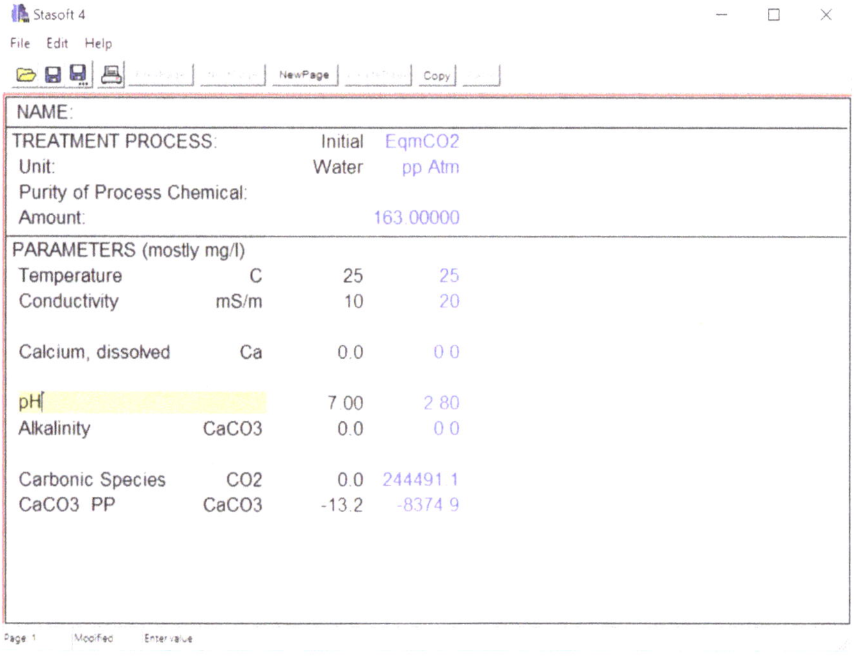

Stasoft 4						— □ ✕

File Edit Help

| NewPage | | Copy | |

NAME:				
TREATMENT PROCESS:		Initial	EqmCO2	
Unit:		Water	pp Atm	
Purity of Process Chemical:				
Amount:			163.00000	
PARAMETERS (mostly mg/l)				
Temperature	C	25	25	
Conductivity	mS/m	10	20	
Calcium, dissolved	Ca	0.0	0.0	
pH		7.00	2.80	
Alkalinity	CaCO3	0.0	0.0	
Carbonic Species	CO2	0.0	244491.1	
CaCO3 PP	CaCO3	-13.2	-8374.9	

Page 1   Modified   Enter value

Fig. 7.8: Solution of Example 7.4 (original Example 5.7c).

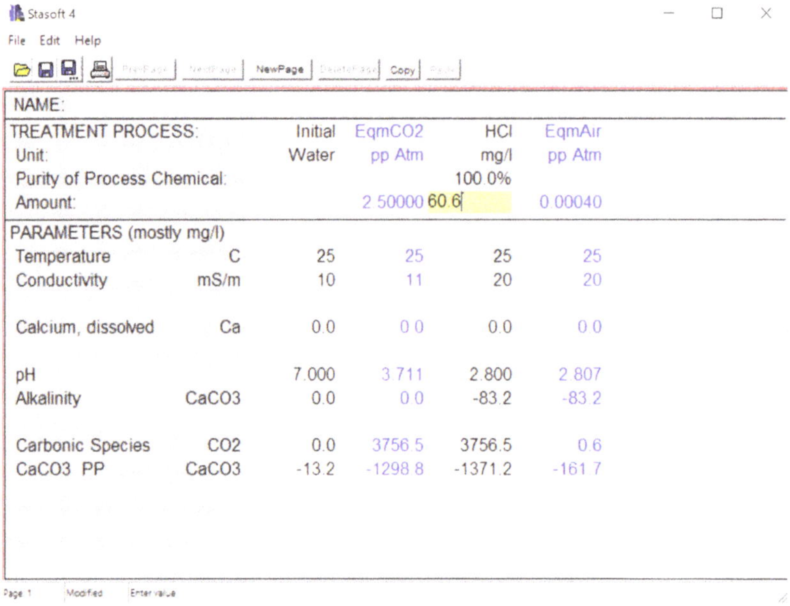

Stasoft 4   — ☐ ✕

File   Edit   Help

🖿 🖫 🖫 🖨 | Prev Page | Next Page | NewPage | Delete Page | Copy | Paste |

NAME:					
TREATMENT PROCESS:		Initial	EqmCO2	HCl	EqmAir
Unit:		Water	pp Atm	mg/l	pp Atm
Purity of Process Chemical:				100.0%	
Amount:			2.50000	60.6|	0.00040
PARAMETERS (mostly mg/l)					
Temperature	C	25	25	25	25
Conductivity	mS/m	10	11	20	20
Calcium, dissolved	Ca	0.0	0.0	0.0	0.0
pH		7.000	3.711	2.800	2.807
Alkalinity	CaCO3	0.0	0.0	-83.2	-83.2
Carbonic Species	CO2	0.0	3756.5	3756.5	0.6
CaCO3 PP	CaCO3	-13.2	-1298.8	-1371.2	-161.7

Page 1   Modified   Enter value

Fig. 7.9: Solution of Example 7.4 (original Example 5.7e).

7.4 Simulation of water treatment processes using the Stasoft4 program

Stasoft4 program enables simulation of equilibrium-based water treatment processes, as illustrated in Examples 7.3 and 7.4. In other words, the water quality when equilibrium is reached can be computed in a straightforward manner. From this definition, it is clear that kinetic considerations cannot be solved by the program. However, from understanding the processes, this problem can sometimes be circumvented, and kinetic data can be entered into the simulation. An example of this is the input of a negative CCPP attained after a certain retention time between the water and calcite. The user must first understand the connection between the retention time of the water with calcite and the resulting CCPP (a connection that can be found empirically). If these data are known, the simulation can be performed in the program. Demonstrations of incorporating this kinetic assumption to Stasoft simulations are given in Chapter 9, Examples 9.2, 9.3 and 9.4. Other examples of manipulating the program to yield results that are based on kinetics are presented in section c of Example 7.4, this section deals with CO_2 vaporization under the limitation that full equilibrium with the atmosphere cannot be achieved at reasonable time scales.

Example 7.5: Evaluating the $CaCO_3$ saturation state (illustrating the use of software for solving Example 6.4)

Calculate the saturation state using the water characterized by:

$$[Ca^{2+}] = 130 \, mg/L \text{ as } CaCO_3, \, Alk = 130 \, mg/L \text{ as } CaCO_3, \, pH = 8.2$$

Solution Insert the given water characteristics as the initial water quality to the Stasoft software (the units of the calcium concentration should be converted to mg/L). No data is given regarding the ionic strength and temperature, therefore the Stasoft default conductivity and temperature of 25 °C are assumed.

Figure 7.10 shows the Stasoft calculation relevant to the given water quality. As the calculated CCPP is 13.6 (bottom line of the first column), i.e., positive, the saturation state is over-saturation. Note that the pH_L calculated manually for this water is 7.75 (Example 6.4). The software can be easily used to calculate the pH of the water following equilibrium with solid $CaCO_3$ (second column in Fig. 7.10). This calculation shows that after precipitating 13.6 mg/L of $CaCO_3$ the water pH is 7.58. Naturally, this pH is not equal to pH_L. However, the pH_L can also be found using Stasoft, by changing the pH value of the given solution (before precipitation, and without changing any other water parameter) until reaching a CCPP value of zero. Note that this pH value too will be different than the one calculated manually, since the ionic strength assumed in the manual calculation was zero.

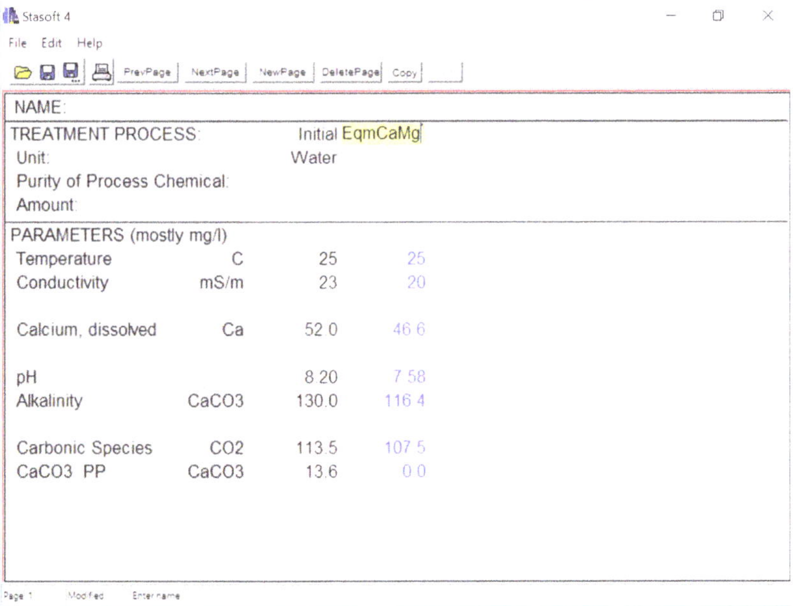

Stasoft 4

File   Edit   Help

PrevPage | NextPage | NewPage | DeletePage | Copy

NAME:			
TREATMENT PROCESS:		Initial EqmCaMg	
Unit:		Water	
Purity of Process Chemical:			
Amount			
PARAMETERS (mostly mg/l)			
Temperature	C	25	25
Conductivity	mS/m	23	20
Calcium, dissolved	Ca	52.0	46.6
pH		8.20	7.58
Alkalinity	CaCO3	130.0	116.4
Carbonic Species	CO2	113.5	107.5
CaCO3 PP	CaCO3	13.6	0.0

Page 1   Modified   Enter name

Fig. 7.10: Solution of Example 7.5 (original Example 6.4).

Example 7.6: Evaluating the effect of exposure to atmospheric CO_2, (illustrating using Stasoft to solve a question based on Example 5.4)
From a water sample analysis, it is obtained:

$$\text{Alkalinity} (H_2CO_3^*) = 80 \text{ mg/L as } CaCO_3, \text{ pH} = 6.3$$

The sample was exposed to the air.

a. What will be the resulting pH if the water reaches equilibrium with atmospheric CO_2? How much CO_2 is transferred and in which direction?

b. Due to kinetic limitation, the water does not reach equilibrium with atmospheric CO_2. Instead, the water $H_2CO_3^*$ concentration is twice higher than at equilibrium with air. What will be the resulting pH?

$H = 30.2 \text{ bar/(mol/L)}$, assume a negligible ionic strength and a temperature of 25 °C.

Solution Insert the given water characteristics as the initial water quality to the Stasoft software.

a. Select "equilibrate with air" as the treatment process. Note that since the software was written in the 1980s the atmospheric CO_2 partial pressure assumed (0.00035 atm) is no longer true. Correct the partial pressure to 0.0004 atm and check the resulting pH, appearing at the second column, see Fig. 7.11. As explained in Chapter 5, for a given alkalinity there is a single pH value at equilibrium with the atmosphere. Therefore, the initial pH does not affect the final pH value.

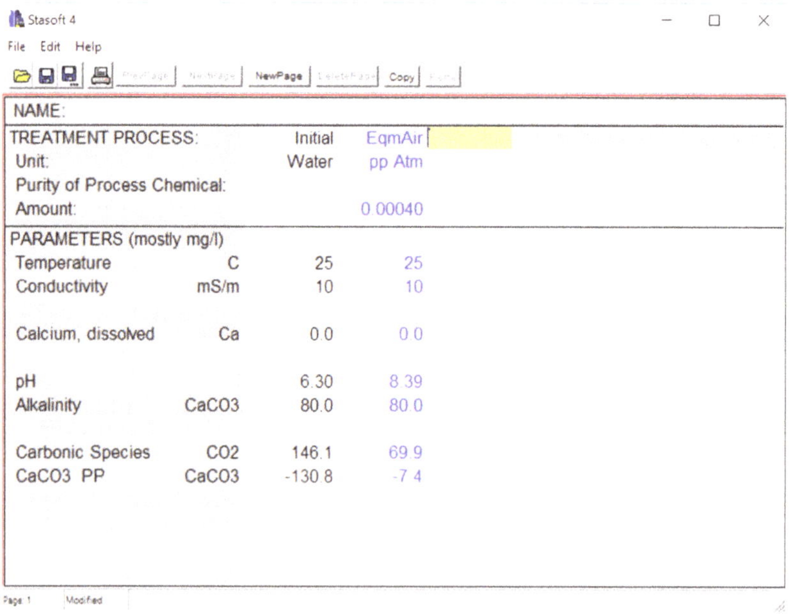

Stasoft 4				— □ ✕
File Edit Help				

NAME:				
TREATMENT PROCESS:		Initial	EqmAir	
Unit:		Water	pp Atm	
Purity of Process Chemical:				
Amount:			0.00040	
PARAMETERS (mostly mg/l)				
Temperature	C	25	25	
Conductivity	mS/m	10	10	
Calcium, dissolved	Ca	0.0	0.0	
pH		6.30	8.39	
Alkalinity	CaCO3	80.0	80.0	
Carbonic Species	CO2	146.1	69.9	
CaCO3 PP	CaCO3	-130.8	-7.4	

Page 1   Modified

Fig. 7.11: Solution of Example 7.6a (based on original Example 5.4).

In order to calculate the amount of CO_2 transferred from the water or into it, one can either follow the change in the C_T, or the Acidity. Since the C_T is one of the parameters appearing in the Stasoft software by default, let us calculate using this parameter:

$$\Delta CO_2 = \Delta C_T = \frac{(69.9 - 146.1)\frac{mg}{L}}{44\frac{mg}{mmol}} = -1.73mM$$

1.73 mM of CO_2 were emitted from the solution.

a. It is given that after exposure to the atmosphere, the $H_2CO_3^*$ concentration is two times higher than at equilibrium with air. Therefore, the water quality is identical to water reaching equilibrium with a gaseous phase containing CO_2 partial pressure two times higher than in the atmosphere, i.e., 0.0008 atm.

Fig. 7.12 shows the water quality in this case. As expected, in this case the final pH is lower, since less CO_2 left the solution.

Stasoft 4				— ☐ ✕
File Edit Help				
🗁 🖫 🖫 🖨	NewPage	Copy		
NAME:				
TREATMENT PROCESS:		Initial	EqmAir	
Unit:		Water	pp Atm	
Purity of Process Chemical:				
Amount:		0.00080		
PARAMETERS (mostly mg/l)				
Temperature	C	25	25	
Conductivity	mS/m	10	10	
Calcium, dissolved	Ca	0.0	0.0	
pH		6.30	8.09	
Alkalinity	CaCO3	80.0	80.0	
Carbonic Species	CO2	146.1	71.0	
CaCO3 PP	CaCO3	-130.8	-8.6	
Page 1 Modified				

Fig. 7.12: Solution of Example 7.6b (based on original Example 5.4).

8 Water softening using the lime-soda ash softening method

8.1 Introduction

Chapter 8 deals with the fundamental question of how to calculate the amount of chemicals required to make a desired change in water quality, or to result in the occurrence of a particular phenomenon. In fact, in the previous chapters, similar simple questions have been discussed: For example, when asking how much acid should be added to the water in order to reduce the pH to a certain value, in which, for example, there would be no precipitation of $CaCO_3$ in pressure lines. The present chapter expands on this by dealing with somewhat more complex questions, dosing several chemicals in parallel to achieve a final water quality characterized by a number of quality parameters. In order to demonstrate the principle, it was chosen to engage in quantitative chemical design of a classical softening method known as lime-soda ash softening.

However, before the method and the accompanying calculations are described, the principles of the change in water quality as a result of the dosage of various chemicals are summarized.

8.2 Deliberate modification of the aqueous-solid equilibrium state characteristics by the addition of chemicals to water

To understand how water quality changes as a result of the dosage of a given chemical, a few steps should be taken: (a) The relevant equations for alkalinity ($H_2CO_3^*$ alk) and acidity (CO_3^{2-} acd) should be written. (b) The rule that a change in one of these parameters occurs only if species that appear in the equation are dosed to the water should be imposed. (c) The new water quality can be calculated based on the new values of the conservative parameters. For example, a change in the carbonic alkalinity (ΔAlk) occurs only if a chemical containing carbonate, bi-carbonate, a strong base (i.e., OH^-), or a strong acid (H^+) is fed into the water. In the following equations, it is assumed that all the chemicals are fed to the water at equivalent-like concentrations (either eq/L or mg/L as $CaCO_3$) and therefore the concentration of carbonate is not multiplied by 2 in the alkalinity equation (the same with the $[H_2CO_3^*]$ concentration).

$$\Delta Alk = [CO_3^{2-}]_{added} + [HCO_3^-]_{added} + [OH^-]_{added} - [H^+]_{added}$$

$$\Delta Acd = [H_2CO_3^*]_{added} + [HCO_3^-]_{added} + [H^+]_{added} - [OH^-]_{added}$$

https://doi.org/10.1515/9783110603958-008

$$\Delta[Ca^{2+}] = [Ca^{2+}]_{added}$$

It is therefore possible to write that:

1. The addition of $CO_{2(g)}$ to water (in units of meq/L or mg/L as $CaCO_3$) results in the following changes in water quality:

$$\Delta[Ca^{2+}] = 0, \quad \Delta Acd = CO_{2\,added}, \quad \Delta Alk = 0$$

2. The addition of $Ca(OH)_2$ to water (in units of meq/L or mg/L as $CaCO_3$) results in the following changes in water quality:

$$\Delta Alk = Ca(OH)_{2\,added}, \quad \Delta Acd = -Ca(OH)_{2\,added}, \quad \Delta[Ca^{2+}] = Ca(OH)_{2\,added}$$

Note that when lime ($Ca(OH)_2$) is added to water, the changes in alkalinity and in calcium (in meq/L units) are the same.

The explanation:

1 mol $Ca(OH)_2 = 1$ mol Ca^{2+} and 2 OH^- mol, which equal 2 equivalents Ca^{2+} and 2 equivalents OH^-.

In other words, 1 mol $Ca(OH)_2 = 2$ equivalents of $Ca(OH)_2$.

Therefore, when 1 equivalent of $Ca(OH)_2$ is added, half a mol $Ca(OH)_2$ or 1 equivalent Ca^{2+} and 1 equivalent OH^- is added.

Similarly, if all concentrations are in units of mg/L as $CaCO_3$, $Ca(OH)_2$ addition of X (mg/L as $CaCO_3$) to water increases alkalinity by X, increases Ca^{2+} by X and decreases acidity by X.

3. Adding Na_2CO_3 to water (in units of meq/L or mg/L as $CaCO_3$) will result in the following changes to the water quality:

$$\Delta Alk = Na_2CO_{3\,added}, \quad \Delta Acd = 0, \quad \Delta[Ca^{2+}] = 0$$

And so on.

Example 8.1: Effect of chemical dosage to water
Analysis of supplied municipal water returned the following results:

$$Alk = 120 \text{ mg/L as } CaCO_3$$

$$[Ca^{2+}] = 100 \text{ mg/L as } CaCO_3$$

$$pH = 9.0, \text{ Temperature} = 20\,°C, \text{ TDS} = 200 \text{ mg/L}$$

Required
a. Determine the precipitation/dissolution potential of the water with respect to $CaCO_{3(s)}$ using Stasoft.
b. Characterize the new state that will be created after the addition of 21 mg/L CO_2 to the water (all concentrations in mg/L as $CaCO_3$)

Solution

a. Insert the given water characteristics as the initial water quality to the Stasoft software (the units of the calcium concentration should be converted to mg/L). According to the bottom row in the main software page (Fig. 8.1, first column), the water CCPP is 22.6 mg/L as $CaCO_3$. That is, the water has a high precipitation potential.

b. A CO_2 dosage of 21 mg/L as $CaCO_3$ (i.e., 9.2 mg CO_2/L) would have no effect on the alkalinity, but would increase the Acd and the C_T by 21 mg/L as $CaCO_3$ (each). Naturally, calcium concentration will remain constant.

   The initial water is at pH 9.0. Therefore, the CO_2 dosage will reduce its pH considerably, and therefore, although the C_T is elevated, the carbonate concentration will be reduced, and consequently the CCPP will be reduced.

   In order to calculate the new CCPP, one can insert the new water quality (i.e., Alk = 120, $[Ca^{2+}]$ = 100 and Acd = 101 + 21 = 122 mg/L as $CaCO_3$), to a new Stasoft page. Alternatively, one can add CO_2 to the original water, as demonstrated in the second column in Fig. 8.1. The resulting CCPP is 5.2.

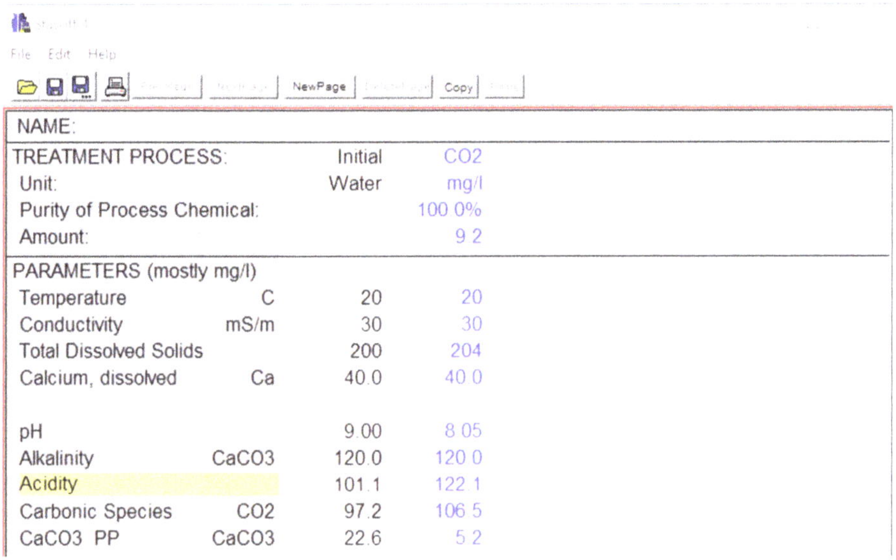

NAME:			
TREATMENT PROCESS:		Initial	CO2
Unit:		Water	mg/l
Purity of Process Chemical:			100 0%
Amount:			9 2
PARAMETERS (mostly mg/l)			
Temperature	C	20	20
Conductivity	mS/m	30	30
Total Dissolved Solids		200	204
Calcium, dissolved	Ca	40.0	40 0
pH		9.00	8 05
Alkalinity	CaCO3	120.0	120 0
Acidity		101.1	122 1
Carbonic Species	CO2	97.2	106 5
CaCO3 PP	CaCO3	22.6	5 2

Fig. 8.1: Solution of Example 8.1, changing water characteristics by adding CO_2.

8.3 Water softening

Water is defined as "soft" or "hard" depending on the concentration of the multivalent metal cations in it. Table 8.1 shows the accepted hardness ranges. These ranges appear in different forms (moving slightly up or down) in various sources in the literature, but the definition in the table is probably the most common.

Table 8.1: Hardness concentrations which correspond to the definition of low, medium, high and very high hardness.

Hardness concentration mg/L as CaCO₃	Definition of hardness
0–60	Soft water
61–120	Moderately hard water
121–180	Hard water
>180	Very hard water

In "natural" water, hardness is almost completely composed of the concentrations of Ca^{2+} and Mg^{2+} ions. Hardness is usually represented in units of mg/L as $CaCO_3$ and sometimes in meq/L units.

Total Hardness $\approx [Ca^{2+}] + [Mg^{2+}]$

The problem associated with hard water is the unwanted precipitation of solids, settling mainly on warm surfaces. As a result of the precipitation and formation of a chemical film (denoted scale, or lime scale), hard water is problematic for use in a variety of industrial processes (boilers, heat exchangers, paper manufacturing processes, glass, electronics, etc.) and also reduces the efficiency of household appliances (washing machine, kettle, dishwasher, solar water heater etc.).

Water softening is defined as a process in which calcium or magnesium ions (or both) are removed so that the hardness of the water decreases. Softening methods should be distinguished from methods based on prevention of crystallization of scale (by sound waves or the activation of a magnetic or electric field) that do not include the actual exclusion of calcium and magnesium ions from the water. Softening of water to obtain the desired hardness concentration for industrial processes depends very much on the type of process the water is designed for, reducing the hardness to a near zero concentration is not uncommon.

In places where water is softened for household supply, it is customary to reduce the concentration to a range of 100–120 mg/L as $CaCO_3$. In many countries, there is no centralized softening of supply water. It should be noted that in many places in the world where softening has been conducted, in recent years, there has been a rethinking due to accumulation of evidence regarding the importance of calcium and magnesium ions in water consumption, especially in the context of heart disease [16, 17].

Water softening can be done in a variety of methods. The most common methods are ion exchange and the "lime-soda" method. A less specialized method is the membrane separation of calcium and magnesium from a certain fraction of the water stream (by RO or NF membranes). In this method, ions are excluded in a non-selective manner. Therefore, in addition to hardness, additional elements are removed from the water. When softening is done using the lime-soda method (see the description of the method in the next section), it is sometimes necessary to stabilize the water after the softening process to prevent them from becoming aggressive to the pipes (dissolving concrete pipes and corrosion to metal pipes).

8.3.1 Softening by lime-soda ash method

The lime-soda ash process is a chemical process designed to partially or completely remove calcium and magnesium ions from water. In this method, although the calcium and magnesium ions are simultaneously removed, chemically these are two different precipitation processes.

Ca^{2+} precipitates as $CaCO_{3(s)}$, while Mg^{2+} precipitates as $Mg(OH)_{2(s)}$. The pH required for the precipitation of $CaCO_3$ (and also of $Mg(OH)_2$) is achieved as a result of the addition of slaked lime $Ca(OH)_2$, which is the cheapest base. As a result of the rise in pH, the bicarbonate ion (HCO_3^-) becomes carbonate (CO_3^{2-}), and when the product of the concentrations Ca^{2+} and CO_3^{2-} is greater than K'$_{sp}$, precipitation begins until equilibrium is achieved at saturation (see Chapter 6).

As stated, magnesium ions precipitate as $Mg(OH)_2$ according to eq. (8.1):

$$Mg(OH)_{2(s)} \leftrightarrow Mg^{2+}_{aq} + 2OH^-_{aq} \qquad K_{sp} = 8.9 \cdot 10^{-12} \qquad (8.1)$$

And not as $MgCO_{3(s)}$ ($K_{sp} = 4 \cdot 10^{-5}$).

Under typical surface water and groundwater qualities, the increase of pH a bit above pH 9.0 (resulting from the addition of $Ca(OH)_2$), calcium ions precipitate as $CaCO_3$. To precipitate the magnesium ions, more $Ca(OH)_2$ must be added to raise the pH to much higher values, up to above pH 11.

Note that precipitation of $Mg(OH)_{2(s)}$ from natural water requires elevation of pH to values higher than those required for $CaCO_{3(s)}$ precipitation. The explanation for this lies in the ions that comprise the precipitant, the composition of the natural water and the solubility constants. For the sake of rough comparison, let us briefly neglect the effect of ionic strength on precipitation reactions and assume that in the initial state, the following concentrations are present in the water: $C_T = 3 \cdot 10^{-3}$ M, $[Mg^{2+}] = [Ca^{2+}] = 10^{-3}$ M. Assume also that the target concentrations after softening are: $[Mg^{2+}] = [Ca^{2+}] = 10^{-5}$ M.

From eq. (8.1), it can be seen that K'$_{sp}(Mg(OH)_2) = [Mg^{2+}]_{eq}[OH^-]^2_{eq} = 8.9 \cdot 10^{-12}$, where the eq subscript represents equilibrium concentrations. Substitute

$[Mg^{2+}]_{eq} = 10^{-5}$ M, to get: $[OH^-]_{eq} = 9.4 \cdot 10^{-4}$ M. Therefore, the pH required for the magnesium to precipitate until reaching the desired concentration is approximately pH 11.

Similarly, $CaCO_3$ precipitation follows the equation: $K_{sp}'(CaCO_3) = [Ca^{2+}]_{eq}$ $[CO_3^{2-}]_{eq} = 8.9 \cdot 10^{-9}$. Therefore, it is obtained that $[CO_3^{2-}]_{eq} = 8.9 \cdot 10^{-4}$ M. For the given C_T, this carbonate concentration will be obtained near pK_2, at pH~9.9.

Considering this explanation, it is possible to understand that after the precipitation of all the magnesium ions, water is left with a relatively high concentration of calcium ions (remember that adding lime to the water to increase the pH until sufficient precipitation of magnesium occurs, also increases the concentration of calcium originated from the lime). Some of the calcium precipitates as $CaCO_3$, but after all the carbonate alkalinity precipitates in this manner, no CO_3^{2-} ions remain in the water and therefore no further precipitation of $CaCO_3$ can occur.

Therefore, to remove the calcium ions remaining in the water, carbonate ions should be added by dosing soda, Na_2CO_3 (Soda ash).

As a final step in the process after precipitating and separating the effluents from the solid phase, CO_2 is added to the water to reduce the pH to an acceptable level for supply water (below pH 9). The addition of CO_2 also allows the increase of C_T, which is required to give the water buffer capacity and to stabilize it chemically.

The softening process reduces the concentration of alkalinity, the concentration of dissolved calcium and magnesium, and the concentration of C_T.

Engineering wise, the purpose of the soda-ash process is to create thermodynamic and kinetic conditions in which the process will "go all the way" during a short retention time and that at the end of the process, the water will be close to saturation. To allow this, the precipitation reactions take place in an environment saturated with nucleation seeds of the precipitating solids. The reaction that takes place on the surface of the nucleation seeds in the reactor is much faster than in clear water with no solid particles. In practice, it is accepted to treat the solution in the reactor as saturated in terms of aqueous-solid phase relations. Another objective of the process is to produce stabilized water, so that no further stabilization processes will be required.

Figure 8.2 shows a schematic description of the lime-soda process.

8.3.2 Basic description of the stages of the lime-soda ash softening process

The purpose of the following description is to explain in general terms the flow chart appearing in Fig. 8.2. The purpose of this description is not to be used as a procedure for designing, and as such hydraulic retention times or sludge ages are excluded from it. The goal of this section is to enable the development of examples that highlight the process planning, and in particular the significance of the

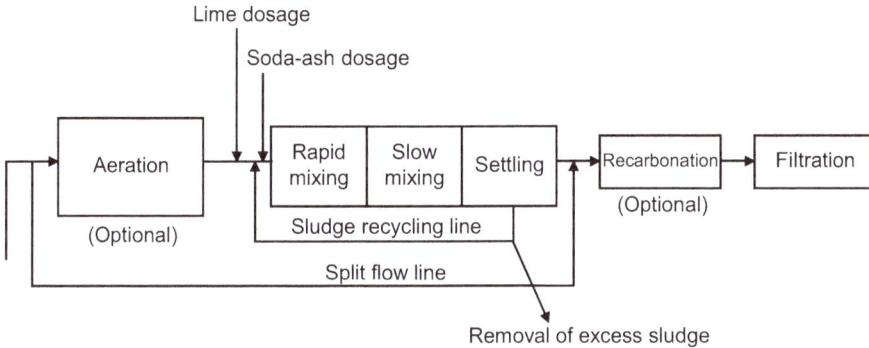

Fig. 8.2: Schematic description of the "lime-soda ash" softening process.

amount of chemicals on changing the properties of water, and how the process engineer can influence the final water characterization by the intelligent dose of chemicals.

The flow of water aimed for softening is sometimes carried over to an aeration reactor. The goal of the aeration is to emit CO_2 to the atmosphere, thereby reduce the acidity and elevate the pH, which can allow for dosing less lime. This step is optional because it is logical to apply it only in cases where the water for treatment is characterized by a significant over-saturation relative to atmospheric $CO_{2(g)}$. After the aeration stage, the water flows to the chemical/physical settling stage, which is divided into three sub-stages: rapid mixing, allowing good contact between the chemicals and the water, slow mixing that allows controllable collisions between particles and the formation of large flakes with good settling properties and a final stage of physical settling. The particles that settled in the third stage are pumped from the sediment, some of which are recirculated back to the entrance of the reactors, where they meet the fresh feed flow, and some are removed from the system as sludge for disposal. Kinetically, it is assumed that in the presence of a high surface area of solids upon which precipitation can occur, the kinetics are very rapid. Thus, immediately after the fresh hard water meets the recirculated solids and the dosed chemicals, the water reaches saturation with respect to the solid $CaCO_3$ and $Mg(OH)_2$. To maintain stable operating conditions, the mass of the particles removed in a day should be the same as the mass of the solids removed from the water in the process. After the separation of the solids in the settling reactor, the softened water flows to a "recarbonation" reactor, where CO_2 is fed into the water in order to increase the concentration of the carbonate system (C_T) and reduce the pH to a level suitable for urban supply (pH <9.0). This step is optional because it is performed only when the pH is too high, and/or the concentration of the carbonate system is too low, and/or the potential of the water to precipitate $CaCO_3$ does not meet the requirements.

The line labeled "split flow" in Fig. 8.2 describes a bypass on the treatment stage: sometimes there is an engineering/economic logic to soften the water until it is completely void of calcium and/or magnesium ions, then dilute the water with a stream that has not been treated at all. It should be noted that in cases where the demand is for the removal of calcium ions alone, it can usually be achieved by the dose of slaked lime alone ($Ca(OH)_2$), no dose of soda (Na_2CO_3) is required. In the final stage of the process, the water is filtered (usually by a deep sand bed) to reduce turbidity caused by fine solids that have not settled.

Example 8.2: Calcium removal from water (Softening) by lime dosage
The following values were obtained from water analysis:

$$Alk = 300 \, mg/L \text{ as } CaCO_3, \, [Ca^{2+}] = 280 \, mg/L \text{ as } CaCO_3, \, [Mg^{2+}] = 41 \, mg/L \text{ as } CaCO_3,$$

$$pH = 7.2, \, Temp = 20 \, °C, \, TDS = 400 \, mg/L.$$

Required
Find the lime ($Ca(OH)_2$) dose required to soften the water to a calcium concentration of 100 mg/L as $CaCO_3$ using the lime-soda method. After separating the water from the solids, find the $Ca(OH)_2$ dose required to raise the CCPP to + 5 mg/L as $CaCO_3$.

Solution
The solution using Stasoft4 is shown in Fig. 8.3. Solution procedure: First select the sub-process involved in the softening process to remove calcium ions, which is dosage of $Ca(OH)_2$. Then, select attaining equilibrium between $CaCO_{3(s)}$ and the aqueous phase, that is, reaching CCPP = 0, recall that due to the presence of nucleation sites, the water is assumed to reach equilibrium with the solid phase. After separating the water from the solids, an additional dosage of $Ca(OH)_2$ is required for stabilizing the water.

The amount of lime required for softening is found by trial and error, conducted until the calcium concentration following equilibrium with the solid phase (i.e., second process column in the Stasoft page) reaches the target concentration, that is, $[Ca^{2+}[= 100$ mg/L as $CaCO_3 = 40$ mg/L. Naturally, the calcium concentration before reaching equilibrium with the solid phase is much higher than the initial concentration.

Figure 8.3 shows that after the addition of 184.8 mg/L $Ca(OH)_2$ and reaching equilibrium with $CaCO_{3(s)}$ and $Mg(OH)_2$ (third column from the left), the water reaches CCPP = 0 (as expected) at a calcium concentration of 40 mg/L as calcium, as required. On the other hand, the magnesium concentration remains constant at these conditions, since the lime dosage is not high enough for magnesium precipitation. It can also be seen that the alkalinity concentration decreased from 300 to 120 mg/L as $CaCO_3$, and that the values of $Acd(CO_3^{2-})$ and C_T also decreased as a result of the addition of the base and precipitation of $CaCO_{3(s)}$, respectively. Technically, the method for finding the required dose of $Ca(OH)_2$ using the software is to manually increase the dose (by trial and error) until the calcium concentration (after reaching equilibrium with the solid phase) reaches 40 mg/L as calcium. Note that the required pH for precipitating calcium does not surpass pH 10. Similarly, in the next step, one should raise the dose of $Ca(OH)_2$ until the calculated CCPP of the water is +5 mg/L as $CaCO_3$. Note that due to the stabilization step, the product water calcium concentration is slightly higher than the target. Thus, in order to comply with the exact requirement of 40 mg/L calcium, one should continue the trial and error procedure to increase the first lime dosage and further soften the

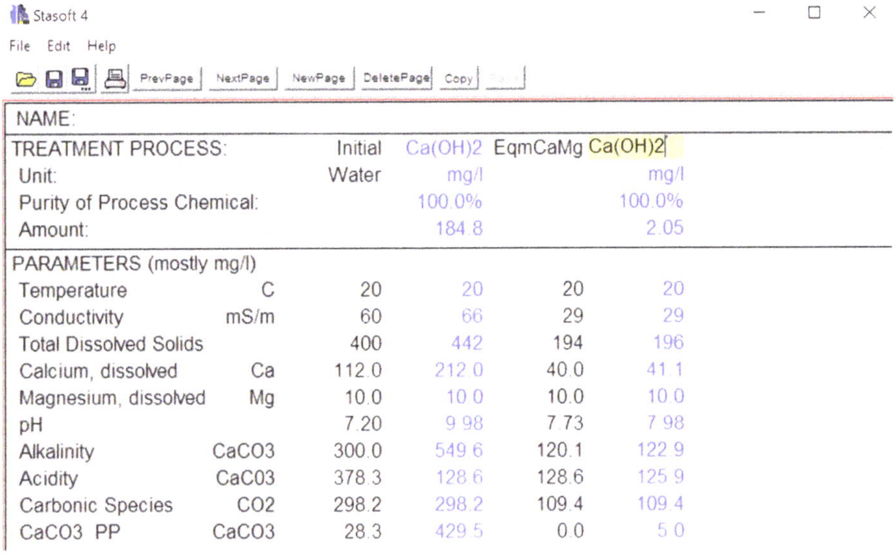

| Stasoft 4 | | | | | | | | — ☐ ✕ |

File   Edit   Help

PrevPage | NextPage | NewPage | DeletePage | Copy

NAME:						
TREATMENT PROCESS:		Initial	Ca(OH)2	EqmCaMg	Ca(OH)2	
Unit:		Water	mg/l		mg/l	
Purity of Process Chemical:			100.0%		100.0%	
Amount:			184.8		2.05	
PARAMETERS (mostly mg/l)						
Temperature	C	20	20	20	20	
Conductivity	mS/m	60	66	29	29	
Total Dissolved Solids		400	442	194	196	
Calcium, dissolved	Ca	112.0	212.0	40.0	41.1	
Magnesium, dissolved	Mg	10.0	10.0	10.0	10.0	
pH		7.20	9.98	7.73	7.98	
Alkalinity	CaCO3	300.0	549.6	120.1	122.9	
Acidity	CaCO3	378.3	128.6	128.6	125.9	
Carbonic Species	CO2	298.2	298.2	109.4	109.4	
CaCO3 PP	CaCO3	28.3	429.5	0.0	5.0	

Fig. 8.3: Solution for Example 8.2.

water. Finally, the required lime dosage for softening is 187.4 mg/L, and the required lime dosage for stabilization remains 2.05 mg/L.

Example 8.3: Concurrent softening of calcium and magnesium ions
A water sample has the following characteristics:

$$\text{Alk} = 110 \text{ mg/L as } CaCO_3, \ [Ca^{2+}] = 210 \text{ mg/L as } CaCO_3, \ [Mg^{2+}] = 40 \text{ mg/L as } CaCO_3$$

$$pH = 7.4$$

Ionic Strength = 0.01 M, Temp = 20 °C.

Required
Calculate the dose of $Ca(OH)_2$ and Na_2CO_3 needed to soften the water to reach calcium at a concentration of 60 mg/L and magnesium at a concentration of 6 mg/L as $CaCO_3$. Assume that the softening is carried out in a saturated reactor in the presence of the solid phase of the two minerals.

Solution
The solution using Stasoft4 is shown in Figure 8.4. Solution procedure: First select the sub-processes involved in the softening process to remove calcium ions: dosages of $Ca(OH)_2$ and Na_2CO_3. Then, select attaining equilibrium between $CaCO_{3(s)}$ and $Mg(OH)_2$ and the aqueous phase.

Technically, the method for finding the required dosages using the software is a two-step procedure: first, one should manually increase the dose (by trial and error) of $Ca(OH)_2$ until the concentration of magnesium (after reaching equilibrium with the solid phase, that is, last column in the Stasoft page) reaches 1.5 mg/L as magnesium. Next, the dosage of Na_2CO_3 is increased until the concentration of calcium (after reaching equilibrium with the solid phase) reaches 24 mg/L.

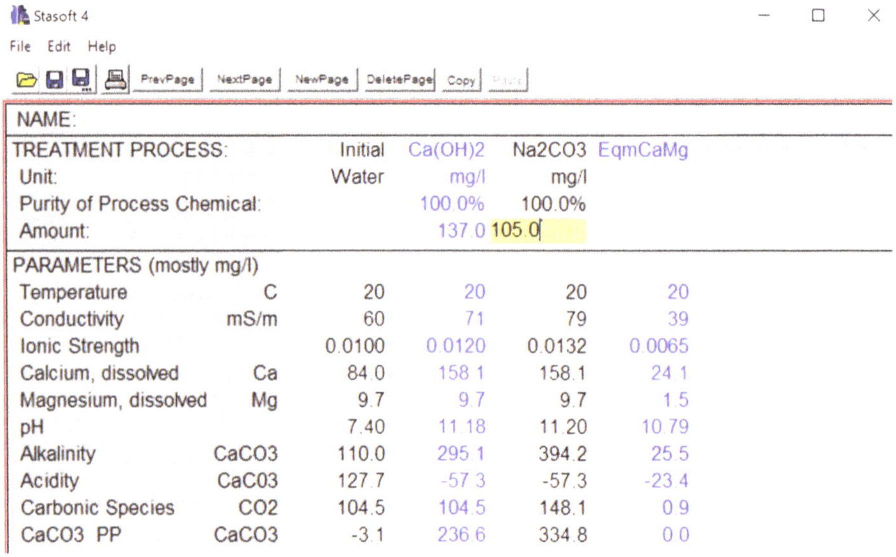

Stasoft 4 — □ ✕

File Edit Help

PrevPage | NextPage | NewPage | DeletePage | Copy | Print

NAME:					
TREATMENT PROCESS:		Initial	Ca(OH)2	Na2CO3	EqmCaMg
Unit:		Water	mg/l	mg/l	
Purity of Process Chemical:			100.0%	100.0%	
Amount:			137.0	105.0	
PARAMETERS (mostly mg/l)					
Temperature	C	20	20	20	20
Conductivity	mS/m	60	71	79	39
Ionic Strength		0.0100	0.0120	0.0132	0.0065
Calcium, dissolved	Ca	84.0	158.1	158.1	24.1
Magnesium, dissolved	Mg	9.7	9.7	9.7	1.5
pH		7.40	11.18	11.20	10.79
Alkalinity	CaCO3	110.0	295.1	394.2	25.5
Acidity	CaCO3	127.7	-57.3	-57.3	-23.4
Carbonic Species	CO2	104.5	104.5	148.1	0.9
CaCO3 PP	CaCO3	-3.1	236.6	334.8	0.0

Fig. 8.4: Stasoft solution for Example 8.3.

Figure 8.4 shows that the required dosages are 137 and 105 mg/L of $Ca(OH)_2$ and Na_2CO_3, respectively.

For didactic purposes, the process is also demonstrated in a step by step manner in Fig. 8.5. For this aim, aqueous-solid equilibrium is assumed twice, once after the lime dosage and second, after the soda dosage. This simulation demonstrates that the dosage of lime indeed leads to the required

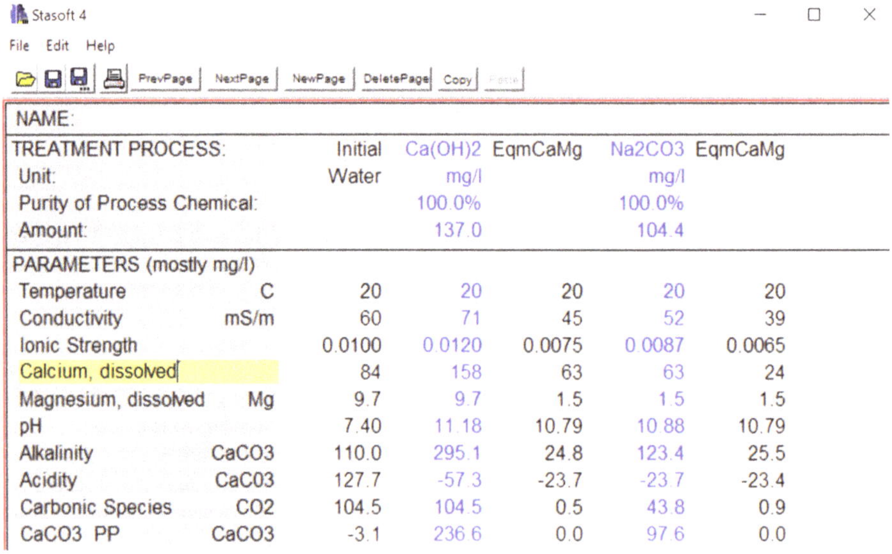

Stasoft 4 — □ ✕

File Edit Help

PrevPage | NextPage | NewPage | DeletePage | Copy | Paste

NAME:							
TREATMENT PROCESS:		Initial	Ca(OH)2	EqmCaMg	Na2CO3	EqmCaMg	
Unit:		Water	mg/l		mg/l		
Purity of Process Chemical:			100.0%		100.0%		
Amount:			137.0		104.4		
PARAMETERS (mostly mg/l)							
Temperature	C	20	20	20	20	20	
Conductivity	mS/m	60	71	45	52	39	
Ionic Strength		0.0100	0.0120	0.0075	0.0087	0.0065	
Calcium, dissolved		84	158	63	63	24	
Magnesium, dissolved	Mg	9.7	9.7	1.5	1.5	1.5	
pH		7.40	11.18	10.79	10.88	10.79	
Alkalinity	CaCO3	110.0	295.1	24.8	123.4	25.5	
Acidity	CaCO3	127.7	-57.3	-23.7	-23.7	-23.4	
Carbonic Species	CO2	104.5	104.5	0.5	43.8	0.9	
CaCO3 PP	CaCO3	-3.1	236.6	0.0	97.6	0.0	

Fig. 8.5: Step by step Stasoft solution for Example 8.3.

magnesium precipitation. As expected, after the addition of lime and precipitation of magnesium (and calcium), the carbonate species are negligible (0.5 mg/L as CO_2). The low alkalinity at this point is solely attributed to the water weak-acid system, that is, the [OH^-]. The excess calcium concentration is $63 - 24 = 39$ mg/L = 1.97 meq/L. Thus, the amount of soda required is 1.97 meq/L = 104.4 mg/L Na_2CO_3. The C_T hardly changes as a result of these two steps (soda dosage and $CaCO_3$ precipitation), since all the soda added is consumed for calcium precipitation.

Note that in this case, since magnesium precipitation is required, the pH is raised above pH 11.

Example 8.4

Refer to the lime-soda process shown in Fig. 8.2:
a. What is the purpose of the "aeration" stage? Why is it presented as optional?
b. In principle, is it appropriate to perform this step for water with the quality detailed below?
c. Assume that the aeration stage is performed on the water characterized below, and it is required to reduce the calcium concentration to 100 mg/L as $CaCO_3$ and the magnesium concentration to 4 mg/L as magnesium. Will the lime ($Ca(OH)_2$) dose required in the process decrease or increase relative to the case where aeration is not performed, or will it remain the same? How will the aeration affect the soda dose (Na_2CO_3), if at all? A well-reasoned answer is required.

Water quality to be softened: Alkalinity (Alk $H_2CO_3^*$) = 230 mg/L as $CaCO_3$, [Ca^{2+}] = 230 mg/L as $CaCO_3$, [Mg^{2+}] = 40 mg/L, pH = 7.0.

Solution
a. The purpose of the aeration stage is to release the CO_2 from the water and thereby reduce the acidity value (Acidity CO_3^{2-}) of the water and in parallel raise the pH value, to reduce the concentration of lime required to raise the pH to the level required for calcium precipitation. The reduction of C_T also reduces the buffer capacity. This too results in a reduction in lime dosage required for attaining a given pH. This step is optional because it makes sense only when the water is significantly oversaturated with regard to $CO_{2(aq)}$.
b. The answer is yes. It is possible to calculate directly and see that the concentration of $CO_{2(aq)}$ is significantly higher than 0.58 mg/L as CO_2, which is approximately the concentration of this species in equilibrium with the atmosphere (assuming the temperature is 20 °C the concentration in the given water is approximately 43 mg/L as CO_2). Another calculation will show that in equilibrium with the atmosphere, the pH will rise to pH 8.79.

    Is it possible to reach a similar answer with no calculations? Again, the answer is yes:
    The relation between the concentrations of HCO_3^- and $H_2CO_3^*$ (i.e., $CO_{2(aq)}$) at pH 7.0 is approximately 4.46 (test by the approximation formula pH=pK+log([HCO_3^-]/[$H_2CO_3^*$])). Since the alkalinity value at pH 7.0 is approximately equal to the concentration of the bicarbonate ion, the concentration of the $CO_{2(aq)}$ according to this approximation is 45.4 mg/L as CO_2 (230/4.46·44/50), which is significantly higher than its concentration at equilibrium with the atmosphere.
c. The answer is that the amount of lime required in the process will decrease. Explanation: To lower the magnesium concentration to 4 mg/L, the pH should be increased to ~pH 10.7 (following the precipitation of magnesium). Aerating the water emits CO_2 as stated, and therefore reduces the acidity ($CO_3^{2-}{}_{acd}$) on the one hand and C_T on the other hand. As a result, the concentration of base dosage (i.e., lime) required to reach the target pH is lower.

As for the soda dose, things are more complex: the correct answer is that the soda dose should not change at all, relative to the situation where aeration is not performed. Explanation: The purpose of the soda dose is to remove the remaining calcium ions present in the water after raising the pH to 11.3 (as stated, following the precipitation the pH decreases to 10.7). In practice, the concentration of soda that should be added to the water in equivalent terms is the same as the excess calcium concentration (i.e., the calcium concentration in the water that is above the required final concentration). In other words, the exact amount of soda added to the water (in equivalent units) precipitates with calcium, as demonstrated in Example 8.3.

Quantitatively, excess calcium ions after pH increase are the difference between alkalinity in the water (after addition of lime and $CaCO_3$ precipitation and initial alkalinity concentration).

From the water characterization, it is also clear that after raising the pH to 11.3 all the carbonate that existed naturally in the water would precipitate with calcium ions as $CaCO_3$. If aeration is applied, the result is a reduction in the concentration of $CO_{2(aq)}$ in the water (with equivalent reduction in C_T value). Since if CO_2 is not removed, it has to be converted to CO_3^{2-} on the way to pH 11, it can be deduced that reduction in lime dose requirement is equivalent to the emitted CO_2:

$$\Delta CO_2 \, (meq/L) = -\Delta Ca(OH)_2 \, (meq/L)$$

The reduction in the required lime dosage (when aeration is performed) is naturally equal to the reduction in excess calcium that should be removed. Finally, this is also identical (in equivalent terms) to the reduction of soda required to precipitate the excess calcium.

In summary, the soda dose remains the same whether aeration is performed or not.

Example 8.5

The following water characteristics are given:

$$H_2CO_3^* \, alk = 170 \, mg/L \text{ as } CaCO_3; [Ca^{2+}] = 80 \, mg/L \text{ as } CaCO_3; pH7.7; \, T = 20\,°C;$$

$$TDS = 400 \, mg/L.$$

a. Calculate the CCPP using Stasoft.
b. Calculate the concentration of soda (Na_2CO_3) to be added to the water to reach equilibrium with $CaCO_3$ (i.e., CCPP = 0).
c. Calculate the concentration of soda to be added to the water to obtain CCPP of 5 mg/L as $CaCO_3$ at the exit from the water treatment facility.

Solution

a. Insert the water quality to Stasoft. The resulting CCPP (first column in Fig. 8.6) is 0.9 mg/L as $CaCO_3 = 0.018$ meq/L.
b. The addition of soda will raise the C_T and the alkalinity, therefore the CCPP will also rise. As the water CCPP is slightly negative, a small dosage of soda is perceived to be enough. Trial and error show that indeed 1.1 mg $Na_2CO_3 = 0.02$ meq/L is required, see Fig. 8.6.
c. Another 6.2 mg/L is required for obtaining a positive CCPP of 5 mg/L.

```
Stasoft 4                                                        —  □  ×
File  Edit  Help
 📂 💾 💾 🖨  PrevPage | NextPage | NewPage | DeletePage | Copy |        |
```

NAME:					
TREATMENT PROCESS:		Initial	Na2CO3	Na2CO3	
Unit:		Water	mg/l	mg/l	
Purity of Process Chemical:			100.0%	100.0%	
Amount:			1.1 6 2		

PARAMETERS (mostly mg/l)

Temperature	C	20	20	20	
Conductivity	mS/m	60	60	60	
Total Dissolved Solids		400	401	405	
Calcium, dissolved	Ca	32.0	32.0	32.0	
pH		7.70	7.73	7.93	
Alkalinity	CaCO3	170.0	171.0	176.8	
Carbonic Species	CO2	155.2	155.7	158.2	
CaCO3 PP	CaCO3	-0.9	0.0	5.0	

Fig. 8.6: Stasoft solution for Example 8.5.

9 Water stabilization and remineralization

9.1 Introduction

One of the fields to which the material discussed in this book is very relevant is the stabilization of soft and low TDS water. Soft and low TDS water exist naturally on all continents, and over the last decade, large volumes of desalinated water have been added, a phenomenon which is expected to become more and more common in the near future. As explained in Chapter 6, water characterized by a negative precipitation potential with respect to $CaCO_{3(s)}$ may be corrosive to both iron and concrete-based pipes. Water low in TDS also has a low buffer capacity, and is thus susceptible to large changes in pH due to reactions that invariably occur in the distribution system, either intentionally (addition of chemicals such as $Cl_{2(g)}$ or F^-) or inadvertently (microbial processes). In addition, water that is very low in dissolved solids has the disadvantage of being tasteless, and often its mineral content is insufficient to support human health needs, particularly with regard to Magnesium and Calcium in the diet. For these reasons, soft water is often required to be stabilized and remineralized. The various processes used to achieve this goal consist of methods which increase the buffering capacity of the water, its hardness concentration (Ca^{2+} alone or both Ca^{2+} and Mg^{2+}) and the pH value, with the purpose of attaining a slightly positive precipitation potential with respect to calcium-carbonate.

9.2 Overview of existing stabilization/remineralization technologies

In general, stabilization/remineralization technologies can be divided into three main groups: (1) processes that are based on direct dosage of soluble chemicals to the water; (2) processes that rely on blending the water with another water source; and (3) processes that rely on dissolution of quarry rocks (typically calcite or dolomite) into the water.

This chapter reviews existing stabilization technologies from these three groups, with emphasis on limitations, advantages and disadvantages, resultant water quality, and potential for compliance with quality regulations. Some of the technologies presented here are commonly used in desalination plants and in other water treatment plants designed for hardening soft water, while others, although being well known, are rarely applied since their disadvantages outweigh their advantages.

https://doi.org/10.1515/9783110603958-009

9.3 Direct dosage of chemicals

Direct chemical addition (also denoted "direct dosage") refers to direct injection of chemicals to the water. The chemicals may be either in a slurry form (usually hydrated lime, $Ca(OH)_2$), dissolved components (e.g., calcium salts), or in a condensed, liquid form of $CO_{2(l)}$ that transforms to $CO_{2(g)}$ and dissolves in the water. Most direct dosage treatments combine the addition of hardness (usually in the form of calcium ions) and alkalinity to the water.

Table 9.1 summarizes the chemicals commonly applied in stabilization units based on the "direct dosage" methods and the associated minerals and other water quality parameters (i.e., C_T and alkalinity) added to the water as a result of the dissolution of 1 mole of each of the chemicals. Evidently, by choosing various chemical types and dosages, a wide range of water qualities can be achieved.

Table 9.1: The increase (in Moles or Equivalents) in Na^+, Cl^-, C_T, alkalinity and Ca^{2+} as a result of the dissolution of 1 mole of each of the chemicals used in the "Direct Dosage" approach.

Chemical dissolved (1 mole)	Added quantity				
	Na^+ Equiv	Cl^- Equiv	C_T mol	Alk Equiv	Ca^{2+} Equiv
CO_2 [a]	0	0	1	0	0
$NaHCO_3$	1	0	1	1	0
Na_2CO_3 [a]	2	0	1	2	0
$Ca(OH)_2$ [a]	0	0	0	2	2
$CaCl_2$ [a]	0	2	0	0	2
$NaOH$	1	0	0	1	0

a – 1 mole of the chemical equals 2 equivalents

Note that direct dosage of each of these chemicals (except lime) can be easily applied as a complementary practice to any other stabilization approach.

9.3.1 Ca(OH)₂ followed by CO₂ addition

Lime ($Ca(OH)_2$) enriches the water with both total hardness and alkalinity (at a 1:1 ratio, in equivalent units), however, carbonate alkalinity is not added, as shown in Table 9.1. As a result, lime does not contribute to the (carbonate) buffering capacity of the water. To efficiently dissolve hydrated lime, the water must be acidified. This is done by dissolving CO_2 into the water prior to the addition of hydrated lime, thereby increasing the carbonate content of the water and its buffering capacity.

Hydrated lime is dosed as a slurry and CO_2 is applied in a condensed, liquid form ($CO_{2(l)}$) that transforms to $CO_{2(g)}$ and $CO_{2(aq)}$ in the water (Fig. 9.1). The method suffers from several drawbacks: the use of hydrated lime slurry is relatively complex from an engineering standpoint [18–21], especially if the water is relatively warm, which reduces the solubility of lime [22]; the method has been reported to suffer from control problems (i.e., difficulties in maintaining consistent pH in the product water) [21]; finally, the approach may raise the water's turbidity to values higher than 5 NTU [21–25]. On the other hand, in this alternative, no unwanted counter ions are added to the water (Table 9.1).

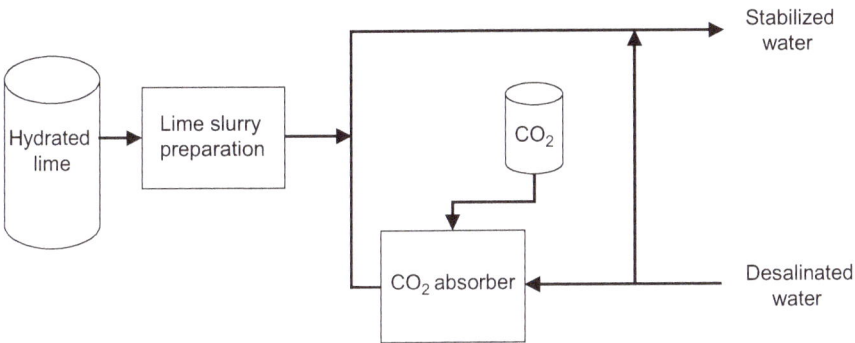

Fig. 9.1: Schematic of the CO_2 lime direct dosage process, modified from Birnhack et al. [26].

9.3.2 Ca(OH)$_2$ and Na$_2$CO$_3$ or Ca(OH)$_2$ and NaHCO$_3$

Using Na_2CO_3 or $NaHCO_3$ as the source of inorganic carbon (C_T) instead of CO_2 (as in the previous alternative) results in an elevated pH value, since these salts (Na_2CO_3 and $NaHCO_3$) consist of the basic species of the carbonate system, while CO_2 is the acidic species (HCO_3^- may be basic or acidic, depending on the initial pH of the water). Consequently, the dissolution potential of lime decreases, and more significantly, even at relatively low alkalinity and Ca^{2+} concentrations (~30 mg/L as $CaCO_3$ and ~10 mg/L, respectively), the resulting pH is excessively high (>pH 10.5), rendering this approach impractical because drinking water cannot typically exceed a pH value of 9.5. Accordingly, Withers [21] stated that this method is more appropriate for raw water which contains a certain initial alkalinity concentration and a relatively high $CO_{2(aq)}$ concentration, which is naturally characterized by relatively low pH values. In such a case, the raw water also contains a substantial amount of buffer capacity as compared to RO permeate, and thus the dissolution capacity of lime increases.

A comparison of these chemical combinations (carbonate salts in addition to lime) to the combination described below (CO_2 based lime dissolution) shows that attaining the same C_T and Ca^{2+} concentration by both methods, results (as can be expected) in lower product water alkalinity if CO_2 is used, as can also be concluded from Table 9.1. Another drawback of the $NaHCO_3$ and Na_2CO_3 based methods is the relatively high Na^+ concentration added to the product water in these alternatives.

9.3.3 CaCl$_2$ and NaHCO$_3$

This process is based on simple dissolution of soluble chemicals and no handling of slurry or gases is required. Hence, from an engineering point of view, it is simpler than the first two options discussed. However, $CaCl_2$ is often a more expensive source of Ca^{2+} than lime and the method involves the introduction of unwanted Cl^- and Na^+ ions to the water. The dissolution of $NaHCO_3$ results in a pH value around 8.3 (the equivalent point of HCO_3^-), and the addition of $CaCl_2$ can lead to practically any required Ca^{2+} concentration and thus also any required CCPP value. However, at low Ca^{2+} concentrations (e.g., $Ca^{2+}<80$ mg/L as $CaCO_3$) resulting from low salt addition, it is required to elevate the pH by adding a base to attain a positive CCPP (or LSI) value.

9.4 Blending of low TDS water and other water sources

Depending upon availability, blending high TDS, high hardness water with soft, low TDS water is a very cheap method, which may be used to increase the concentration of desired ions in the latter. However, such blending may also increase the concentration of undesired species. The concentrations of all the introduced salts are a function of the blended water composition and the dilution fraction, making the control over product water quality limited. More specifically, when this technique is used to remineralize soft water, only the concentration of a single component can be adjusted to its required value, while all other components found in the source water would be added to the desalinated water to a value that is an unavoidable outcome of the chosen component concentration target. As a result, this method is not recommended for water intended for domestic use [24] or agricultural use [27]. In case the soft water is destined for irrigation, blending with seawater or brackish water may result in both an environmental and economic negative impact. Nevertheless, the scientific literature continues to address the option of blending, either as a sole remineralization technique, for example, when groundwater is blended with reverse osmosis permeate [28–31], or as a complementary alternative [18, 30]. Since it is often recognized that blending alone introduces mainly hardness and TDS to the water, in places where blending is practiced as a stabilization approach, in many cases it would be followed by pH correction, for example, in the City of Abu Dhabi , which receives all its water from a desalination plant, where the stabilization approach is to blend the RO permeate with seawater at a ratio of 1:500 followed by final pH elevation with NaOH dosage [32]. Such a combination of blending followed by pH control is also commonly applied in BW desalination plants in the USA [33].

Generally speaking, the effect of blending with brackish water is a function of the composition of the specific BW coupled with the target concentration of the chosen component. Table 9.2 lists the composition of various BWs used as source water in desalination plants. Using Table 9.2 one can easily calculate the outcome of the preferred blending strategy.

Table 9.2: Relevant water quality parameters of seven BWs used in desalination plants.

Water source	TDS mg/L	$[Mg^{2+}]$ mg/L	$[Ca^{2+}]$ mg/L	$[SO_4^{2-}]$ mg/L	$[HCO_3^-]$ mg/L
Saja, Sharjah Emirates[*]	3216	48.6	44.8	550	NA
Martin County, Florida[**]	3664	132	179	384	146
Sarasota County, Florida[***]	1180	70	166	609	144
Port Hueneme, California[**]	1320	58	175	670	260
Indian Wells Valley, California[***]	1630	49	164	570	370
Colorado River Water[***]	1021	38	104	342	160
El Paso Water Utilities[***]	3170	38	176	301	75
Maagan Michael, Israel	3900	153	201	251	372

*source: from Almulla et al. [34]
**source: from Greenlee et al. [35]
***source: from National Research Council [36]

9.4.1 Blending case study

The city of Eilat (Israel) receives more than 90% of its water from three desalination plants: Sabha A and Sabha B, which desalinate brackish groundwater, and Sabha C which is fed with a mixture of water from the Red Sea and the brine of Sabha A and Sabha B [37]. Thus, the quality of the desalinated water has a great impact on the distribution system, public health e.g., [38–42] and the crops and lawns irrigated using this water e.g., [43–45]. Since the water at the outlet of the desalination plants is soft and corrosive, improving it through stabilization is clearly required. The following scheme is implemented in the Eilat-Sabha plants: the 51,000 m^3 of desalinated water produced per day are blended with 3,000 m^3 of brackish water; that is, a mixing ratio of 5.9%. Following the mixing, the pH and alkalinity (and consequently also the CCPP) of the desalinated water are increased via NaOH dosage [37].

The brackish ground water blended with the raw desalinated water is hard: average dissolved Ca^{2+} and Mg^{2+} of 240 mg/L and 140 mg/L, respectively, and is characterized by a moderate alkalinity concentration of 180 mg/L as $CaCO_3$. Considering that the desalination plant permeate has low Ca^{2+}, Mg^{2+} and alkalinity concentrations, the water quality achieved as a result of the mixing strategy is adequately hard (above 70 mg/L as $CaCO_3$ of hardness), but has a relatively low alkalinity concentration. The subsequent addition of NaOH raises the alkalinity value. However, since the buffer capacity of the blended solution is low, small addition of the strong base (NaOH) results in a steep pH elevation, limiting the NaOH addition. Thus, only a small NaOH dosage is possible and the resultant alkalinity and CCPP is, in the authors' opinion, too low. This problem exemplifies the main drawback of blending as a post treatment strategy to attain a set of water quality criteria.

9.5 Post treatment methods based on (quarry) calcite dissolution

Unlike direct dosage, calcite (and dolomite) dissolution processes are conducted in reactors, in which the retention time is in the order of minutes (typically >15 min). These reactors are often mistakenly called "filter beds" although their exclusive purpose is the introduction of Ca^{2+} and CO_3^{2-} (and also Mg^{2+} in case dolomite is dissolved) to the water through dissolution, and not filtration. To enable rapid dissolution of a high concentration of $CaCO_3$, the pH value must be reduced before the desalinated water is introduced into the dissolution reactor. It should also be stressed that in any dissolution reactor, thermodynamic equilibrium cannot practically be attained, due to kinetic limitations. In other words, the CCPP of the water leaving the calcite dissolution reactor is always slightly negative. Thus, although the dissolution of $CaCO_3$ elevates the pH (to around 6.5), it must be further increased, both for achieving a more appropriate pH value for drinking water and for elevating the $CaCO_3$ dissolution potential above zero; that is, to produce water that is chemically stable within the distribution system.

9.6 Acidic chemical agents used to enhance calcite dissolution

Two acidic substances are typically used to lower pH, thereby increasing the $CaCO_3$ dissolution potential: H_2SO_4 and $CO_{2(g)}$. The H_2SO_4-based dissolution process was chosen, for example, as the stabilization method that is applied in the Ashkelon (~140 Mm^3/y) and Palmachim (~90 Mm^3/y) desalination plants in Israel [46]. The main advantage of this approach is the high calcium carbonate dissolution potential (CCDP, which is equal to -CCPP) values that can be attained when a strong acid is introduced to the water and the rapid dissolution rate of calcite at these conditions, it is possible to dissolve a significant amount of calcite in the water that passes through the reactor, and let the majority of the flow bypass the reactor, a fact that renders the reactor considerably cheaper [47]. The percentage of treated water out of the total desalinated water flow rate is denoted as "%split flow" in this book (see Fig. 9.2, for example).

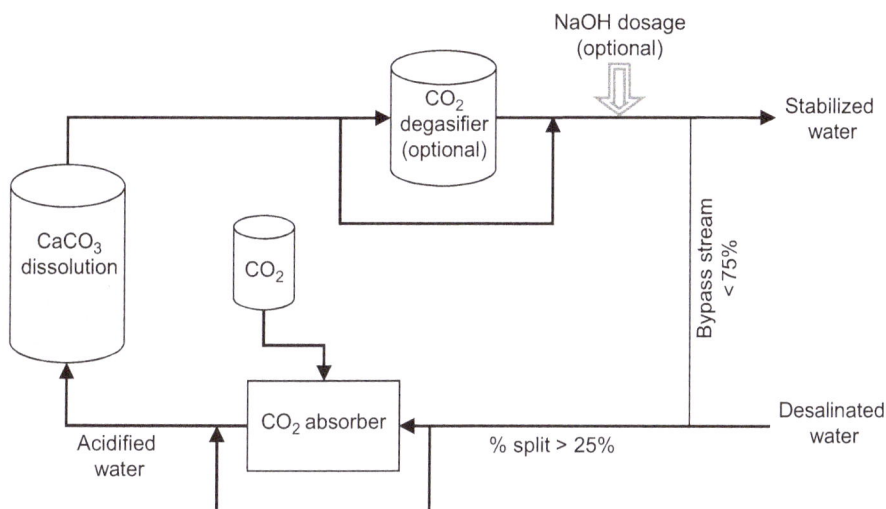

Fig. 9.2: Schematic of the CO_2-based calcite dissolution process. Final pH adjustment can be achieved either by CO_2 stripping or by NaOH dosage.

From the water quality point of view, the main difference between using H_2SO_4 and CO_2 is that the H_2SO_4-based calcite dissolution process results in a dissolved calcium to alkalinity concentration ratio that is always equal to, or higher than 2 to 1 (in equivalent units) while the alternative process, that is, the CO_2-based calcite dissolution, results in a ratio of approximately 1 to 1. This can be observed, for example, in the actual water quality produced in post-treatment in plants that apply this method [29, 48]. The 2 to 1 ratio is explained by considering the data shown in Table 9.3, which shows the changes in alkalinity, Acidity and Ca^{2+} concentrations as a result of applying the H_2SO_4-based calcite dissolution process. The process ends when the pH is raised (by NaOH addition) to around 8.3; that is, close to the HCO_3^- equivalent point. At this point, the alkalinity value exactly equals the acidity value (see Chapter 3). Substituting the expressions for the final alkalinity and acidity (given in the last row of Table 9.3) yields:

$$Alk_{final} = Acd_{final} \Rightarrow \quad y + z - x = x - z$$

Thus,

$$y = 2(x - z)$$

Table 9.3: Alterations in water quality parameters as a result of the steps in the H_2SO_4-based calcite dissolution process.

Chemical added in the process	dosage	Alteration in component value		
	Equiv/L	Alk Equiv/L	Acd Equiv/L	$[Ca^{2+}]$ Equiv/L
H_2SO_4	x	−x	+x	0
$CaCO_3$	y	+y	0	+y
NaOH to ~pH 8.3	z	+z	−z	0
Product water		y + z − x	x − z	y

Substituting the expression for y into the final alkalinity yields:

$$Alk_{final} = y − (x − z) = 2(x − z) − (x − z) = x − z = 0.5y$$

That is, the final alkalinity equals exactly half the Ca^{2+} concentration.

The 1:1 ratio attained in the CO_2-based process can be simply concluded from the data given in the first row of Table 9.1 and the second row of Table 9.3.

Note that the dissolution rate of calcite at pH < 5.5 is significantly more rapid than the rate observed at higher pH values (or in other words at lower CCDP values) (e.g., [49, 50]). As a result, in H_2SO_4-based dissolution reactors, ~75% of the calcite, is dissolved in the first ~15 cm of the reactor [51]. At this point, once the majority of calcite has dissolved, and the pH has been raised above pH 4.5, the $CO_{2(aq)}$ concentration may become very high (for %split of 20%, $CO_{2(aq)}$ concentration may be >200 mg/L, for example), and in fact, acid/base conditions closely resemble those prevailing in a CO_2-based dissolution reactor.

Calcite dissolution enhanced by a combined dosage of CO_2 and H_2SO_4 is also possible. In this case, the resultant water quality (e.g., from the alkalinity to Ca^{2+} ratio and the CO_2 concentration standpoints) will fall in between the qualities obtained when either CO_2 or H_2SO_4 are applied as the acidifying agents.

9.7 Final pH adjustment

Elevation of the pH value in the water flowing out of the calcite dissolution reactor is usually carried out by a controlled NaOH dosage. However, in some cases, it can be also achieved by controlled CO_2 stripping. pH elevation by CO_2 stripping can be practiced only in cases where a significant CO_2 super-saturation exists with respect to atmospheric $CO_{2(g)}$ at the outlet of the dissolution reactor, since stripping is based on emitting excess CO_2 and approaching aqueous–gas phase equilibrium with respect to CO_2. As a result, this approach can be applied only when CO_2 is used as the acidifying agent. The maximal final pH value that can be theoretically achieved in this method is primarily a function of the extent of supersaturation, that is, the concentration of CO_2 that can be emitted before equilibrium is attained. In practice, equilibrium can never be attained because of kinetic limitations, and the final distance from equilibrium can be controlled by operational parameters such as the applied air flow-rate, the hydraulic retention time (HRT) and the average air bubbles diameter.

Often, following the stripping step, the pH and CCPP values are still not high enough. In such cases, further pH elevation can be realized by NaOH dosage, as practiced, for example, in Kuwait [52] and Qatar [48].

9.8 Unintentional $CO_{2(g)}$ emission during calcite dissolution

CO_2-based $CaCO_3$ dissolution reactors are normally operated at pressures higher than atmospheric (i.e., sealed reactors). As a result, loss of CO_2 to the atmosphere is minimal in these reactors. On the other hand, H_2SO_4-based $CaCO_3$ dissolution reactors are often operated in a fashion that is open to the atmosphere, since such configurations are less costly. In these reactors, after a significant amount of calcite had dissolved, the water is characterized by low pH values (~pH 5) and a relatively high C_T concentration, which comprises almost entirely of $CO_{2(aq)}$. In other words, the water is highly supersaturated with respect to atmospheric CO_2. At the outlet of the reactor, the C_T concentration is higher, but the pH is also higher. Consequently, the dissolved CO_2 (or $H_2CO_3^*$) concentration is lower. Nevertheless, CO_2 supersaturation still prevails. According to this description, it is clear that either up-flow or down-flow operation of H_2SO_4-based $CaCO_3$ dissolution in an open reactor can result in CO_2 loss to the atmosphere. It is worthy to note that CO_2 emission does not affect the Ca^{2+} or alkalinity concentrations; on the other hand, it reduces the C_T and Acidity values, and increases pH. As a result, a decreased dosage of NaOH is required to reach the same final pH (~8.3), and thus an alkalinity elevation resultant of NaOH dosage is reduced. In this case, the ratio between dissolved Ca^{2+} and alkalinity concentrations becomes higher than 2, as observed in the water quality produced, for example, in the Ashkelon desalination plant : alkalinity = 45–50 mg/L as $CaCO_3$, $[Ca^{2+}]$ = 90–110 mg/L as $CaCO_3$ [29].

9.9 Dolomite dissolution as means of supplying Ca^{2+}, Mg^{2+} and carbonate alkalinity

Dissolution of dolomite in a packed bed reactor as means of treating soft water is frequently suggested in the literature. However, to the writers' knowledge, its implementation is hardly documented. A few works that examined the performance of dolomite dissolution reactors in the context of post treatment of desalinated water [53, 54] found it impractical and expensive.

Birnhack et al. [53] examined the dissolution of dolomite rock using three different up-flow velocities. Summary of the results is shown in Fig. 9.3, indicating that 80% of the total Mg^{2+} mass added to the water dissolves in the first few cm of the dissolution reactor. The pH at which the Mg^{2+} concentration leveled out under all the operational conditions tested was approximately pH 5.5. However, although the Mg^{2+}-containing part of the rock (i.e., dolomite mineral) stopped dissolving at ~pH 5.5, the calcite part continued to dissolve at higher pH values. Nevertheless, since calcite constituted only a small fraction of the rock, the Ca^{2+} concentration in the water increased relatively gradually. As long as the calcite part of the rock dissolved, the alkalinity value also increased. The pH, on the other hand, increased very gradually in the upper part of the reactor, not only because the rock dissolved slowly but also because the buffer capacity of the carbonate system at this pH range is high. The alkalinity and pH values obtained in the effluent in this alternative were very low, as also manifested by the very negative CCPP value in the effluent: between −200 mg/L as $CaCO_3$ (at 10 m/h) and −170 mg/L as $CaCO_3$ (at 4 m/h).

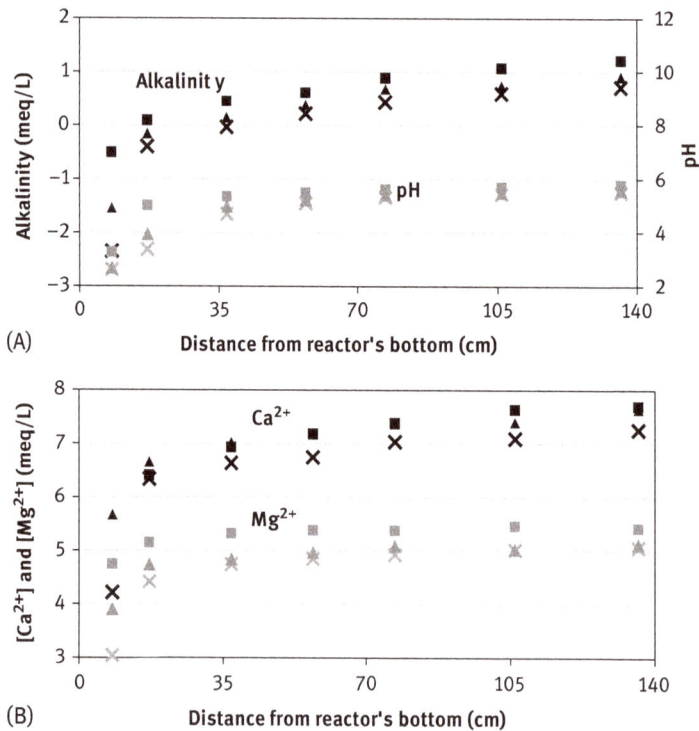

(A)

(B)

Fig. 9.3: Results of dolomite dissolution. pH of inlet solution = 2.00. Concentrations of alkalinity (black), pH (gray), Ca^{2+} (black) and Mg^{2+} (gray) (Fig. 9.3 (A) and (B), respectively) at three flow velocities: 10, 7 and 4 m/h (×, ▲ and ■, respectively). Modified from Birnhack et al. [53].

The differences between the water qualities obtained with the three flow velocities were not significant.

To overcome the problem of low pH and alkalinity, NaOH can be theoretically dosed, but such an approach is neither technically nor economically feasible. It is technically infeasible for three reasons: (1) The buffer capacity of the reactor's effluent is low. Thus, the addition of NaOH results in a significant pH increase, and the upper pH limit (pH 8.5) is exceeded before the alkalinity threshold (80 mg/L as CaCO$_3$) is attained. According to Birnhack et al. [53] attaining Mg^{2+} concentration of 12 mg/L and alkalinity of 80 mg/L as CaCO$_3$ results in a pH of 10.27; (2) As a result of the previous conclusion, when NaOH is dosed to attain the alkalinity threshold, the CCPP value becomes excessively high (62.1 mg/L as CaCO$_3$, at the mentioned conditions); (3) In case a requirement of 24 mg/L Mg^{2+} is set, the %split flow is increased and the pH and CCPP values of the effluent are very low. If the pH value is to be raised to supply water that is not very negative with regard to CaCO$_3$ precipitation potential, a very high NaOH dosage would be required.

The described drawbacks lead to the conclusion that dolomite dissolution *per se* cannot be applied in cases where the water must comply with the Israeli criteria set (or similar). In this regard, the reader is invited to also read in [54]. Therefore, the following dissolution alternative was suggested in order to overcome the drawbacks associated with the dissolution of dolomite (mainly caused by the very low CCPP).

In the suggested process, illustrated in Fig. 9.4, acidified water is pumped through dolomite and calcite reactors, operated in series. As in the conventional calcite dissolution process, NaOH is dosed to the treated water for final adjustment of alkalinity, pH and CCPP. The logic behind this alternative is that the effluents of the dolomite dissolution reactor, which are rich in CO$_{2(aq)}$, can be used in order to further dissolve calcite and by this increase the final alkalinity, pH and CCPP values, without the need for a high NaOH dosage. Such a strategy is preferred over pure dolomite dissolution from both the water quality and cost-effectiveness standpoints.

The results of feeding a calcite reactor with the effluents of a dolomite reactor are shown in Fig. 9.5. The results show that (1) The pH and C$_T$ values in the effluent of the dolomite reactor, which were found unsuitable for further dolomite dissolution (Fig. 9.3 and right hand side of Fig. 9.5), are appropriate for dissolving further several meq/L of calcite; (2) The CCPP values at the outlet of the combined dolomite-calcite reactor were between −105 mg/L as CaCO$_3$ (upflow velocity = 10 m/h) and −50 mg/L as CaCO$_3$ (4 m/h) (results not shown). That is, the CCPP values were higher than in the case where only dolomite was dissolved; (3) Using this process, it is possible to attain the following water quality: [Mg^{2+}] = 12 mg/L, [Ca^{2+}] = 120 mg/L as CaCO$_3$, alkalinity = 75 mg/L as CaCO$_3$, pH = 8.17 and CCPP = 3.1 mg/L as CaCO$_3$.

To date, this process was examined only at the laboratory scale (see [53]).

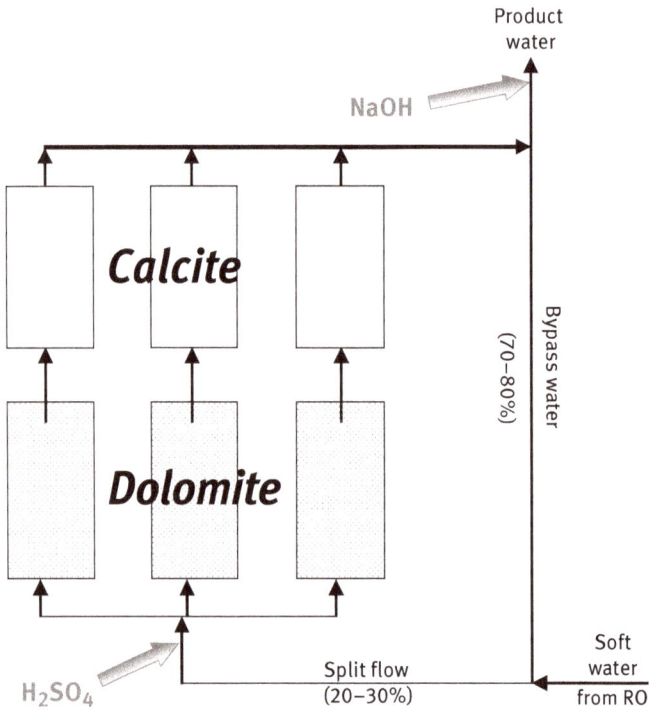

Fig. 9.4: Schematic of an in-series operation of dolomite and calcite dissolution reactors. **Source:** Birnhack et al. [53].

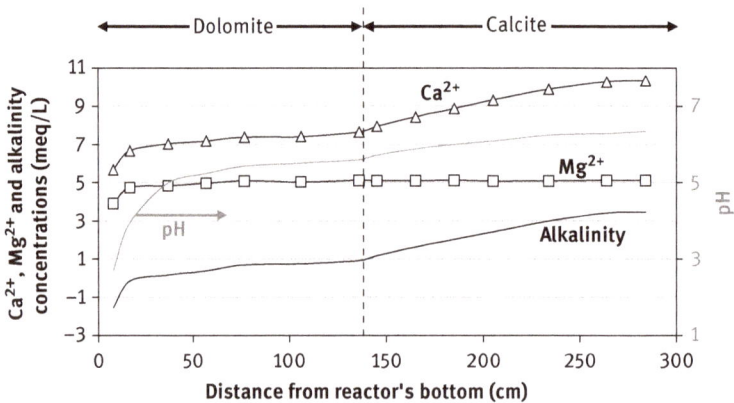

Fig. 9.5: Dissolution of dolomite and calcite in series. pH of inlet solution = 2.00, upflow velocity = 7 m/h. Concentrations of Ca^{2+}, Mg^{2+} and of alkalinity and pH (triangles, squares, black line and gray line, respectively). **Source:** Birnhack et al. [53]

9.10 Design of stabilization/remineralization processes

The most common stabilization/remineralization process is based on a combined dosage of lime $(Ca(OH)_2)$ and $CO_{2(g)}$. The popularity of this process stems from its relative ease of implementation. On the downside, one can list the cost and availability of $CO_{2(g)}$ and a relatively emanating turbidity from incomplete lime dissolution. Furthermore, typically in this alternative all the water flow is treated. The following example demonstrates the chemical dosages required in this alternative.

Example 9.1 Determine the chemical dosages of lime and CO_2 required to attain the following set of criteria. Assume that the initial water is desalinated water with TDS = 20 mg/L (all composed of Cl^- and Na^+). Water temperature = 20 °C.

Quality criteria required (according to [55]): Alk ≥ 80 mg/L as $CaCO_3$; $[Ca^{2+}]$ ≥ 80 mg/L as $CaCO_3$; CCPP ≥ 1 mg/L as $CaCO_3$. The pH value should be within the requirements for drinking water; that is, $6.5 < pH < 9.5$.

Solution For solving the problem, let us use the program Stasoft4. Figure 9.6 shows the initial characteristics of the water, and the changes occurring in the water as a result of the dosage of lime and $CO_{2(g)}$. Since the $[Ca^{2+}]$ requirement is >80 mg/L as $CaCO_3$ (0.8 mM), the required dosage of lime should be identical (i.e., 59.2 mg/L as $Ca(OH)_2$). The next step is to add $CO_{2(g)}$ such that the CCPP value will equal 1.0 mg/L as $CaCO_3$. This is achieved by the addition of 70.95 mg/L of $CO_{2(g)}$. At this point, the pH stabilizes at pH 8.06.

Note the one to one ratio between the $[Ca^{2+}]$ and alkalinity values attained in this technique (in mg/L as $CaCO_3$ or mM units).

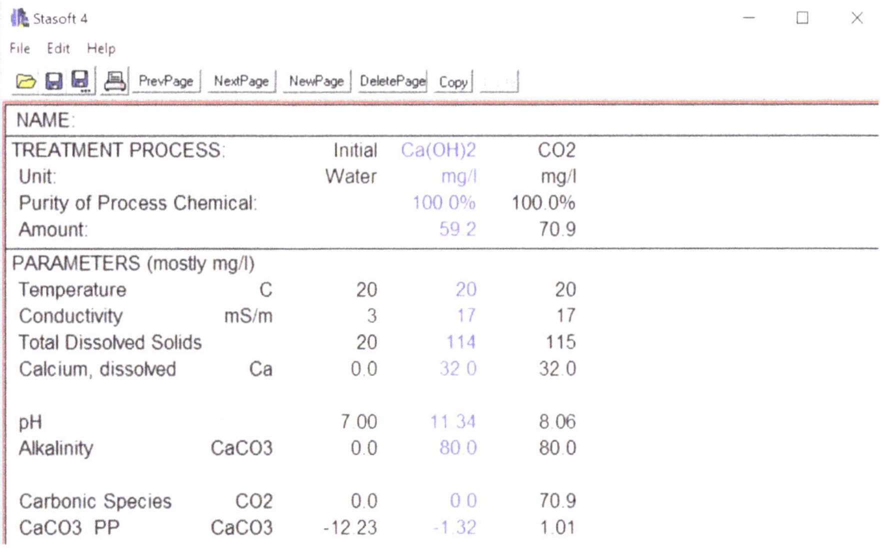

Stasoft 4				— ☐ ✕

File   Edit   Help

📂 💾 💾 🖨 PrevPage | NextPage | NewPage | DeletePage | Copy |

NAME:				
TREATMENT PROCESS:		Initial	Ca(OH)2	CO2
Unit:		Water	mg/l	mg/l
Purity of Process Chemical:			100.0%	100.0%
Amount:			59.2	70.9
PARAMETERS (mostly mg/l)				
Temperature	C	20	20	20
Conductivity	mS/m	3	17	17
Total Dissolved Solids		20	114	115
Calcium, dissolved	Ca	0.0	32.0	32.0
pH		7.00	11.34	8.06
Alkalinity	CaCO3	0.0	80.0	80.0
Carbonic Species	CO2	0.0	0.0	70.9
CaCO3 PP	CaCO3	-12.23	-1.32	1.01

Fig. 9.6: Simulative Stasoft4 solution for Example 9.1.

Another common way of adding carbonate alkalinity and Ca^{2+} ions to water is through the dissolution of calcite (quarried $CaCO_3$) into the water, followed by NaOH dosage aimed at attaining the required CCPP value and increasing pH. To enhance the dissolution potential, either H_2SO_4 or $CO_{2(g)}$ is added prior to pumping the water through the $CaCO_{3(s)}$ bed. Operational conditions typically consist of empty bed retention time (EBRT) higher than 15 min and $CaCO_3$ bead size of 1–4 mm. Examples 9.2 and 9.3 address the general design of these processes. Figure 9.7 is presented to depict the general case of calcite dissolution using both acid types.

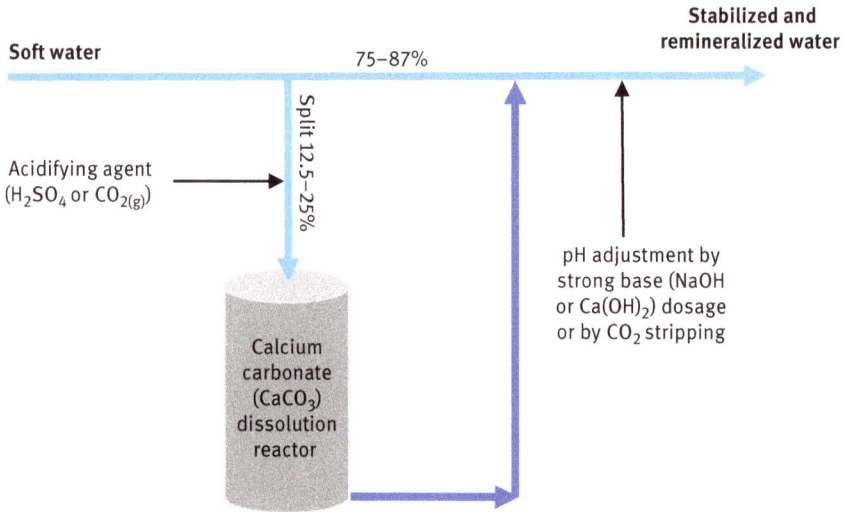

Fig. 9.7: Typical calcite dissolution setup using either $CO_{2(g)}$ or H_2SO_4 to enhance dissolution potential of the water pumped into the calcite bed.

Example 9.2 Provide the dosages of $CaCO_3$, H_2SO_4 and strong base required to attain the following quality criteria using the calcite dissolution remineralization technique described in Fig. 9.7 (split ratio = 25%). Quality criteria required: Alk ≥ 50 mg/L as $CaCO_3$, CCPP ≥ 2 mg/L as $CaCO_3$. Explain the ratio attained in this method between $[Ca^{2+}]$ and Alk in the product water. Assume that the initial water is desalinated water with TDS = 20 mg/L (comprising solely of Cl^- and Na^+). Use NaOH as the strong base. Water temperature = 20 °C.

Solution The following Stasoft4 screens demonstrate the solution to the problem. The design sequence starts with a dosage of H_2SO_4 to provide a dissolution potential that is slightly higher than the required one, which in this case is ~400 mg/L as $CaCO_3$. This figure (400 mg/L as $CaCO_3$) is derived from the 2 to 1 ratio expected between $[Ca^{2+}]$ and Alk in the product water in this process (see Table 9.3). Since the required alkalinity in the product water is 50 mg/L and the stream split ratio is 25%, $50 \cdot 1/0.25 \cdot 2 = 400$ mg/L as $CaCO_3$ of Ca^{2+}. In calcite dissolution reactors operated at hydraulic retention times higher than 15 min, it is common to assume that the dissolution potential will be almost materialized and the CCPP at the outlet of the dissolution reactor will be around −25 mg/L as $CaCO_3$. Accordingly, the solution technique in Stasoft4 is to iteratively add H_2SO_4 to a $CaCO_3$ dosage of 400 mg/L until the CCPP at the outlet of this step settles at −25 mg/L as $CaCO_3$. As shown in Fig. 9.8, this requires dosing 241.1 mg/L of H_2SO_4. The next step is to dilute the treated

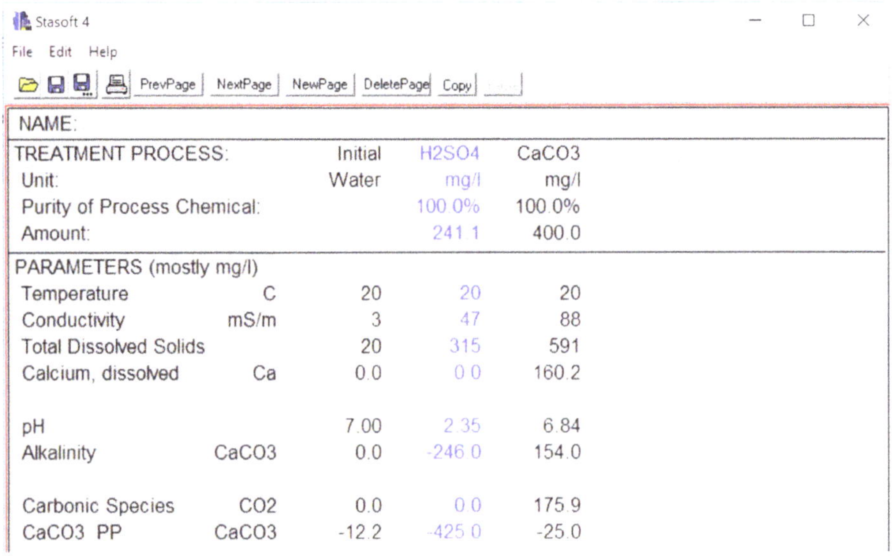

Stasoft 4				— ☐ ✕

File   Edit   Help

PrevPage | NextPage | NewPage | DeletePage | Copy

NAME:					
TREATMENT PROCESS:		Initial	H2SO4	CaCO3	
Unit:		Water	mg/l	mg/l	
Purity of Process Chemical:			100.0%	100.0%	
Amount:			241.1	400.0	
PARAMETERS (mostly mg/l)					
Temperature	C	20	20	20	
Conductivity	mS/m	3	47	88	
Total Dissolved Solids		20	315	591	
Calcium, dissolved	Ca	0.0	0.0	160.2	
pH		7.00	2.35	6.84	
Alkalinity	CaCO3	0.0	-246.0	154.0	
Carbonic Species	CO2	0.0	0.0	175.9	
CaCO3 PP	CaCO3	-12.2	-425.0	-25.0	

Fig. 9.8: Simulative Stasoft4 solution for Example 9.2, part a.

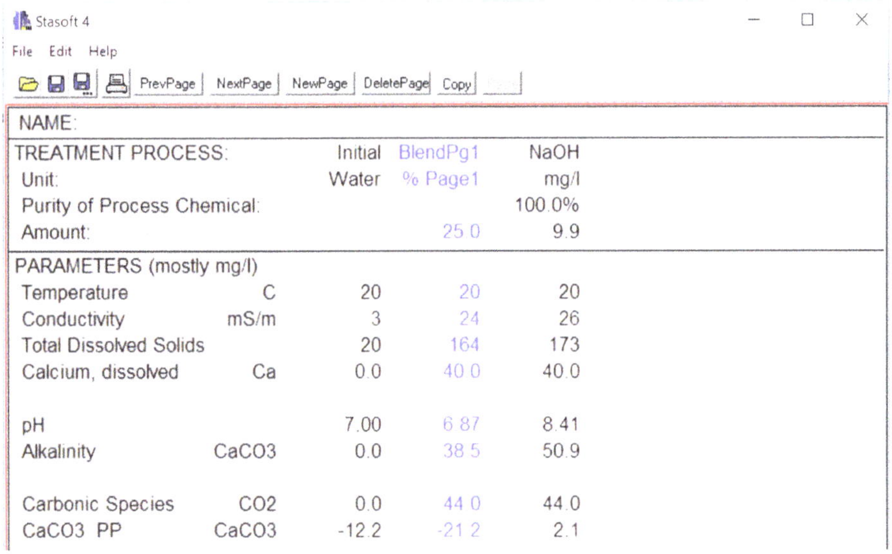

Stasoft 4				— ☐ ✕

File   Edit   Help

PrevPage | NextPage | NewPage | DeletePage | Copy

NAME:					
TREATMENT PROCESS:		Initial	BlendPg1	NaOH	
Unit:		Water	% Page1	mg/l	
Purity of Process Chemical:				100.0%	
Amount:			25.0	9.9	
PARAMETERS (mostly mg/l)					
Temperature	C	20	20	20	
Conductivity	mS/m	3	24	26	
Total Dissolved Solids		20	164	173	
Calcium, dissolved	Ca	0.0	40.0	40.0	
pH		7.00	6.87	8.41	
Alkalinity	CaCO3	0.0	38.5	50.9	
Carbonic Species	CO2	0.0	44.0	44.0	
CaCO3 PP	CaCO3	-12.2	-21.2	2.1	

Fig. 9.9: Simulative Stasoft4 solution for Example 9.2, part b.

stream at a 4 to 1 ratio. This is done by clicking on "NewPage" and defining in it the untreated water characteristics (in this case water that contain TDS = 20 mg/L and nothing more).

The next calculation step is depicted in Fig. 9.9. The first step is to blend the water from Page 1 at a ratio of 4 to 1 (i.e., 25% split). Thereafter, a strong base is added (NaOH) to adjust the CCPP value to >2 mg/L as $CaCO_3$ and concurrently raise the pH value. The chemical dosages required in Example 9.2 are thus:

$[H_2SO_4] = 241.1 \cdot 0.55 = 60.28$ mg acid per 1 liter of product water.

$[CaCO_3] = 400$ mg $CaCO_3$ per 1 liter of product water.

$[NaOH] = 9.9$ mg base per 1 liter of product water.

As expected, the ratio between the Ca^{2+} concentration (100 mg/L as $CaCO_3$) and the Alk concentration in the product water approaches 2 to 1.

Example 9.3 Provide the dosages of $CaCO_3$, CO_2 and NaOH required to attain the following quality criteria using the calcite dissolution remineralization technique described in Fig. 9.7 (split ratio = 32%). Quality criteria required (based on [53]): $[Ca^{2+}] \geq 80$ mg/L as $CaCO_3$, CCPP ≥ 3 mg/L as $CaCO_3$. Explain the ratio attained in this method between $[Ca^{2+}]$ and Alk in the product water. Assume that the initial water is desalinated water with TDS = 20 mg/L (comprising solely of Cl^- and Na^+) and that at the outlet of the reactor, the water is characterized by CCPP = −25 mg/L as $CaCO_3$. Water temperature = 20 °C.

Solution Figures 9.10 and 9.11 simulate the required dosages. It is required to arrive at a $[Ca^{2+}]$ value that equals 80 mg/L in the product water hence it is required to dissolve $80/0.32 = 250$ mg/L of $CaCO_3$. The CO_2 dosage is determined in a trial and error fashion so that the CCPP after the dissolution of 250 mg/L of $CaCO_3$ would be −25 mg/L. In this case, the dosage is 174.7 mgCO_2/L in the treatment line and 55.9 mgCO_2/L of product water (after blending with the untreated desalinated water stream). Note that since the CO_2 dosage does not affect the alkalinity value and the $CaCO_3$ dissolution results in a 1 to 1 ratio between Ca^{2+} and Alk, the water leaving the calcite dissolution reactor has a $[Ca^{2+}]$ to Alk ratio of unity (in units of eq/L or mg/L as $CaCO_3$). However, since a strong base must be added in order to increase CCPP to positive territory, the Alk to $[Ca^{2+}]$ ratio in the product water is higher than 1 to 1 (in this case 1.28:1). The amount of NaOH that needs to be dosed to the water to attain CCPP of 3 mg/L as $CaCO_3$ is shown on Fig. 9.11. The resulting pH value (pH 8.1) is within the required range for drinking water supply.

Stasoft 4				— ☐ ✕
File Edit Help				

NAME:				
TREATMENT PROCESS:		Initial	CO2	CaCO3
Unit:		Water	mg/l	mg/l
Purity of Process Chemical:			100.0%	100.0%
Amount:			174.7	250.0
PARAMETERS (mostly mg/l)				
Temperature	C	20	20	20
Conductivity	mS/m	3	3	47
Total Dissolved Solids		20	22	312
Calcium, dissolved	Ca	0.0	0.0	100.1
pH		7.00	4.39	6.86
Alkalinity	CaCO3	0.0	0.0	250.0
Carbonic Species	CO2	0.0	174.7	284.6
CaCO3 PP	CaCO3	-12.2	-275.0	-25.0

Fig. 9.10: Simulative Stasoft4 solution to Example 9.3: The calcite dissolution line.

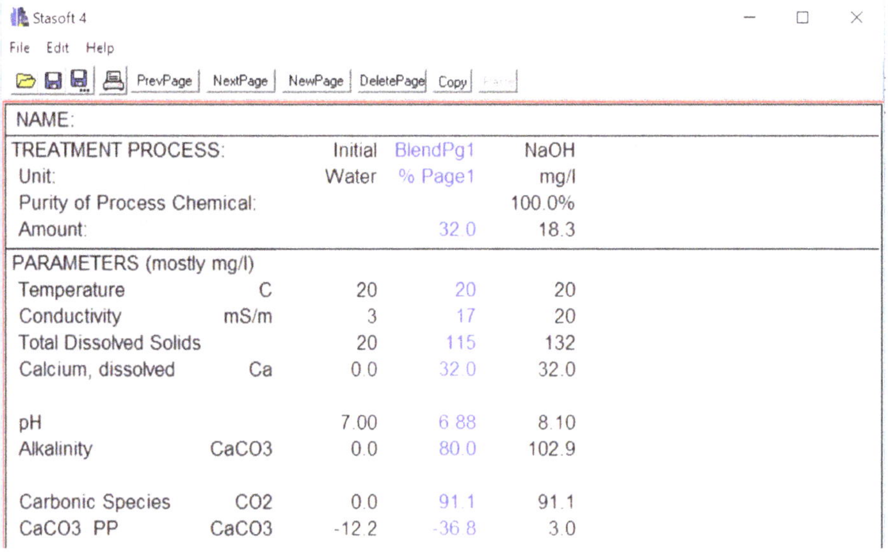

```
 Stasoft 4                                                    —  □  ×
File  Edit  Help
   🖿 🖫 🖫 🖴  PrevPage | NextPage | NewPage | DeletePage | Copy |

NAME:
TREATMENT PROCESS:              Initial  BlendPg1    NaOH
Unit:                            Water   % Page1      mg/l
Purity of Process Chemical:                          100.0%
Amount:                                   32.0       18.3

PARAMETERS (mostly mg/l)
  Temperature            C          20       20        20
  Conductivity         mS/m          3       17        20
  Total Dissolved Solids           20      115       132
  Calcium, dissolved     Ca        0.0     32.0      32.0

  pH                              7.00     6.88      8.10
  Alkalinity           CaCO3       0.0     80.0     102.9

  Carbonic Species      CO2        0.0     91.1      91.1
  CaCO3  PP           CaCO3      -12.2    -36.8       3.0
```

Fig. 9.11: Simulative Stasoft4 solution to Example 9.3, part b: blending of the two streams followed by NaOH dosage.

Example 9.4 Since NaOH is a costly chemical and it also has the disadvantage of adding Na^+ ions to the water, $Ca(OH)_2$ may be used to partially or completely replace it. The drawbacks associated with this alternative is that $Ca(OH)_2$ does not dissolve as well as NaOH at ~pH 8 and thus turbidity may develop in the product water. Another issue that may arise is heterogenic precipitation of $CaCO_3$ at the point of dosage on the surface area of the $Ca(OH)_2$ that have not dissolved. Nevertheless, let us now simulate the dosages required when $CaCO_3$ is dissolved using CO_2 and $Ca(OH)_2$ replacing NaOH as the strong base. The final water quality criteria is as in Example 9.3.

Solution In this case, since both alkalinity and calcium are added to the water when lime is dosed, $CaCO_3$ dissolution can be reduced. Fig. 9.12 depicts one of the possible simulative solutions to the $CaCO_3$ step. Because less $[Ca^{2+}]$ is required both the $CaCO_3$ and CO_2 dosages are lower than in the case NaOH is used (Example 9.3). Figure 9.13 presents the simulation of the second step (blending and lime dosage). It can be seen that the $Ca(OH)_2$ dosage is much lower (in M terms) than the required NaOH dosage and since it is also much cheaper, the advantages are clear. Having said that, most of the applied calcite dissolution processes use NaOH because of the high turbidity values emanating from the use of lime as a strong base for drinking water.

Another option for raising the pH as part of the CO_2-enhanced $CaCO_3$ dissolution process is to strip CO_2 from the blended stream. In such a case, the Acd concentration is reduced while neither the alkalinity nor the Ca^{2+} concentration change. This option has the advantage that it saves on the strong base dosage and in case NaOH is used, also

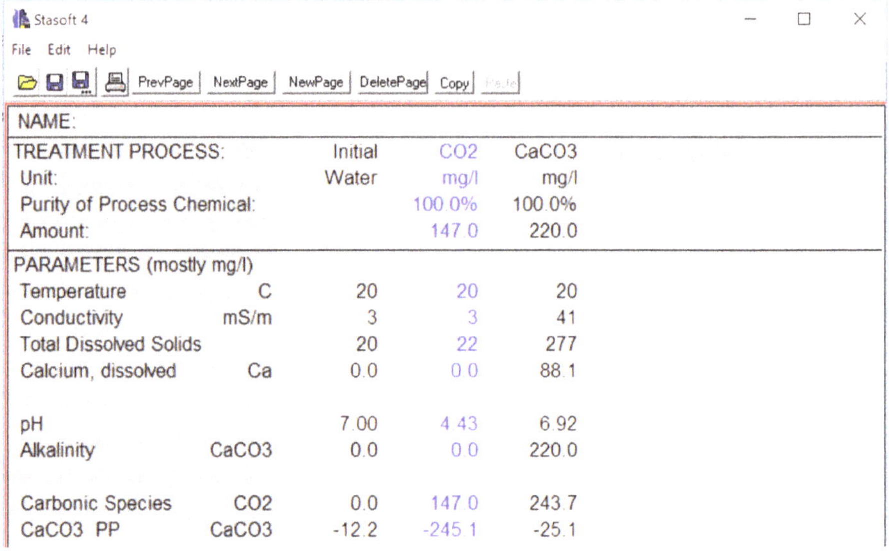

Stasoft 4 — □ ✕

File   Edit   Help

PrevPage | NextPage | NewPage | DeletePage | Copy | Paste

NAME:				
TREATMENT PROCESS:		Initial	CO2	CaCO3
Unit:		Water	mg/l	mg/l
Purity of Process Chemical:			100.0%	100.0%
Amount:			147.0	220.0
PARAMETERS (mostly mg/l)				
Temperature	C	20	20	20
Conductivity	mS/m	3	3	41
Total Dissolved Solids		20	22	277
Calcium, dissolved	Ca	0.0	0.0	88.1
pH		7.00	4.43	6.92
Alkalinity	CaCO3	0.0	0.0	220.0
Carbonic Species	CO2	0.0	147.0	243.7
CaCO3 PP	CaCO3	-12.2	-245.1	-25.1

Fig. 9.12: Simulative Stasoft4 solution to Example 9.4: the calcite dissolution step.

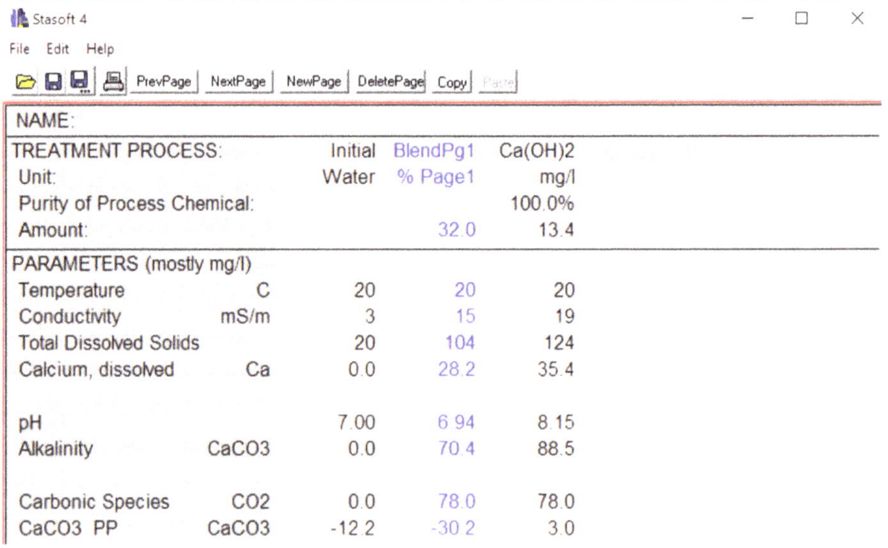

Stasoft 4 — □ ✕

File   Edit   Help

PrevPage | NextPage | NewPage | DeletePage | Copy | Paste

NAME:				
TREATMENT PROCESS:		Initial	BlendPg1	Ca(OH)2
Unit:		Water	% Page1	mg/l
Purity of Process Chemical:				100.0%
Amount:			32.0	13.4
PARAMETERS (mostly mg/l)				
Temperature	C	20	20	20
Conductivity	mS/m	3	15	19
Total Dissolved Solids		20	104	124
Calcium, dissolved	Ca	0.0	28.2	35.4
pH		7.00	6.94	8.15
Alkalinity	CaCO3	0.0	70.4	88.5
Carbonic Species	CO2	0.0	78.0	78.0
CaCO3 PP	CaCO3	-12.2	-30.2	3.0

Fig. 9.13: Simulative Stasoft4 solution to Example 9.4, part b: blending of the two streams followed by Ca(OH)$_2$ dosage.

reduces the detrimental Na^+ concentration in the water. The disadvantages are the large footprint of the stripping reactors and the somewhat problematic control over the final pH value.

Example 9.5 Provide the dosages of $CaCO_3$, CO_2 and the amount of CO_2 that needs to be stripped from the blended stream to attain the following quality criteria using the calcite dissolution remineralization technique described in Fig. 9.7 (split ratio = 32%). Quality criteria required: $[Ca^{2+}] \geq 80$ mg/L as $CaCO_3$, CCPP ≥ 3 mg/L as $CaCO_3$.

Solution The first step in this simulation is similar to the solution appearing in Fig. 9.10. Figure 9.14 shows the CO_2 stripping step. CO_2 stripping is applied in Stasoft4 simply by assigning a negative value to the CO_2 dosage. It can be seen that by stripping 21 mgCO_2/L from the blended stream the CCPP is set at 3 mg/L as $CaCO_3$ and the pH at 8.28, as required.

Stasoft 4						— ☐ ✕
File Edit Help						

☐ 🖫 🖫 🖨 | PrevPage | NextPage | NewPage | DeletePage | Copy |

NAME:					
TREATMENT PROCESS:		Initial	BlendPg1	CO2	
Unit:		Water	% Page1	mg/l	
Purity of Process Chemical:				100.0%	
Amount:			32.0	-21.1|	
PARAMETERS (mostly mg/l)					
Temperature	C	20	20	20	
Conductivity	mS/m	3	17	17	
Total Dissolved Solids		20	115	114	
Calcium, dissolved	Ca	0.0	32.0	32.0	
pH		7.00	6.88	8.28	
Alkalinity	CaCO3	0.0	80.0	80.0	
Carbonic Species	CO2	0.0	91.1	70.0	
CaCO3 PP	CaCO3	-12.2	-36.8	3.0	

Fig. 9.14: Simulative Stasoft4 solution to Example 9.5, blending of the two streams followed by $CO_{2\,(g)}$ stripping.

In this case, the ratio between Alk and $[Ca^{2+}]$ is exactly 1 to 1 (in meq/L or mg/L as $CaCO_3$ units).

10 Problems and solutions

The 21 problems (and solutions) presented in this chapter aim at encompassing the material covered in this book. They may be used for rehearsing and sharpening the knowledge of the reader, preparing toward exams or as a tool for assisting in the general design of commonly encountered processes in the water and wastewater treatment field.

Some constants and data frequently used in the solutions are summarized here:

Carbonic system equilibrium constants: $pK_{c1} = 6.3$; $pK_{c2} = 10.3$

Phosphoric system equilibrium constants: $pK_{p1} = 2.12$; $pK_{p2} = 7.2$; $pK_{p3} = 12.35$

Sulfide system equilibrium constants: $pK_{s1} = 7$; $pK_{s2} = 12.9$

Calcite solubility constant: $K_{sp}(CaCO_{3\,(s)}) = 10^{-8.48}$

Partial pressure of carbon dioxide in the atmosphere: $P_{p(CO_2)} = 0.0004\,\text{bar}$

Partial pressure of oxygen in the atmosphere: $P_{p(O_2)} = 0.21\,\text{bar}$

Henry constants:

$$H_{O_2} = 730\ \text{bar/M}$$

$$H_{CO_2} = 28.4\ \text{atm/M}$$

$$H_{NH_3} = 0.017\ \text{bar/M}$$

Question 1

The following groundwater characteristics are given: Alkalinity $(H_2CO_3^*) = 250$ mg/L as $CaCO_3$, pH = 6.7, $[Fe^{2+}] = 22.3$ mg/L, $[Mn^{2+}] = 25$ mg/L.

It is known that in aqueous environments divalent iron (Fe^{2+}) is oxidized to trivalent iron and precipitates as $Fe(OH)_{3(s)}$, divalent manganese is oxidized to manganese dioxide, MnO_2. Assume that the oxidant is dissolved oxygen $(O_{2(aq)})$ and that the described process goes all the way (i.e., all the iron and manganese are oxidized, and thereafter all the iron precipitates).

a. Write the balanced oxidation-reduction equations.

b. Calculate the alkalinity of the water after the oxidation and precipitation took place.

c. Develop a parametric equation for calculating the pH of the water following the process. Substitute all the required parameters, in the correct units. Qualitatively explain how you would expect the pH to be affected by the processes (will it considerably/moderately increase/decrease? will it remain constant?)?

d. It is known that at pH values below 4.5, the oxidation rate of Fe^{2+} is considerably reduced. What is the minimum required strong acid dosage, in order to reduce the pH of the initial solution from pH 6.7 to pH 4.5, to slow down the process?

e. Assume that the initial water (before the redox reaction takes place) is at equilibrium with atmospheric oxygen. Is the initial oxygen concentration in the water sufficient for complete oxidation of both iron and manganese?

---

https://doi.org/10.1515/9783110603958-010

! **Solution**

a. Half-cell redox equations:

$$Fe^{2+} + 3H_2O \rightarrow Fe(OH)_{3(s)} + 3H^+ + e^-$$

$$Mn^{2+} + 2H_2O \rightarrow MnO_2 + 4H^+ + 2e^-$$

$$O_2 + 4e^- + 4H^+ \rightarrow 2H_2O$$

Balanced redox equations:

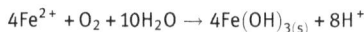

$$4Fe^{2+} + O_2 + 10H_2O \rightarrow 4Fe(OH)_{3(s)} + 8H^+$$

$$2Mn^{2+} + O_2 + 2H_2O \rightarrow 2MnO_2 + 4H^+$$

b.
$$\Delta Alk_{(for\ 1\ mol\ Fe(II)\ oxidized\ and\ precipitated)} = -8/4 = 2\ eq\ of\ alk\ consumed$$

$$\Delta Alk_{(for\ 1\ mol\ Mn(II)\ oxidized)} = -4/2 = 2\ eq\ of\ alk\ consumed$$

$$Alk\ (new) = Alk\ (old) - (2\cdot\Delta Fe\ (II) + 2\cdot\Delta Mn(II)) = 250/50 - 2\cdot22.3/55.85 - 2\cdot25/55 = 3.3\ meq/L$$

c.
$$C_T = (250/50000 - [OH^-] + [H^+]) \cdot ([H^+]^2 + K_1[H^+] + K_1K_2) / (2K_1K_2 + [H^+]\ K_1)$$

where $[H^+] = 10^{-6.7}$

The new pH can be found from (C_T constant):

$$Alk(new) = 3.3 meq/l = (2K_1K_2 + [H^+]\ K_1)C_T/([H^+]^2 + K_1\ [H^+] + K_1K_2) + [OH^-] - [H^+]$$

The new alkalinity is 165 mg/L as $CaCO_3$, that is, about 85 mg/L as $CaCO_3$ lower than the original alkalinity. This change is relatively substantial, but there is still enough alkalinity left in the solution that can stabilize the pH, at the discussed pH range. The original pH was 6.7, just above the low pK of the carbonate system, making for a high buffering capacity. The new pH is thus expected to be only moderately lower than 6.7.

d. There are two ways to solve such a question. The first one is based on the assumption that at pH 4.5, the water is at its $H_2CO_3^*$ equivalence point, and therefore Alk ($H_2CO_3^*$) → 0. Therefore, the minimum acid dosage equals the original alkalinity = 250/50, that is, 5 meq/L of strong acid. Clearly, this is an approximated value because the exact $H_2CO_3^*$ equivalence point was not determined. Alternatively, one can calculate the required strong acid dosage, based on the following equations:

$$Acid\ dose = -\Delta Alkalinity.$$
$$C_T\ final = C_T initial$$

$$Alk\ H_2CO_{3\ final}^* = [HCO_3^-] + 2[CO_3^{2-}] + [OH^-] - [H^+]$$

The final alkalinity (Alk = 3.5 mg/L as $CaCO_3$) is calculated by substituting in the explicit terms of the carbonate system species, and $pH_{final} = 4.5$. As expected, this alkalinity is almost equal to zero. The acid dosage calculated in this way is 4.93 mN.

e.
$$P_{O_2(g)} = 0.21 bar; \qquad H_{O_2} = 730\left(\frac{bar\cdot L_l}{mol}\right)$$

$$H = (O_2)_g/(O_2)_{aq} \rightarrow$$

$$(O_2)_{aq} = (O_2)_g/H = 0.21/730 = 0.28\ mM$$

Fe(II) oxidation will consume: 0.25*22.3/55.85 = 0.1 mM
Mn(II) oxidation will consume: 0.5*25/55 = 0.227 mM
Total O_2 consumption = 0.327 mM > 0.28 mM.
Conclusion: there is not enough $O_{2(aq)}$ for complete oxidation.

Question 2

The following water quality is given: alkalinity ($H_2CO_3^*$, NH_4^+) = 200 mg/L as $CaCO_3$, pH = 11.4, [NH_3] = 14 mg/L as N, [Mg^{2+}] = 15 mg/L, [Ca^{2+}] = 10 mg/L

Consider $Ca(OH)_2$ dosage of 26.3 mg/L as $CaCO_3$, resulting in a decrease in the magnesium concentration to 0.2 mg/L and a decrease in the calcium concentration to 5.4 mg/L, determine:
a.   What is the alkalinity of the water at the end of the process?
b.   Half a liter of the original water (before the lime dosage) comes in contact with clean air (containing only N_2, O_2 and CO_2), until gas-liquid equilibrium is attained. The acidity (Acidity (CO_3^{-2})) at equilibrium is −38.2 mg/L as $CaCO_3$. Assume a closed system. Determine how much CO_2 and how much NH_3 were transferred between the phases (i.e., stripped or dissolved). Determine the alkalinity in the aqueous phase.

Solution

a.   The changes in the values of three conservative parameters (including alkalinity) within the process should be calculated:

$$[Ca]_f = [Ca]_i + [Ca(OH)_2]_{dosed} - [CaCO_3]_{precipitated}$$

$$[Mg]_f = [Mg]_i - [Mg(OH)_2]_{precipitated}$$

$$Alk_f = Alk_i + [Ca(OH)_2]_{dosed} - [CaCO_3]_{precipitated} - [Mg(OH)_2]_{precipitated}$$

3 equations with 3 unknowns

$$Alk_f(mg/L\ CaCO_3) = 200 + 26.3 - [(10 - 5.4)/20 * 50 + 26.3] - (15 - 0.2)/12.15 * 50$$
$$= 127.5\ mg/L\ as\ CaCO_3$$

b.   The initial carbonate alkalinity can be calculated from the initial alkalinity:

$$Alk\ (H_2CO_3^*, NH_4^+) = Alk(H_2CO_3^*) + [NH_3] = Alk\ (H_2CO_3^*) + 1\ meq/L = 4\ meq/L$$

Therefore, Alk ($H_2CO_3^*$) = 3 meq/L
At equilibrium with the atmosphere, all the ammonia (practically) will be stripped out, thus:

$$N_T = 0$$

Therefore, $\Delta Alk_T = -\Delta Alk(NH_4^+) = -\Delta N_T$
Original $N_T = [NH_3]*(K_N + (H^+))/K_N = 1.007\ mM$

$$\Delta Alk_T = -1.007\ meq/L$$

$$Alk_{T\ final} = 4 - 1.007 = 3\ meq/L$$

$$\Delta N = -1.007\ mM \cdot 0.5\ L = -0.5\ mmol.$$

0.5 mmol of NH_3 were transferred to the atmosphere.

$$Acd_i = (1 + 2 \cdot 10^{-pH}/K_1)/(1 + 2 \cdot K_2/10^{-pH}) * (Alk - K_w/10^{-pH} + 10^{-pH}) + 10^{-pH} - K_w/10^{-pH}$$
$$= -2.48 \text{ meq/L}$$

$$\Delta Acd = Acd_f - Acd_i = -38.2/50 + 2.48 = 1.7 \text{ meq/L}$$

$$CO_{2(aq)\,stripped} = \Delta Acd = 1.7 \text{ meq/L, or } 1.7(meq/L)/2(meq/mmol) = 0.85 \text{ mM}$$

0.425 mol of CO_2 were stripped out of the 0.5 L.

Question 3

Two solutions are given: (1) 1 L of water in which 2 mM $H_2PO_4^-$ were dissolved (distilled water to which NaH_2PO_4 was added) and an unknown $[Ca^{2+}]$ concentration; (2) 1 L in which 10 mM of CO_3^{2-} were dissolved (distilled water to which Na_2CO_3 was added).

Given: $pK(H_3PO_4/H_2PO_4^-) = 2.1$, $pK(H_2PO_4^-/HPO_4^{-2}) = 7.2$, $pK(HPO_4^{-2}/PO_4^{-3}) = 12.7$
pK_{sp} ($CaCO_3$) = 8.05. Ignore ionic strength and temperature effects.

a. How many ml of the carbonate solution should be added to the phosphate solution in order to get a pH of 10.3?
b. Assume that the calcium carbonate precipitation potential (CCPP) of the mixture attained in the previous section (a) is 0. Determine the calcium concentration in the phosphate solution

---

Solution

a. Solution 1: $Alk(H_2CO_3^*, H_2PO_4^-) = 0$, $P_T = 2$ mM
    Solution 2: $Alk(H_2CO_3^*, H_2PO_4^-) = 2 \cdot (CO_3^{2-})$ dosed $= 20$ meq/L, $C_T = 10$ mM
    This is a mixing problem. Represent the volume of the carbonate solution added as x liters.
    In the mixed solution:

$$Alk\,(H_2CO_3^*, H_2PO_4^-)\,mix\,(meq/L) = [0\,(meq/L) \cdot 1\,(L) + 20\,(meq/L) \cdot x]/(1+x)(L) = 20x/(1+x)$$

$$Alk\,(H_2CO_3^*, H_2PO_4^-)\,mix = -[H_3PO_4] + [HPO_4^{2-}] + 2[PO_4^{3-}] + [HCO_3^-] + 2[CO_3^{2-}] + [OH^-] - [H^+]$$

$$@\,pH = 10.3\,(=pK_{C2}): [HCO_3^-] = [CO_3^{2-}] = 0.5C_T, [HPO_4^{2-}] \sim P_T, \text{and} [H^+] \ll [OH^-]$$

$$Alk\,mix\,(meq/L) \approx [HPO_4^{2-}] + 2[PO_4^{3-}] + 0.5C_{Tmix} + 2 \cdot 0.5C_{Tmix} + [OH^-]$$

After neglecting:

$$Alk\,mix\,(meq/L) \approx P_{Tmix} + 1.5C_{Tmix} + [OH^-]$$

$$C_T(mix, mM) = [0 \cdot 1 + 10 \cdot x]/(1+x) = 10x/(1+x)$$

$$P_T(mix, mM) = [2 \cdot 1 + 0 \cdot x]/(1+x) = 2/(1+x)$$

$$Alk\,(H_2CO_3^*, H_2PO_4^-)(mix) = 20x/(1+x) \approx 2/(1+x) + 1.5 \cdot 10x/(1+x) + (10^{-3.7} \cdot 1000_{meq/eq})$$

$$x = 0.458\,L$$

b.

$$CCPP = 0 \rightarrow [Ca^{2+}][CO_3^{2-}] = K_{sp}$$

$$[Ca^{2+}]_{mix} = K_{sp}/[CO_3^{2-}]\,mix = 10^{-8.05}/0.5\,C_{Tmix} = 10^{-8.05}/(0.5 \cdot 3.14 * 10^{-3}) = 5.67 \cdot 10^{-6}$$

$$[Ca^{2+}]_{mix}(mM) = ([Ca^{2+}]_{solution\,1} \cdot 1 + [Ca^{2+}]_{solution\,2} \cdot 0.458)/(1 + 0.458)$$

$$[Ca^{2+}]_{solution\,1} = [Ca^{2+}]_{mix} \cdot 1.458 = 8.3 \cdot 10^{-6}\,M$$

Question 4

The permeate stream of desalination processes is typically very soft and contains a negligible carbonate alkalinity. Two common methods for improving the quality of the water are: (1) CO_2 dosage combined with $Ca(OH)_2$ dissolution; (2) H_2SO_4 dosage followed by $CaCO_3$ dissolution. In both approaches, the acid is required in order to drive the dissolution reaction.

a. Fill in the following table by substituting in the change in conservative parameters due to the addition of 1 mM of the listed chemicals.

Parameter changed	Component dosed			
	CO_2	$Ca(OH)_2$	H_2SO_4	$CaCO_3$
ΔAlkalinity (meq/L)				
ΔAcidity (meq/L)				
$\Delta[Ca^{2+}]$ (meq/L)				
ΔC_T (mM)				

b. Considering that the chemical dosages in both methods are such that the final Ca^{2+} concentrations in the water are identical, which method would result in a higher alkalinity concentration in the product water?

c. Consider the dissolution of $Ca(OH)_2$: why using a strong acid such as H_2SO_4 is not recommended in this case?

d. In the method in which $CaCO_3$ is dissolved, often the pH is reduced to 2.1 by dosing H_2SO_4 and thereafter a large amount of calcite is dissolved. Calculate the <u>maximum concentration</u> of calcite that can be dissolved in this case, while avoiding the precipitation of $CaSO_{4(s)}$, which is described by the following reaction and equilibrium constant:

$$K_{sp}(CaSO_4) = 2.4 \cdot 10^{-5} \quad Ca^{2+}_{(aq)} + SO^{2-}_{4(aq)} \leftrightarrow CaSO_{4(s)}$$

In this question, neglect the effects of ionic strength and temperature.

Solution

a.

Parameter changed	Component dosed			
	CO_2	$Ca(OH)_2$	H_2SO_4	$CaCO_3$
ΔAlkalinity (meq/L)	0	2	−2	+2
ΔAcidity (meq/L)	+2	−2	+2	0
$\Delta[Ca^{2+}]$ (meq/L)	0	0	0	2
ΔC_T (mM)	+1	0	0	+1

b. CO_2 dosage combined with $Ca(OH)_2$ dissolution would result in higher alkalinity. This can be explained as follows: while the ratio of Alkalinity to Ca^{2+} (eq to eq) added as a result of $Ca(OH)_2$ and $CaCO_3$ dosages is identical (1:1), H_2SO_4 dosage reduces the alkalinity and CO_2 does not effect it.

As a result, at the outlet of the H_2SO_4 enhanced $CaCO_3$ dissolution reactor, the Alkalinity to Ca^{2+} ratio will be 0.5 to 1 while in the $Ca(OH)_2/CO_2$ method it will be exactly 1 to 1.

c.   In this case, since no C_T is added to the water, the buffer capacity will be too low and so will the CCPP.

d.
$$pH = 2.1 \Rightarrow [H^+] = 10^{-2.1}M \Rightarrow H_2SO_4 \text{ dose} = 10^{-2.1}/2 = 3.97\,mM$$

that is, the SO_4^{2-} concentration will be 3.97 mM.

$$Ksp(CaSO_4) = 2.4 \cdot 10^{-5} = [Ca^{2+}][SO_4^{2-}]$$

$$\Rightarrow [Ca^{2+}] = 2.4 \cdot 10^{-5}/(3.97 * 10^{-3}) = 0.006M$$

Therefore, the maximum concentration of calcite that can be dissolved is 6 mM.

Question 5

Consider $1\,m^3$ of air, polluted with ammonia ($NH_{3\,(g)}$) at a partial pressure of 0.02 atm. Four alternative solutions are considered for absorbing the ammonia from the air: **(1)** $1\,m^3$ of pure water, to which NaOH was added to reach pH 11; **(2)** $0.5\,m^3$ of distilled (pure) water; **(3)** $0.3\,m^3$ of pure water, to which HCl was added to reach pH 3.9; and **(4)** $0.3\,m^3$ water with $0.1\,M\,H_3PO_4$.
   Assume that the entire process takes place at 25 °C and ignore ionic strength effects.
   $K_H\,(NH_3) = 61\,M/atm$ (at 25 °C).

a.   Which solution is the most suitable for absorbing the ammonia? that is, equilibrium with which solution would result in the lowest ammonia concentration in the air? Explain.

b.   Write a parametric equation for calculating the pH of solution (4) after it has reached equilibrium with the ammonia polluted air. Assume that at equilibrium, the partial pressure of NH_3 in the headspace of the reactor is 0.002 atm.

c.   Solution (3) was contacted with ammonia polluted air until the pH of the solution was raised to 4.2. Write a parametric equation to calculate the N_T concentration of the solution at this point.

d.   The solution from section C was brought to equilibrium with clean air (i.e., partial pressure of ammonia = 0). What would be the alkalinity value of the solution at equilibrium?

e.   Write a parametric equation(s) for calculating the pH of a mixture of solution (1) and solution (4). Make sure you write all the known parameters and all the unknowns.

Solution

a.   The fourth alternative is characterized by a low pH (therefore, the NH_3 absorbed will react with H^+ to form NH_4^+, thus the NH_3 concentration will remain low even after a considerable NH_3 absorption). In addition, the fourth solution has a high buffer capacity, since it is a weak acid with two highly buffered pH areas, located many pH units below pH 9.25 (the NH_4^+/NH_3 pK).

b.   Since the partial pressure of $NH_{3(g)}$ is known, one can calculate $[NH_{3(aq)}]$ using Henry's law :

$$[NH_{3(aq)}] = K_H \cdot 0.02$$

Since this is a monoprotic acid system, the acidic species is represented as follows:

$$[NH_4^+] = \frac{[H^+][NH_3]}{K'_N} = \frac{10^{-pH}K_H 0.02}{K'_N}$$

The initial acidity (acidity (PO_4^{3-}, NH_3)) = 0.3 eq/L, and the adsorption of NH_3 does not change this value.

$$\text{Acidity}\,(PO_4^{\,3-}, NH_3) = 3[H_3PO_4] + 2[H_2PO_4^{\,-}] + [HPO_4^{\,2-}] + [NH_4^{\,+}] + [H^+] - [OH^-] = 0.3$$

In order to solve for pH, the following explicit equation (after substituting 0.1 M for P_T and the term for ammonium concentration) should be solved:

$$0.3 = \frac{0.1 \cdot (3 \cdot 10^{-3pH} + 2K'_{P_1}10^{-2pH} + K'_{P_1}K'_{P_2}10^{-pH})}{K'_{P_1}K'_{P_2}K'_{P_3} + K'_{P_1}K'_{P_2}10^{-pH} + K'_{P_1}10^{-2pH} + 10^{-3pH}} + \frac{K_H 0.02 \cdot 10^{-pH}}{K'_N} + \frac{K'_w}{10^{-pH}} - 10^{-pH}$$

c. Before the adsorption of ammonia, the alkalinity of the solution was simply

$$\text{Alk}_{\text{initial}} = -[H^+]_{\text{initial}} = -10^{-3.9}\,\text{eq/L}$$

Since NH_3 at this pH range acts as a strong base to all intents and purposes, then the addition of NH_3 (which equals the addition of N_T) adds N_T eq/L to the alkalinity value. Thus at pH 4.2:

$$\text{alkalinity}\,(NH_4^{\,+})_f = \text{Alk}_{\text{initial}} + N_T = -10^{-3.9} + N_T$$

In addition, the following equation for alkalinity can be written:

$$\text{alkalinity}\,(NH_4^{\,+})_f = ([NH_3] + [OH^-] - [H^+])_{\text{at pH 4.2}}$$

Equating the two terms and substituting into the new equation the explicit term for ammonia and pH = 4.2, results in:

$$-10^{-3.9} + N_T = \frac{N_T \cdot 10^{-9.25}}{10^{-9.25} + 10^{-4.2}} + \frac{K'_w}{10^{-4.2}} - 10^{-4.2}$$

d. At equilibrium with clean air $N_T = 0$. In other words, the ammonia stripped out of the solution is equal to N_T. thus, Δalkalinity due to equilibrium with clean air = $-N_T$
   At equilibrium with clean air: $\text{Alk} = -[H^+]_{\text{initial}} + N_T - N_T = -10^{-3.9}$

e. First define and calculate the conservative parameters in the mixed solution. Only one weak-acid system is present in the mixture: the phosphate weak-acid system. Therefore, two conservative parameters should be determined. The easiest to determine are the alkalinity(H_3PO_4) and P_T.
   In solution (1): alkalinity(H_3PO_4)$_1$ = (OH^-) = 10^{-3} N, and $P_{T,\,1} = 0$.
   In solution (4): alkalinity(H_3PO_4)$_4$ = 0, and $P_{T,\,4} = 0.1$ M.
   Let us define the alkalinity of the mixture as alk (H_3PO_4)$_{\text{mix}}$ and the total phosphate concentration of the mixture ($(0.1 \cdot 0.3 + 0)/(1 + 0.3) = 0.023$ M) as $P_{T\,\text{mix}}$.
   Accordingly:

$$\text{alk}(H_3PO_4)_{\text{mix}} = 3[PO_4^{\,3-}] + 2[HPO_4^{\,2-}] + [H_2PO_4^{\,-}] + [OH^-] - [H^+] = (0 + 10^{-3})/(0.3 + 1)$$
$$= 7.7 \cdot 10^{-4}\,\text{eq/L}$$

$$7.7 \cdot 10^{-4} = \frac{0.023 \cdot (3 \cdot 10^{-3pH} + 2K'_{P_1}10^{-2pH} + K'_{P_1}K'_{P_2}10^{-pH})}{K'_{P_1}K'_{P_2}K'_{P_3} + K'_{P_1}K'_{P_2}10^{-pH} + K'_{P_1}10^{-2pH} + 10^{-3pH}} + \frac{K'_w}{10^{-pH}} - 10^{-pH}$$

? **Question 6**

High salinity groundwater that is characterized by $Acd(CO_3^{2-}) = 650\ mg/L$ as $CaCO_3$, pH 6.7, TDS=4000 mg/L and Temperature=25 °C is desalinated using an RO membrane with 100% salt rejection and practically no rejection of $CO_{2(aq)}$ (i.e., assume equal dissolved carbon dioxide concentrations in all streams: feed, brine and permeate).

The permeate water from the RO stage is characterized by TDS = 10 mg/L and pH 4.27. This water is pumped within a very short residence time through a column that contains granular $Mg_3(PO_4)_2$. It is known that the pH of the water flowing out of the column (denoted "column discharge") is pH 5.0.

a.   What are the magnesium and P_T concentrations in the column discharge? Write detailed alkalinity and acidity expressions. Neglect species only if their contribution to the final value is lower than 1%.

b.   What is the final alkalinity of the solution? Define the proper alkalinity expression.

c.   The column discharge is now brought to equilibrium with the atmosphere. Write a short qualitative explanation for the change in the water pH caused by the exposure to the atmosphere, and add a parametric equation for the calculation of pH.

! **Solution**

a.   Convert carbonate system constants according to the given ionic strength (TDS = 4000 mg/L):
$pK'_1 = 6.18$, $pK'_2 = 9.95$, $pK'_w = 13.88$

At pH 6.7, the water subsystem can be neglected. Since $Acd(CO_3^{2-})$ is known, C_T in the saline groundwater can be calculated from

$$Acd\ (CO_3^{2-}) = 2[H_2CO_3^*] + [HCO_3^-] - [OH^-] + (H^+)$$

$$13 \cdot 10^{-3} = \frac{(2(H^+)^2 + K'_1(H^+)) \cdot C_T}{(H^+)^2 + K'_1(H^+) + K'_1K'_2} \rightarrow C_T = 10.55 \cdot 10^{-3}\ M$$

From C_T and pH, the $CO_{2(aq)}$ concentration can be calculated (in its manifestation as $H_2CO_3^*$):

$$[H_2CO_3^*]_{Feed} = [H_2CO_3^*]_{Permeate} = \frac{(H^+)^2 \cdot C_T}{(H^+)^2 + K'_1(H^+) + K'_1K'_2} = 2.45 \cdot 10^{-3}\ M$$

Since the $H_2CO_3^*$ concentration is assumed to be equal in all the streams, one can calculate the C_T in the permeate stream from the knowledge of the given pH (pH 4.27) in the permeate (note that for this stream the apparent equilibrium constants are equal to the thermodynamic ones):

$$2.45 \cdot 10^{-3}\ M = \frac{(H^+)^2 C_T}{(H^+)^2 + K'_1(H^+) + K'_1K'_2} \Rightarrow C_{T(permeate)} = 2.47 \cdot 10^{-3}\ M$$

The acidity with respect to the most basic species of the phosphate system will not change due to the dissolution of $Mg_3(PO_4)_2$.

Let us calculate the initial acidity (before the phosphate dissolution):

$$Acid(CO_3^{2-}, PO_4^{3-}) = 2[H_2CO_3^*] + [HCO_3^-] + 3[H_3PO_4] + 2[H_2PO_4^-] + [HPO_4^{2-}] - [OH^-] + [H^+]$$

At this stage, $P_T = 0$. By inserting the known pH and C_T values, $Acid\ (CO_3^{2-}, PO_4^{3-})_{initial} =$

Acid $(CO_3^{2-}) = 4.97 \cdot 10^{-3}$ eq/L is obtained. This is also the final acidity value (after the phosphate dissolution).

Since the pH after the dissolution is known (pH = 5.0), from the knowledge of the acidity and C_T, the new P_T (after the dissolution) can be calculated using the acidity equation.

$$4.97 \cdot 10^{-3} = \frac{(2(H^+)^2 + K'_{c1}(H^+)) \cdot C_T}{(H^+)^2 + K'_{c1}(H^+) + K'_{c1}K'_{c2}} + \frac{\left(3(H^+)^3 + 2K'_{p1}(H^+)^2 + (H^+)K'_{p1}K'_{p2}\right) \cdot P_T}{(H^+)^3 + K'_{p1}(H^+)^2 + K'_{p1}K'_{p2}(H^+) + K'_{p1}K'_{p2}K'_{p3}}$$

$$+ [H^+] - \frac{K'_w}{[H^+]} \to P_T = 6.95 \cdot 10^{-5} M$$

P_T is the amount of phosphate added to solution due to the dissolution of $Mg_3(PO_4)_2$. Thus:

$$[Mg^{2+}] = 1.5 P_T = 1.04 \cdot 10^{-4} M$$

b. The initial alkalinity is zero, since the desalinated water does not contain alkalinity species and CO_2 addition does not change the $H_2CO_3^*$ alkalinity value.

As a result of the dissolution of $Mg_3(PO_4)_2$, the alkalinity ($H_2CO_3^*$, H_3PO_4) value will increase at a ratio of 3 to 1 for each mole of PO_4^{3-} added to the water, while P_T is equal to the overall phosphate added. Hence:

$$Alk\,(H_2CO_3^*.H_3PO_4) = 2[CO_3^{2-}] + [HCO_3^-] + 3[PO_4^{3-}] + 2[H_2PO_4^-] + [HPO_4^{2-}] + [OH^-] - [H^+]$$

$$= 3 \cdot P_T = 2.09 \cdot 10^{-4} eq/L$$

c. Let us first calculate the supersaturation state of $CO_{2(aq)}$ in the column discharge:

$$[H_2CO_3^*]_{Product} = \frac{(H^+)^2 \cdot C_T}{(H^+)^2 + K'_1(H^+) + K'_1K'_2} = 2.35 \cdot 10^{-3} M$$

$$[H_2CO_3^*]_{Equlibrium} = K_H \cdot P_p^{CO2} = 1.36 \cdot 10^{-5} M$$

Clearly, the $CO_{2(aq)}$ concentration is supersaturated relative to the atmosphere and thus $CO_{2(g)}$ is expected to leave the solution upon exposure to the atmosphere thereby resulting in pH increase.

Since the alkalinity value does not change upon CO_2 stripping, the following parametric equation can be used to calculate the new pH:

$$Alk(H_2CO_3^*, H_3PO_4) = \frac{2P_pK_HK'_{c1}K'_{c2}}{(H^+)^2} + \frac{P_pK_HK'_{c1}}{(H^+)} + \frac{\left(3K'_{p1}K'_{p2}K'_{p3} + 2K'_{p1}(H^+)^2 + (H^+)K'_{p1}K'_{p2}\right) \cdot P_T}{(H^+)^3 + K'_{p1}(H^+)^2 + K'_{p1}K'_{p2}(H^+) + K'_{p1}K'_{p2}K'_{p3}} - [H^+] - \frac{K'_w}{[H^+]}$$

Question 7

In order to supply the required magnesium dose in the post-treatment stage at desalination plants, it was proposed to dissolve dolomite ($CaMg(CO_3)_2$), instead of calcite rock. The suggested process is shown in Fig. 10.1. The required water quality at the outlet of the post-treatment stage (i.e., stream #3 in Fig. 10.1) is (all concentrations are expressed in mg/L as $CaCO_3$):

$$H_2CO_3^*alk > 80, [Ca^{2+}] > 80, [Mg^{2+}] > 80, CCPP > 0 \text{ and } pH < 8.5.$$

NaHCO₃

Product water

③

②

CaMg(CO₃)₂(s) dissolution reactor

By pass water 70–80%

H₂SO₄

①

% Split (20–30)

Soft water from RO

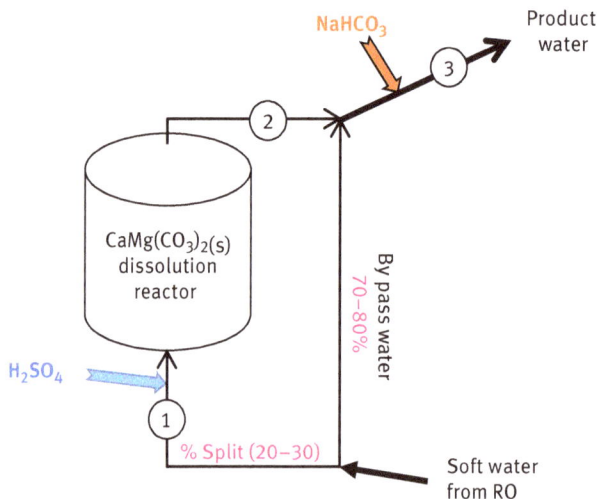

Fig. 10.1: A scheme of the dolomite dissolution process described in question 7. The labels represent the stream numbers used in the question.

a. Assume that only 25% of the feed flow enters the dolomite dissolution reactor, 452 mg/L of sulfuric acid (H_2SO_4) is fed to the reactor suction line (i.e., to stream #1), and the total hardness (combined concentration of calcium and magnesium) in the product water (i.e., in stream #3) is 160 mg/L as $CaCO_3$. What are the values of $H_2CO_3^*$ alkalinity and CO_3^{2-} acidity at the dolomite dissolution reactor <u>discharge</u> (i.e., at stream #2)? Assume that the discharge is in a closed pipe with no exposure to the atmosphere.

b. Given that the pH in the <u>reactor discharge</u> (stream #2) is lower than pH 6. What is the saturation state of $CaCO_3$ at this point (positive or negative CCPP)? Neglect ionic strength and temperature effects.

c. How much sodium bicarbonate ($NaHCO_3$ in mg/L) should be added to the product water, in order to comply with the alkalinity requirement?

Solution

a. From the dolomite dissolution reaction, one can conclude that in molar terms, the concentrations of Ca^{2+} and Mg^{2+} added to the water due to the dolomite dissolution are equal, and C_T is doubled.

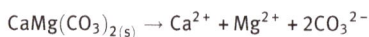

$$CaMg(CO_3)_{2(s)} \rightarrow Ca^{2+} + Mg^{2+} + 2CO_3^{2-}$$

Considering the 25% flow split (4:1 ratio between product flow rate and the dolomite dissolution reactor discharge flow rate), one can calculate the CO_3^{2-} dosage as follows:

$$[CO_3^{2-}]_{dosed} = [Mg^{+2}]_{dosed} + [Ca^{+2}]_{dosed} = TH_{dosed} = 4TH_{product} = 4\frac{160 \left(\frac{mg}{L} \text{ as } CaCO_3\right)}{50,000 \frac{mg}{eq}} = 12.8 \cdot 10^{-3}N$$

The alkalinity ($H_2CO_3^*$) at the entrance to the dolomite dissolution reactor is (as a function of the dosed H_2SO_4 concentration):

$$Alk_i = 0 - 2\Delta H_2SO_4 = 0 - 2\frac{eq}{mol}\cdot\frac{452\left(\frac{mg}{L}\right)}{98,000\left(\frac{mg}{mol}\right)} = -9.22\cdot 10^{-3}\ eq/L$$

Following the dolomite dissolution, the alkalinity increases at 1:1 ratio with the equivalents of $(CO_3^{2-})_{dosed}$:

$$Alk_{(dolomite\ discharge)} = Alk_i + [CO_3^{2-}]_{dosed} = -9.22\cdot 10^{-3} + 12.8\cdot 10^{-3} = 3.58\cdot 10^{-3} eq/L$$

The value of acidity (CO_3^{2-}) at the inlet to the dolomite dissolution reactor is equal to the H_2SO_4 addition, that is, $9.22\cdot 10^{-3}$ eq/L. Since CO_3^{2-} is the reference species, this acidity value does not change due to the dolomite dissolution. Thus, acidity $(CO_3^{2-})_{dolomite\ discharge} = 9.22\cdot 10^{-3}$ eq/L.

b. Let us check the conditions at pH 6. The Ca^{2+} concentration is known (80 mg/L as $CaCO_3$). The C_T is twice the molar concentration of calcium, or half the $(CO_3^{2-})_{dose}$ (expressed in equivalent units), that is, $C_{T\ dolomite\ discharge} = 0.5\ mol/eq \cdot (CO_3^{2-})_{dosed} = 6.4$ mM. Calculate first the CO_3^{2-} concentration (note, the carbonate concentration is not equal to the carbonate dose) and the quotient (Q) for the $CaCO_3$ precipitation reaction:

$$[CO_3^{2-}] = \frac{K_1'K_2'\cdot C_T}{(H^+)^2 + K_1'(H^+) + K_1'K_2'} = 1.07\cdot 10^{-7},\quad Q = [Ca^{2+}][CO_3^{2-}] = 3.2\cdot 10^{-3}\cdot 1.07\cdot 10^{-7}$$

$$= 10^{-9.46} < 10^{-8.48} = K_{sp}$$

Therefore, CCPP is negative.

c. The addition of $NaHCO_3$ adds to the alkalinity $(H_2CO_3^*)$ value at a 1 to 1 ratio (eq to eq):

$$Alk_{product} = Alk_{dolomite\ discharge}\cdot 0.25 + (NaHCO_3)_{dose} = 80\frac{mg}{L}\ as\ CaCO_3 \div 50000\frac{mg}{eq} \rightarrow (NaHCO_3)_{dose}$$

$$= 7.075\cdot 10^{-4} M = 59.43\frac{mg}{L}$$

Question 8

The following water quality is given: $H_2CO_3^*$ alkalinity = 150 mg/L as $CaCO_3$, pH = 7.00, $[NO_3^-]$ = 28 mg/L as N. It is known that solid nanometer-sized particles of elemental iron $(Fe_{(s)})$ can reduce nitrate (NO_3^-) to ammonium (NH_4^+) in an aqueous solution, while iron is oxidized to divalent iron (ferrous), Fe^{2+}.

a. Write the balanced oxidation-reduction equation.
b. Calculate how many grams of solid iron should be added to 100 L of water (at the given quality) to reduce the concentration of nitrate such that it will meet drinking water regulations, that is, 70 mg/L as NO_3. Assume that all the iron reacts with the nitrate and no other reactions occur.
c. What is the alkalinity $(H_2CO_3^*, NH_4^+)$ of the water at the end of the reaction?
d. If the pH must be kept fixed during the reaction, how much acid/base is needed to be added to the water?
e. This reaction is known to be faster at lower pH values. Assuming that this is a closed system, how much acid is needed to reduce the pH of the original water (before the reaction with the iron) to pH = 3? Give your answer in meq/L units.

! **Solution**

a. The balanced two half reactions:

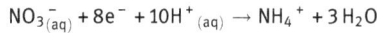

$$Fe_{(s)} \rightarrow Fe^{2+} + 2e^-$$

$$NO_{3(aq)}^- + 8e^- + 10H^+_{(aq)} \rightarrow NH_4^+ + 3H_2O$$

In order to balance the number of electrons released (first half reaction, oxidation of iron) and the number of electrons accepted (second half reaction, reduction of nitrate), the first half reaction should be multiplied by 4. After multiplication, summing up the two half reactions, one gets:

$$4Fe_{(s)} + NO_{3(aq)}^- + 10H^+_{(aq)} \rightarrow NH_4^+ + 4Fe^{2+} + 3H_2O$$

b. The amount of nitrate that should be reduced:

$$\Delta NO_3^- = \frac{28\frac{mg}{L} \text{ as N}}{14\frac{mgN}{mmol}} - \frac{70\frac{mg}{L} NO_3}{62\frac{mgNO_3}{mmol}} = 0.87mM$$

The $Fe_{(s)}$: NO_3^- stoichiometric ratio is 4:1, therefore,

$$Fe_{(s)dose} = 100L \cdot 0.87\frac{mmol}{L} \cdot 4\frac{mol \text{ } Fe}{mol \text{ } NO_3} \cdot 55.85\frac{g \text{ } Fe}{mol \text{ } Fe} = 19.5 \text{ } g \text{ } Fe_{(s)}$$

c. According to the balanced redox reaction, any mole of nitrate reduced results in an increase of 10 equivalents of alkalinity due to the consumption of H^+. The release of ammonium does not affect the alkalinity ($H_2CO_3^*$, NH_4^+).

d.
$$\Delta Alk(H_2CO_3^*, \text{ } NH_4^+) = 10\frac{eq \text{ } H^+}{mol \text{ } NO_3^-} \cdot NO_{3 \text{ } reduced}^- = 8.7mN$$

$$Alk_{new} = 3 + 8.7 = 11.7 \text{ } mN$$

e.
$$acid_{dose} = \Delta alkalinity = 8.7mN$$

f. Initial state: Alk = 3 mN, pH = 7.0.
Final state: Alk = 3 − acid$_{dose}$, pH = 3.0
Principally, in order to find the acid dose one needs to find the initial C_T. The C_T does not change as a result of the acid dosage. Therefore, $C_{T \text{ } final} = C_{T \text{ } initial}$. Given the final C_T and pH one can calculate the final alkalinity. The change in alkalinity is equal to the acid dose. However, following the acid dose, the pH is such that the concentrations of all carbonate species appearing in the alkalinity equation can be neglected. Therefore, the question can be solved without calculating the C_T.

$$Alk(H_2CO_3^*)_{final, \text{ at pH3}} = 2[CO_3^{2-}] + [HCO_3^-] + [OH^-] - [H^+] \approx -[H^+] = -10^{-3}N$$

$$acid_{dose} = \Delta alkalinity = 3 - (-1) = 4mN$$

? **Question 9**

Read carefully and circle the most correct answer. Please explain your answer.

1. The following water quality is given: pH = 8.4, S_T = 1 mM, Alkalinity $_{H_3PO_4, \text{ } H_2S}$ = 2.5$\frac{meq}{L}$. It is

known that $H_2S_{(g)}$ evaporated to the air from this water. To quantify how much H_2S is emitted you can:

a. Calculate the difference between the concentration of $H_2S_{(aq)}$ before and after evaporation.

b. Assume that the evaporation is negligible because at pH 8.4 $H_2S_{(aq)}$ concentration is negligible.

c. Calculate the difference between Alkalinity $_{H_3PO_4, H_2S}$ before and after evaporation.

d. Calculate S_T after the evaporation by measuring the new pH and comparing to the original S_T. Argument: the correct answer is d, since Alk and P_T have not changed. pH measurement and known Alk and P_T are enough for S_T calculation.

2. Rainwater is at equilibrium with atmospheric CO_2. Dosing 2 meq/L of H_2SO_4 to it, will inevitably lead to:
   a. A significant decline in pH because the buffer capacity of the water is low.
   b. A small decline in pH because the buffer capacity of the water is high.
   c. A significant decline in pH only if the initial pH is greater than 6.3.
   d. An identical decrease in the pH as a result of dosing 4 mM of HCl.
   Argument: the correct answer is a, since C_T of rain water is low, pH is ~5.5, so the buffer capacity is low.

3. Water with CCPP = 0 is given. Which of the following arguments is true:
   a. Addition of $CaCl_2$ at a concentration of 10 mg/L as $CaCO_3$ will bring the CCPP to 10 mg/L as $CaCO_3$.
   b. Addition of HCl will result in a LSI < 0.
   c. Heating the water will not change the CCPP.
   d. There is not enough data to calculate the LSI.
   Argument: the correct answer is b, since acid dosage reduces CCPP. At CCPP≤0 LSI is also negative

4. Lime-Soda softening is based on the following principles:
   a. pH elevation by the addition of $Ca(OH)_2$ will cause the required magnesium precipitation without the precipitation of calcium.
   b. Dosing $NaHCO_3$ is aimed to precipitate $MgCO_3$.
   c. pH elevation by the addition of $Ca(OH)_2$ may cause the required magnesium precipitation, but water may be left with excess concentration of calcium.
   d. Before removing solids from the water, carbon dioxide should be added in order to lower the pH below 9.5.
   Argument: the correct answer is c. the pH required for $Mg(OH)_2$ precipitation is higher than that required for $CaCO_3$ precipitation, therefore excess calcium remains in the solution.

5. Assuming a closed system, the addition of a strong acid results in:
   a. Alkalinity reduction, regardless of the reference species, and also reduction of CCPP.
   b. Alkalinity reduction, only if the buffer capacity is low.
   c. Higher buffer capacity if the initial pH (i.e., prior to acid addition) was 8.3.
   d. Higher NH_4^+ concentration.
   Argument: the correct answer is a. The alkalinity is reduced regardless of the pH or the buffer capacity since it is a conservative parameter. Answer c is incorrect since reduction of pH from 8.3 to, for example, below pH4 decreases the buffer capacity. Answer d is incorrect since for an initial pH much lower than the pK, further reducing the pH hardly affects the NH_4^+ concentration.

? **Question 10**

Read carefully and circle the most correct answer. Please explain your answer

1. Ionic strength of water has influence on:
    a. The solubility of divalent based salts more than on monovalent based salts.
    b. Thermodynamic equilibrium constants, excluding Henry's constant.
    c. TDS of the water, and therefore solubility of all salts.
    d. Apparent equilibrium constants, especially Henry's constant.

    Argument: the correct answer is a. For monovalent based salts: $K'_{sp} = \frac{K_{sp}}{\gamma_m^2}$, while for divalent based salts: $K'_{sp} = \frac{K_{sp}}{\gamma_d^2}$, and since $\gamma_d < \gamma_m$ the monovalent K'$_{sp}$ is less influenced by ionic strength than the divalent one.

2. When two different solutions with known pH, containing only the carbonic system, are mixed in a 1:1 ratio, the additional required parameters for calculating the pH of the mixture are: (more than one option is possible)
    a. Alkalinity, ionic strength and temperature of each solution.
    b. Alkalinity of each solution.
    c. Ionic strength of each solution.
    d. C_T, ionic strength and temperature of each solution.
    e. Alkalinity, TDS and temperature of each solution.

    Argument: the correct answers are a, d and e. The mixture contains a single weak-acid system. Therefore, two independent parameters are required, on top of ionic strength (or TDS) and temperature. pH is given, the second parameter can be alkalinity or C_T.

3. Post-treatment of desalinated water can include dissolution of $CaCO_3$, followed by pH adjustment. The purpose of the pH adjustment is to:
    a. Increase the C_T, in order to increase water buffer capacity.
    b. Reach a positive CCPP, in order to protect the water distribution pipes.
    c. Add sodium to the water, in order to improve their taste.
    d. Increase $CaCO_3$ dissolution rate.

    Argument: the correct answer is b. at the outlet of the dissolution reactor the CCPP is always negative (due to kinetic limitations). Dosing a strong base elevates the pH, alkalinity and CCPP. C_T is unaffected by NaOH addition. A positive CCPP is required for minimizing corrosion.

4. 107mg NH_4Cl were added to 1 L of distilled water. The resulting solution is (ignore ionic strength effects):
    a. At pH < 9.25
    b. At the equivalent point of NH_3.
    c. Under-saturated with respect to the NH_3 partial pressure in clean atmosphere.
    d. With LSI > 0

    Argument: the correct answer is a. The resulting solution is at the ammonium equivalence point, which is at pH ≤ 9.25, that is, below the pK.

? **Question 11**

Given are three methods to change the CCPP value of a sample of water. Explain <u>qualitatively</u> how each method is expected to affect the CCPP value of the solution given below.

Water quality: Alkalinity ($H_2CO_3^*$) = 150 mg/L as $CaCO_3$, $[Ca^{2+}]$ = 50 mg/L, pH = 7.8, TDS = 600 mg/L.

The methods are:

a. Exchanging the Ca^{2+} concentration with Mg^{2+} using an ion exchanger (resulting in a reduced Ca^{2+} concentration and an equivalent elevation of the Mg^{2+} concentration, while all other water quality parameters remain constant).
b. Equilibrating the solution with clean air.
c. Dissolving a considerable amount of NaCl in the solution.

Solution

a. CCPP will be reduced due to reduction in the Ca^{2+} concentration.
b. Equilibrating with air will result in CO_2 stripping (calculate $[H_2CO_3^*]$ at equilibrium with the atmosphere and in the given solution and show that the $CO_{2(aq)}$ is supersaturated relative to the $CO_{2(aq)}$ concentration in equilibrium with the atmosphere). Thus, pH will increase. Accordingly $[CO_3^{2-}]$ will rise and CCPP will increase.
c. Dissolving NaCl will elevate the ionic strength. Thus γ will decrease (always: $0 < \gamma < 1$) and since $K'sp = Ksp/\gamma_d^2$, at higher ionic strength $K'sp$ will increase, while the quotient, Q, will remain constant. Consequently, CCPP will decrease.

Question 12

The following results were obtained from groundwater analysis. This groundwater was allowed to percolate through a calcite rock layer, thus it can be assumed that it is at equilibrium with $CaCO_{3(s)}$ (i.e., CCPP = 0).

pH = 7.4; TDS = 1000 mg/L; Temp = 25 °C; $[Ca^{2+}]$ = 60 mg/L;

a. Calculate the CO_3^{-2} concentration in the water
b. What is the $H_2CO_3^*$ concentration in the water?
c. It was decided to pump this water to the distribution system, following a stay in a water reservoir. What is the problem that can arise with the pumping equipment and the distribution system as a result of the water's exposure to the atmosphere? Explain, and base your answers on calculations
d. In order to overcome the problem in Section c, it was decided to dose the water (before it is exposed to the atmosphere) with a 3 molar HCl solution in order to decrease the water's pH to 6.0. Calculate the required acid volume (provide the answer in units of liter acid per cubic meter of water)
e. A dosage of HCl at high concentrations is usually an unwanted solution, why is this so (give at least two reasons?)
f. Accordingly, propose a different approach for solving the problem appearing in Section c, explain the proposed treatment in general terms (2–3 sentences) and point out which chemicals (if at all) are to be dosed into the water. There is no need for calculations.

$$\gamma_m = 0.86, \ \gamma_d = 0.54$$

Solution

$$K'_{C1} = \frac{K_{C1}}{\gamma_m}, \ K'_{C2} = \frac{\gamma_m}{\gamma_d} \cdot K_{C2}$$

a. For CCPP = 0, $Q = [Ca^{2+}][CO_3^{2-}] = K'_{sp} = \frac{Ksp}{\gamma_d^2}$

Carbonate concentration can be isolated and calculated:

$$[CO_3^{2-}] = \frac{K_{sp}}{\gamma_d^2[Ca^{2+}]} = \frac{10^{-8.48}}{0.54^2 \cdot 1.5 \cdot 10^{-3}} = 7.57 \cdot 10^{-6} M$$

b. The water's C_T can be calculated from the known carbonate concentration and pH. The following equation should be used, after converting the equilibrium constants:

$$[CO_3^{2-}] = \frac{K'_1 K'_2 C_T}{(H^+)^2 + K'_1(H^+) + K'_1 K'_2}$$

$$= \frac{10^{-6.28} \cdot 10^{-10.19} C_T}{10^{-7.4 \cdot 2} + 10^{-6.28} \cdot 10^{-7.4} + 10^{-6.28} \cdot 10^{-10.19}} \Rightarrow C_T = 5.03 \cdot 10^{-3} M$$

Now, the explicit term for $H_2CO_3^*$ is used, with pH 7.4 and the calculated C_T, to get:

$$[H_2CO_3^*] = \frac{(H^+)^2 C_T}{(H^+)^2 + K'_1(H^+) + K'_1 K'_2} = 3.54 \cdot 10^{-4} M$$

c. From Henry's law, at equilibrium with the atmosphere $H_2CO_3^*$ concentration is,

$$[H_2CO_3^*]_{Equlibrium} = K_H \cdot P_p^{CO2} = 1.36 \cdot 10^{-5} M$$

While in the given water, $H_2CO_3^*$ concentration is considerably higher. Thus, due to the exposure to the atmosphere, a significant CO_2 stripping might take place, which will result in a significant pH elevation and therefore also a significant increase in CCPP. Eventually, this might lead to considerable precipitation and damage to pumps, valves, etc.

d.
$$acid_{dose} = \Delta alkalinity$$

The initial alkalinity can be calculated based on pH 7.4 and $C_T = 5.03$ mM.

$$Alk (H_2CO_3^*) \, initial = [HCO_3^-] + 2[CO_3^{2-}] + [OH^-] - [H^+] - 4.68 \, mN.$$

After the dosage of the acid, the final alkalinity can be calculated based on pH 6.0 and the same C_T.

$$Alk (H_2CO_3^*) \, final = 1.73 \, mN.$$

Therefore, the required acid dosage is 2.95 mN. Using a 3 M HCl, the required dose is 3.28 mM/3M = 0.98 ml to each liter of water.

e. A dosage of HCl at a high concentration is usually an unwanted solution since it adds chloride ions at considerable concentrations thereby possibly resulting in aggressive (corrosive) water. In addition, supplying water at pH 6 is undesirable.

f. Instead of dosing acid, several solutions can be pointed out to prevent equipment blocking, following the exposure to the atmosphere. The first and most trivial is to prevent the exposure to the atmosphere, in case it is possible. However, when it is not a practical option, other solutions are, for example (a) to soften the water, using an ion exchanger or the lime-soda process, to reduce the CCPP, even after exposure to atmosphere and pH elevation; (b) to dose an acid after exposure to the atmosphere. At this stage, the water buffer capacity is reduced, and a slight acid dosage is enough for attaining a small, positive CCPP.

Question 13

It is known that reverse osmosis membranes hardly reject $CO_{2(aq)}$, which is a small uncharged molecule. Therefore, the $CO_{2(aq)}$ concentration measured at both ends of the membrane (feed water, concentrate and the permeate) is practically identical.

Calculate the approximate C_T value expected in the concentrate of the desalination plant treating the groundwater characterized below, and with a recovery-ratio of 90% (i.e., for every m^3 of raw water, 900 L pass the membrane).

For the purpose of the calculation, assume that all ions are fully rejected by the membrane.

$H_2CO_3^*$ alk = 250 mg/L as $CaCO_3$; TDS = 1500 mg/L; Temperature = 25 °C; pH 6.8; Ca^{2+} = 100 mg/L

Solution For 90 % recovery ratio, and complete rejection of all ions, the concentrations of each ion in the concentrate are ten times higher than in the feed.

First, the apparent equilibrium constants should be calculated, for the feed (TDS = 1500 mg/L) and for the concentrate (TDS = 10·1500 = 15,000 mg/L).

For the feed: $I_{feed} = 2.5 \cdot 10^{-5} \cdot 1500 = 0.038M$; $\gamma_{m\ feed} = 0.83$; $\gamma_{d\ feed} = 0.48$;

$pK_{1\ feed} = 6.27$; $pK_{2\ feed} = 10.09$

For the concentrate: $I_{conc} = 2.5 \cdot 10^{-5} \cdot 15000 = 0.375M$; $\gamma_{m\ conc} = 0.70$;

$\gamma_{d\ conc} = 0.24$; $pK_{1\ conc} = 6.19$; $pK_{2\ conc} = 9.86$

The C_T of the groundwater can be calculated based on the known Alk (5 mN) and pH (pH 6.8), using the alkalinity equation:

$$\text{Alk}\ (H_2CO_3^*)\ \text{feed} = [HCO_3^-] + 2[CO_3^{2-}] + [OH^-] - [H^+] = \frac{2K'_1K'_2 \cdot C_T + K'_1(H^+)C_T}{(H^+)^2 + K'_1(H^+) + K'_1K'_2} + [H^+] - \frac{K'_w}{[H^+]}$$

$$\Rightarrow C_{Tfeed} = 6.57\ \text{mM}.$$

Based on this C_T, one can calculate the concentration of $H_2CO_3^*$ of the feed:

$$[H_2CO_3^*] = \frac{(H^+)^2 C_T}{(H^+)^2 + K'_1(H^+) + K'_1K'_2} = 1.5 \cdot 10^{-3}M$$

And since $C_T = [H_2CO_3^*] + [HCO_3^-] + [CO_3^{-2}]$, the sum of the charged species of the carbonate system is

$$[HCO_3^-] + [CO_3^{-2}] = C_T - [H_2CO_3^*] = 6.57 - 1.49 = 5.08 mM$$

These species are the ones rejected by the membrane. Their concentration in the concentrate is ten times higher, therefore, $[HCO_3^-]_{conc} + [CO_3^{-2}]_{conc} = 50.8$ mM, while $[H_2CO_3^*]_{conc} = 1.49$ mM.

$$C_{T\ conc} = 50.8 + 1.49 = 52.29\ \text{mM}$$

Question 14

Corrosion of iron is an oxidation-reduction reaction, in which iron ($Fe_{(s)}$) dissolves in water. Assume that $O_{2(aq)}$ is the electron acceptor, and the end product of iron oxidation is dissolved ferric (Fe^{3+}).

a. Balance the oxidation-reduction reaction.

b. The following water quality enters the distribution system: Alk = 50 mg/L as $CaCO_3$; pH = 8.3. Assume that due to the corrosion of the pipe 14 mg/L of dissolved Fe are added to the water. What will be the alkalinity of the water after the corrosion takes place. How will the pH be affected (will it increase or decrease)? Write a parametric equation to calculate the resulting pH, in which the pH is the only unknown; substitute in the equation all the known values.

! **Solution**

a.   The balanced two half reactions:

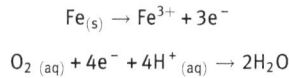

$$Fe_{(s)} \rightarrow Fe^{3+} + 3e^-$$

$$O_{2\ (aq)} + 4e^- + 4H^+_{\ (aq)} \rightarrow 2H_2O$$

To balance the number of electrons released (first half reaction, oxidation of iron) and the number of electrons accepted (second half reaction, reduction of oxygen), the first and second reactions should be multiplied by 4 and 3, respectively. After multiplication, summing up the two half reactions, one gets:

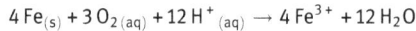

$$4\,Fe_{(s)} + 3\,O_{2\,(aq)} + 12\,H^+_{\ (aq)} \rightarrow 4\,Fe^{3+} + 12\,H_2O$$

b.   It is given that 14 mg/L of Fe are released to the water as a result of the reaction balanced in section a of this example.

$$\Delta Fe^{3+} = 14\,\frac{mg}{L} = 0.25mM$$

According to stoichiometric relations given in the balanced reaction, the associated uptake of hydronium is:

$$\Delta H^+ = -3\Delta Fe^{3+} = -0.75mM = -\Delta Alk$$

And the alkalinity is thus raised by 0.75 meq/L:

$$Alk_{new} = Alk_{old} + 0.75mN = \frac{50\,\frac{mg}{L}\ as\ CaCO_3}{50\,\frac{7mg\ CaCO_3}{meq}} + 0.75mN = 1.75mN$$

The pH will increase, since alkalinity increases and C_T remains constant.

In order to calculate the new pH, one should first calculate the initial C_T, based on the initial Alk and pH, as follows:

pH 8.3 is the equivalence point of bicarbonate. Therefore, Alk $(HCO_3^-) = Acd\ (HCO_3^-) = 0$ (see the links between the various acid-base terms given in Chapter 3). Using equations 3.38 and 3.39, one can conclude that at pH 8.3 Alk $= Acd = C_T$. Therefore, $C_T = 1\,mM$.

After the corrosion takes place, Alk $= 1.75mN$, $C_T = 1\,mM$. The new pH can be found by substituting these values into the alkalinity equation:

$$Alk\ (H_2CO_3^*) = 2[CO_3^{2-}] + [HCO_3^-] + [OH^-] - [H^+] =$$

$$= \frac{2K'_1K'_2 \cdot C_T + K'_1(H^+)C_T}{(H^+)^2 + K'_1(H^+) + K'_1K'_2} + \frac{K_w}{(H^+)} - (H^+)$$

After solving the above equation, one gets pH=10.3.

Note that since the initial buffer capacity of the water was very low, the consumption of alkalinity had a significant impact on the pH.

Question 15

The process shown in the scheme below (Fig. 10.2), aims at removing $CO_{2\,(aq)}$ from stream #1 by stripping it to the gas phase, and utilizing the emitted CO_2 in a second stream (stream #2 in the figure below). The goal of the process is to dissolve $CaMg(CO_3)_2$ in stream #2 by reducing its precipitation potential (PP) with respect to $CaMg(CO_3)_2$.

Fill in the missing values in the scheme (Fig. 10.2). Show calculations or explain shortly how you evaluated each value, under the following assumptions:

a. Acid is added to stream #1 in order to convert all the species of the carbonate system to $CO_{2\,(aq)}$, to facilitate efficient gas stripping.
b. In the degasifier: assume that 67% of the C_T from stream #1 is stripped.
c. Assume that the ratio between the flow rates of stream #1 and the air is 2:1.
d. Assume that the ratio between the flow rates of stream #2 and the air is 1:1.
e. Assume that only 70% of the CO_2 in the air leaving the degasifier is absorbed by stream #2. When calculating the concentration of $CO_{2\,(g)}$ in the air, assume that there was no $CO_{2(g)}$ in the air before it was introduced into the degasifier.
f. 92 mg of $CaMg(CO_3)_2$ dissolve in each liter of stream #2 that enters the dissolution reactor.
g. Neglect effects of temperature and ionic strength on the process.

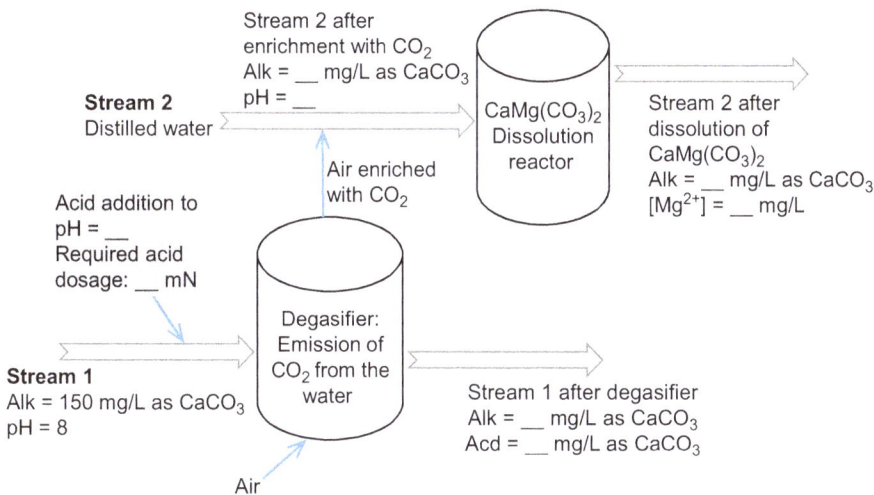

Fig. 10.2: Scheme of the process described in question 15.

Solution

Stream 1: According to the first assumption, after the acid addition, the water is at the equivalence point of $H_2CO_3^*$. Therefore, following the acid addition, the water is at pH~4.5, Alk = 0.

$$HCl_{dose} = Alk_{initial} = 3\,mN.$$

Stream 1 after degasifier: alkalinity does not change as a result of CO_2 stripping, and acidity is reduced by the amount of CO_2 stripped (in equivalent units), and it is elevated by the acid dosage:

$$Alk = 0.$$

$$Acd = Acd_{initial} + HCl_{dose} - 2 \cdot CO_{2 \text{ stripped}}$$

The initial acidity can be calculated by substituting Alk = $3 \cdot 10^{-3}$ and pH = 8 into eq. (3.40):

$$CO_{3\ acd}^{2-} = \frac{1 + 2 \cdot 10^{-pH}/K'_{C1}}{1 + 2K'_{C2}/10^{-pH}} \cdot \left\langle H_2CO_3^{*}{}_{alk} - \frac{K'_w}{10^{-pH}} + 10^{-pH} \right\rangle + 10^{-pH} - \frac{K'_w}{10^{-pH}}$$

To get initial Acd = $3.1 \cdot 10^{-3}$.

To calculate how much CO_2 was emitted in the degasifier, one should calculate the initial C_T. Based on eq 3.37:

$$C_{T\ initial} = 0.5(Alk + Acd) = 0.5(3 + 3.1) = 3.05 mM.$$

Based on assumption #2, $CO_{2 \text{ stripped}} = 0.67 \cdot C_{T\ initial}$

Now, let us substitute these values to calculate the acidity following the degasifier:

$$Acd = Acd_{initial} + HCl_{dose} - 2 \cdot CO_{2 \text{ stripped}} = 3.1 + 3 - 2 \cdot 0.67 \cdot 3.05 = 2.01 mN.$$

Stream #2 after CO_2 enrichment:

Alkalinity does not change as a result of CO_2 enrichment: Alk = 0.

The flow rate ratio between stream #1 and stream #2 is 2 : 1 (assumptions #3 and #4), therefore the concentration of CO_2 after the enrichment is (according to assumption #5):

$2 \cdot 0.7 \cdot CO_{2 \text{ stripped}} = 2.86$ mM

This water is also at the equivalence point of $H_2CO_3^{*}$, thus

$[H^+] = [HCO_3^-]$

Substituting into this equation, the explicit term for bicarbonate (eq. (2.31)) and $C_T = 2.86mM$, one gets pH = 4.47.

Stream 2 after dolomite dissolution:

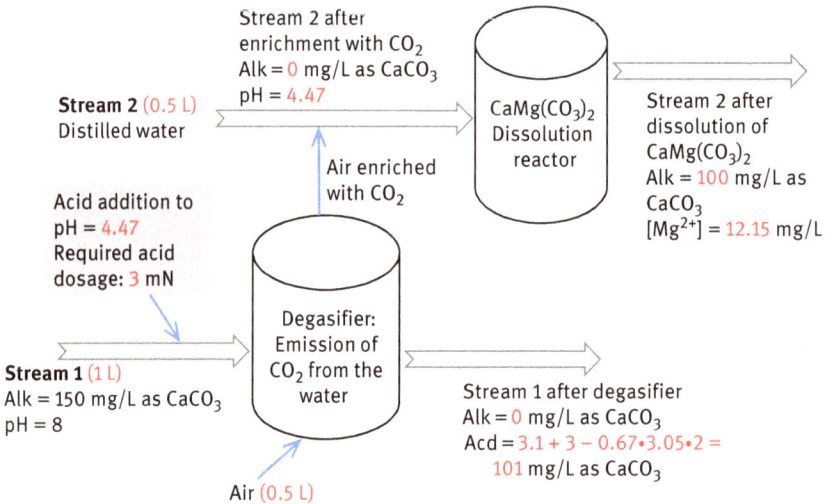

Fig. 10.3: Solution of question 15.

The number of moles of dolomite dosed to the solution is:

$$\text{dolomite}_{dose} = \frac{92 \text{ mg CaMg(CO}_3)_2}{M_w(\text{CaMg(CO}_3)_2) \cdot L} = \frac{92 \text{ mg CaMg(CO}_3)_2}{184.3 \frac{mg}{mmol} \cdot L} = 0.5mM$$

The increase in alkalinity (in normal units) is twice the dose of carbonate (in molar units), which is twice the dose of dolomite. Therefore:

$$\text{Alk}_{new} = \text{Alk}_{old} + 0.5mM \text{ dolomite} \cdot 2 \frac{\text{mol CO}_3^{2-}}{\text{mol dolomite}} \cdot 2 \frac{eq}{\text{mol CO}_3^{2-}} = 0 + 2mN = 2mN$$

The magnesium added to the water is the same as the dolomite:

$$[Mg^{2+}] = \text{dolomite}_{dose} = 0.5mM \cdot 24.3 \frac{g \text{ Mg}}{\text{mol Mg}} = 12.15 \frac{mg}{L}$$

Question 16

?

Analysis of sulfate concentration is based on the precipitation reaction of barium sulfate, creating $BaSO_{4(S)}$ (named barite) and its settling. The following water quality is given:

Alkalinity $(H_2CO_3^*) = 180$ mg/L as $CaCO_3$, $[Ca^{2+}] = 60$ mg/L, pH = 7.8, TDS = 650 mg/L,

$$[SO_4{}^{2-}] = 40 \text{ mg/L as S,}$$

$$\gamma_m = 0.879, \ \gamma_d = 0.597, \ \gamma_t = 0.313, \ Ba_{MW} = 137.32 \text{ g/mol}, \ Cl_{MW} = 35.453 \text{ g/mol}, \ S_{MW} = 32 \text{ g/mol},$$

$$O_{MW} = 16 \text{ g/mol}, \ Ca_{MW} = 40 \text{ g/mol}$$

Solubility constant of $BaSO_4$: $K_{sp} (BaSO_4) = 1.1 \cdot 10^{-10}$

a. Determine how many mg of $BaCl_2$ should be added to 200 ml of water at the given quality for completely precipitating the sulfate. That is, calculate the dose required for 99.9% removal, while assuming immediate precipitation of the added barium with the sulfate.

   In sections b and c, assume $BaSO_{4(s)}$ precipitation does not occur.

b. 2 mM of barium $(BaCl_2)$ is added to the water. Is there a propensity for barium precipitating as $BaCO_3$? Base your answer on calculations.

   Solubility constant of $BaCO_3$: $K_{sp} (BaCO_3) = 5.1 \cdot 10^{-9}$

c. What is the minimum required dose of $BaCl_2$ to precipitate $BaCO_3$ at the given water quality. In other words, how much barium needs to be dosed in order to produce a PP of zero relative to $BaCO_3$?

Solution

!

a. Assuming that aqueous-solid equilibrium is attained: $Q = K'_{sp}$. One should first convert the thermodynamic constant according to the ionic strength and convert the concentrations to molar units:

$$K_{sp} = (Ba^{2+})(SO_4^{2-}) = \gamma_d[Ba^{2+}]\gamma_d[SO_4^{2-}] = \gamma_d{}^2 K'_{sp}$$

$$K'_{sp} = \frac{K_{sp}}{\gamma_d{}^2} = \frac{1.1 \cdot 10^{-10}}{0.597^2} = 3.1 \cdot 10^{-10}$$

$$[SO_4^{2-}] = \frac{40mg \text{ S/L} \cdot 10^{-3}}{32mgS/mmol} = 1.25 \cdot 10^{-3}M$$

At the given water sulfate concentration, precipitation will start at the following barium concentration:

$$Q = [Ba^{2+}][SO_4^{2-}] = K'_{sp}$$

$$[Ba^{2+}] = \frac{K'_{sp}}{[SO_4^{2-}]} = \frac{3.1 \cdot 10^{-10}}{1.25 \cdot 10^{-3}} = 2.47 \cdot 10^{-7} M$$

In order to reduce the sulfate concentration by three orders of magnitude, that is, 99.9% removal, the barium concentration in the water should be:

$$[Ba^{2+}]_f = \frac{K'_{sp}}{[SO_4^{2-}]_f} = \frac{3.1 \cdot 10^{-10}}{1.25 \cdot 10^{-6}} = 2.47 \cdot 10^{-4} M$$

Since this concentration is found at 0.2 L, the barium dose required for this is

$$Ba^{2+}_{dose, 1} = 2.47 \cdot 10^{-4} M \cdot 0.2L = 4.9 \cdot 10^{-5} mol$$

In addition, the amount of barium precipitating with sulfate is

$$Ba^{2+}_{dose, 2} = 1.25 \cdot 10^{-3} M \cdot 0.2L \cdot 0.999 = 2.5 \cdot 10^{-4} mol$$

Altogether, the required barium dose is,

$$Ba^{2+}_{dose} = 2.5 \cdot 10^{-4} + 0.49 \cdot 10^{-4} = 3.0 \cdot 10^{-4} mol$$

$$BaCl_{2dose} = 3.0 \cdot 10^{-4} mol \cdot (137.32 + 2 \cdot 35.45)g/mol = 0.062g$$

b.

$$[Ba^{2+}] = 2mM$$

Given the water alkalinity (3.6 mN) and pH, one can find the water acidity using eq. (3.40), after converting the constants according to the given TDS:

$$CO_{3acd}^{2-} = \frac{1 + 2 \cdot 10^{-pH}/K'_{C1}}{1 + 2K'_{C2}/10^{-pH}} \cdot \left(H_2CO_{3alk}^* - \frac{K'_w}{10^{-pH}} + 10^{-pH} \right) + 10^{-pH} - \frac{K'_w}{10^{-pH}} = 3.8 \text{ mN}$$

Using the relation Acd (CO_3^{2-}) + Alk $(H_2CO_3^*) = 2C_T$ (eq. (3.37)), one can calculate the C_T to be 3.7 mM.

Using eq. (2.32), the carbonate concentration is found to be

$$[CO_3^{2-}] = \frac{K'_{a1}K'_{a2}C_T}{(H^+)^2 + K'_{a1}(H^+) + K'_{a1}K'_{a2}} = 1.66 \cdot 10^{-5} M$$

where, $K'_{a1} = \frac{K_{a1}}{\gamma_m}$ and $K'_{a2} = \frac{\gamma_m K_{a2}}{\gamma_D}$.

Now compare the concentration product, Q, with the apparent equilibrium constant:

$$Q = [Ba^{2+}][CO_3^{2-}] = 2 \cdot 10^{-3} \cdot 1.66 \cdot 10^{-5} = 10^{-7.5}$$

$$K'_{sp} = \frac{K_{sp}}{\gamma_d^2} = \frac{5.1 \cdot 10^{-9}}{0.597^2} = 10^{-7.8}$$

Therefore, $Q > K'_{sp}$ and the precipitation propensity is positive.

c. At PP of zero relative to $BaCO_3$ the following equation holds: $Q = K'_{sp}$.

$$[Ba^{2+}]_{min} = \frac{K'_{sp}}{[CO_3^{2-}]} = \frac{10^{-7.8}}{1.66 \cdot 10^{-5}} = 8.62 \cdot 10^{-4} M$$

The minimum barium chloride dose required is thus

$$BaCl_{2dose} = 8.62 \cdot 10^{-4} M \cdot 0.2L \cdot 208.22 \frac{g\ BaCl_2}{mol} = 3.59 \cdot 10^{-2} g\ BaCl_2$$

Question 17

Desalination plants are required to remove boron from their permeate stream. In the following table, the quality of two water sources of a desalination plant are given. The flow rates of the two sources are identical. Solve the following section while neglecting the influence of ionic strength. Additional data: the pH of the mixture is $pH_{mix} = 8.5$; Water source #2: alk $(HCO_3^-) = 0$.

a. Fill in all the blanks in the table.
b. What will be the total alkalinity of the mixture after dosing 0.3 milimolar of sodium hydroxide (NaOH) to the mixture.
c. What is the pH in which the concentration of the basic species of the borate system (that is $B(OH)_4^-$) comprises 95% of the total boric system concentration?
d. CO_2 aeration is applied on water source #1 in order to raise its pH, until reaching equilibrium with the atmosphere. Write a parametric equation for calculating the pH.

Water	B_T mM	C_T mM	Alk $_{H_2CO_3}$* mN
Source 1	1		
Source 2	0	3	
Mixture		2.5	

Solution

a. The following parameters are additive and their concentrations can be calculated based on weighted average: BT, CT and all alkalinity and acidity terms. Since the flow rates of the two sources are identical, a simple average is applied:

$$B_{T\ mix} = \frac{B_{T1} + B_{T2}}{2} = \frac{1+0}{2} = 0.5 mM$$

$$C_{T\ mix} = \frac{C_{T1} + C_{T2}}{2} \Rightarrow C_{T1} = 2C_{T\ mix} - C_{T2} = 2mM$$

Regarding the mixture, based on the calculated C_T and given pH (pH 8.5), one can calculate the alkalinity, by substituting into the equation the explicit terms for the concentrations:

$$Alk_{mix} (H_2CO_3^*) = 2[CO_3^{2-}] + [HCO_3^-] + [OH^-] - [H^+] = 2.45 mN$$

Regarding source #2, it is given that alk $(HCO_3^-) = 0$, thus the water is at the equivalence point of bicarbonate (by definition), and this means that pH is equal to 8.3. In addition, it is given that $C_T = 3mM$. The alkalinity can be calculated in the same manner:

$$Alk_2 (H_2CO_3^*) = 2[CO_3^{2-}] + [HCO_3^-] + [OH^-] - [H^+] = 3mN$$

The alkalinity of source 1 can now be calculated from the average:

$$Alk_1 = 2Alk_{mix} - Alk_2 = 1.9 \, mN$$

Water	B_T	C_T	Alk $_{H2CO3*}$
	mM	mM	mN
Source 1	1	2	1.9
Source 2	0	3	3
Mixture	0.5	2.5	2.45

b. The total alkalinity relates to the carbonate and boric acid systems. Let us first calculate the total alkalinity before the addition of NaOH:

$$Alk_{mix} (H_2CO_3^*, B(OH)_3) = 2[CO_3^{2-}] + [HCO_3^-] + [OH^-] - [H^+] + [B(OH)_4^-]$$
$$= Alk_{mix} (H_2CO_3^*) + [B(OH)_4^-]_{mix} = 2.45mN + 7.55 \cdot 10^{-5} = 2.52mN$$

The addition of 0.3 mM of NaOH, being a strong monoprotic base, adds 0.3 mN to the alkalinity: $Alk_{mix} (H_2CO_3^*, B(OH)_3)_{final} = 2.52 + 0.3 = 2.82 \, mN$

c.
$$[B(OH)_4^-] = 0.95 \, B_T = \frac{K_B B_T}{K_B + [H^+]}$$

pH = 10.528. That is, at a pH value which is more than one pH unit above the pK of the system, the basic species comprise 95% of the system species.

d. When aerating the water, CO_2 is either stripped or dissolved into the water. In both cases, the alkalinity remains constant. At equilibrium with the gaseous phase, the carbonate system species are a function of the pH and the CO_2 partial pressure, therefore the alkalinity equation can be written as follows:

$$Alk_1 (H_2CO_3^*, B(OH)_3) = 2[CO_3^{2-}] + [HCO_3^-] + [OH^-] - [H^+] + [B(OH)_4^-] =$$
$$1.9 \cdot 10^{-3} = 2\frac{K_{c1}K_{c2}P_{p(CO_2)}}{H_{CO_2}[H^+]^2} + \frac{K_{c1}P_{p(CO_2)}}{H_{CO_2}[H^+]} + \frac{K_w}{[H^+]} - [H^+] + \frac{K_B B_T}{K_B + [H^+]}$$

To solve for pH, one should substitute the values of the carbonate and boric acid system constants, the Henry constant for CO_2 ($H_{CO_2} = 28.4 \, atm/M$) and the atmospheric CO_2 partial pressure ($P_{p(CO_2)} = 0.0004 \, bar$).

?

Question 18

The alkalinity value of a water sample was measured by titration to pH 4.5. The volume of the strong acid (HCl) titrated to the endpoint was 3.0 ml and its concentration 0.05 N. The volume of the sample was 20 ml, and the pH of the original sample was 9.2. The original water sample consisted of the carbonate system at an unknown concentration and the sulfide di-protic weak-acid system (pKs given below) at total concentration of $S_T = 32$ mg/L as S.

What is the contribution of each weak-acid system to the alkalinity (with reference species HCO_3^- and HS^-) value? Give the answer in meq/L or mg/L as $CaCO_3$ units.

$$pK(H_2S/HS^-) = 7.0; \ pK(HS^-/S^{2-}) = 13.65$$

Solution

Unit conversion: $S_T = \dfrac{32\frac{mg}{L}asS}{32\frac{mgS}{mmol}} = 1\,mM.$

The alkalinity measured was alkalinity with the reference species $H_2CO_3^*$ and H_2S, and it was found to be alkalinity $(H_2CO_3^*, H_2S) = \frac{3\,ml \cdot 0.05\,N}{20\,ml} = 7.5\,mN.$

$$\text{alkalinity}\,(H_2CO_3^*, H_2S) = 2[CO_3^{2-}] + [HCO_3^-] + 2[S^{2-}] + [HS^-] + [OH^-] - [H^+]$$

$$= \text{alkalinity}\,(H_2CO_3^*) + \frac{(2(H^+)^2 + K_{S1}(H^+)) \cdot S_T}{(H^{+2} + K_{S1}(H^+) + K_{S1}K_{S2}}$$

Substituting all the known values,

$$7.5\,mN = \text{alkalinity}\,(H_2CO_3^*) \ + \ \frac{(2 \cdot 10^{-2 \cdot 9.2} + 10^{-7}10^{-9.2}) \cdot 1}{10^{-2 \cdot 9.2} + 10^{-7}10^{-9.2} + 10^{-13.65}10^{-9.2}}$$

$$\Rightarrow \text{alkalinity}\,(H_2CO_3^*) = 7.5 - 1.02 = 6.48\,mN$$

$$\text{alkalinity}\,(H_2CO_3^*) = 2[CO_3^{2-}] + [HCO_3^-] + [OH^-] - [H^+] = 6.48\,mN$$

given the pH, one can compute the C_T using, first eq. (3.40):

$$CO_3{}^{2-}_{acd} = \frac{1 + 2 \cdot 10^{-pH}/K'_{C1}}{1 + 2K'_{C2}/10^{-pH}} \cdot \left\langle H_2CO_3{}^*_{alk} - \frac{K'_w}{10^{-pH}} + 10^{-pH} \right\rangle + 10^{-pH} - \frac{K'_w}{10^{-pH}}$$

Acd = 5.58 mN.

Using the link Acd + Alk = $2C_T$, one gets $C_T = 6.03$ mM.

Now, one can substitute into the following equation (after using the explicit terms for each species concentration) the C_T, S_T and pH, and solve:

$$\text{alkalinity}\,(HCO_3^-, HS^-) = [CO_3^{2-}] - [H_2CO_3^*] + [S^{2-}] + [H_2S] + [OH^-] - [H^+] = -0.46\,mN$$

?

Question 19

Tap water analysis shows the temperature to be 25 °C, total dissolved solids (TDS) 200 mg/L, alkalinity (H_2CO_3) 150 mg/L as $CaCO_3$, Calcium 100 mg/L as $CaCO_3$, Magnesium 25.1 mg/L as $CaCO_3$ and pH = 7.8.

a. Determine the water saturation state with respect to $CaCO_3$ and to $MgCO_3$.
b. You are requested to reduce the water hardness to $[Ca^{2+}] = 15$ mg/L and $[Mg^{2+}] = 6.1$ mg/L using the lime-soda treatment. Which chemical(s) should be used?

c. Do you recommend aerating the water before it enters the reactor? Explain based on calculations.

d. It was decided to aerate the water, before it enters the reactor. It is known that due to kinetic limitations, the water does not reach equilibrium with the partial pressure of atmospheric carbon dioxide, rather with double that pressure.

    I.   Write a parametric equation for calculating the resultant water pH after the aeration. Then, substitute in the equation, all the available information, in their correct units.

    II.   Explain qualitatively whether you expect the pH value to increase or decrease.

Solution

a. Let us first correct the equilibrium constants according to the given TDS:

$$I = 2.5 \cdot 10^{-5} \cdot TDS = 2.5 \cdot 10^{-5} \cdot 200 = 0.005M$$

$$\gamma_m = 0.926, \quad \gamma_d = 0.736$$

$$Ksp_{CaCO_3} = (CO_3^{2-})(Ca^{2+}) = \gamma_d[CO_3^{2-}]\gamma_d[Ca^{2+}] \Rightarrow Ksp'_{CaCO_3} = \frac{Ksp}{\gamma_d^2} = \frac{10^{-8.48}}{0.736^2} = 10^{-8.21}$$

$$Ksp_{MgCO_3} = (CO_3^{2-})(Mg^{2+}) = \gamma_d[CO_3^{2-}]\gamma_d[Mg^{2+}] \Rightarrow Ksp'_{MgCO_3} = \frac{Ksp}{\gamma_d^2} = \frac{10^{-5.16}}{0.736^2} = 10^{-4.89}$$

$$K_1 = \frac{(H^+)(HCO_3^-)}{(H_2CO_3^*)} = \gamma_m K' \Rightarrow K'_1 = \frac{1}{0.926}10^{-6.3} = 10^{-6.27}$$

$$K_2 = \frac{(H^+)(CO_3^{2-})}{(HCO_3^-)} = \frac{\gamma_d}{\gamma_m}K'_2 \Rightarrow K'_2 = \frac{0.926}{0.736}10^{-10.3} = 10^{-10.20}$$

The saturation state can be evaluated using the LSI index. When used to evaluate the saturation state of carbonate minerals other than calcite, it should be modified.

$$Alkalinity = \frac{150\frac{mg}{L} \text{ as } CaCO_3}{50000\frac{mgCaCO_3}{eq}} = 3 \cdot 10^{-3}\frac{eq}{L}$$

$$[Ca^{+2}] = \frac{100\frac{mg}{L} \text{ as } CaCO_3}{50000\frac{mgCaCO_3}{eq}2\frac{eq}{mol}} = 10^{-3}M, \quad [Mg^{+2}] = \frac{25.1\frac{mg}{L} \text{ as } CaCO_3}{50000\frac{mgCaCO_3}{eq}2\frac{eq}{mol}} = 2.51 \cdot 10^{-4}M$$

pH 7.8

Langeleir index: $LSI = pH - pH_L$

$$pH_L = pK'_{c_2} - pK'_{sp} - \log(Alk) - \log[Ca^{2+}]$$

For calcite:

$$pH_L = pK'_{c_2} - pK'_{sp} - \log(Alk) - \log[Ca^{2+}] = 10.20 - 8.21 - \log(3 \cdot 10^{-3}) - \log(10^{-3}) = 7.5$$

$$\rightarrow LSI = 7.8 - 7.5 = 0.3 > 0$$

A positive value represents over saturation.

For magnesite, $MgCO_3$:

$$pH_L = pK'_{C_2} - pK'_{sp} - \log(Alk) - \log[Ca^{2+}] = 10.20 - 4.89 - \log(3 \cdot 10^{-3}) - \log(2.5 \cdot 10^{-4})$$

$$= 11.43 \rightarrow LSI = 7.8 - 11.43 = -3.6 < 0$$

A negative value represents under saturation.

b. The final required concentrations are: $[Ca^{2+}] = 15$ mg/L = 0.375 mM and $[Mg^{2+}] = 6.1$ mg/L = 0.251 mM. Therefore, only calcium should be removed from the solution. In the lime-soda process, lime is added to reach the required pH for precipitating $CaCO_3$ and $Mg(OH)_2$. In our case, $Mg(OH)_2$ precipitation is not required. Therefore, no excess calcium will be present in the solution following the precipitation of $CaCO_3$ to the required calcium concentration. To conclude, only lime should be dosed to the water.

c. In this case, aeration is beneficial if the initial solution is oversaturated with respect to atmospheric CO_2. Aeration would increase the pH and decrease the required lime addition, which is the only chemical required. Let us first calculate $[H_2CO_3^*]$ in the solution:
Using eq. (3.40), the acidity in the water can be found:

$$CO_3^{2-}{}_{acd} = \frac{1 + 2 \cdot 10^{-pH}/K'_{C1}}{1 + 2K'_{C2}/{10^{-pH}}} \cdot \left(H_2CO_3^*{}_{alk} - \frac{K'_w}{10^{-pH}} + 10^{-pH} \right) + 10^{-pH} - \frac{K'_w}{10^{-pH}}$$

Acid = 3.17 mN.

$$C_T = (Alk + Acd)/2 = 3.08 \text{ mM.}$$

$$[H_2CO_3^*] = \frac{(H^+)^2 C_T}{(H^+)^2 + (H^+)K_1 + K_1 K_2} = 9.79 \cdot 10^{-5} M$$

CO_2 at equilibrium with the atmosphere is

$$[H_2CO_3^*]_{eq} = Pp \cdot K_H = \frac{0.0004 \text{ bar}}{28.8 \text{ bar}/M} = 1.39 \cdot 10^{-5} M$$

Thus, the CO_2 concentration in the solution is much higher than at equilibrium and aeration is recommended.

d.

I. The alkalinity does not change when CO_2 is stripped. At equilibrium with gaseous phase, the following expressions for carbonic system species should be used:

$$Alk_{H_2CO_3^*} = \frac{2K'_1 K'_2 \frac{P_{CO_2(g)}}{H_{CO_2}}}{[H^+]^2} + \frac{K'_1 \frac{P_{CO_2(g)}}{H_{CO_2}}}{[H^+]} + \frac{K'_w}{[H^+]} - [H^+]$$

The equilibrium constants are substituted, as well as the known alkalinity and the partial pressure (i.e., $P_p(CO_2) = 2 * 0.0004 = 0.0008$ bar):

$$3 \cdot 10^{-3} = \frac{2 \cdot 10^{-10.39} \cdot 10^{-6.26} \cdot \frac{0.0004 \cdot 2}{28.8}}{[H^+]^2} + \frac{10^{-6.26} \cdot \frac{0.0004 \cdot 2}{28.8}}{[H^+]} + \frac{10^{-14}}{[H^+]} - [H^+]$$

II. The pH value is expected to increase due to aeration, since CO_2 is the acidic species, and the initial pH is above the first pK (i.e., CO_2 is not dominant). Stripping of the acidic species will cause the species to shift toward the more acidic ones, and the consumption of protons.

Question 20

A biological treatment plant removes ammonia from wastewater. The wastewater flow rate is 100 m^3/day. The total ammoniacal nitrogen ($[NH_3] + [NH_4^+]$) of the incoming wastewater is 42 mg/L as N, and the total ammoniacal nitrogen in the plant effluent is 7 mg/L as N. Ammonia removal is performed by nitrifying bacteria. It is known that in a nitrification process, the ammonia is oxidized to nitrate (NO_3^-), and the dissolved oxygen is reduced (it is the electron acceptor). The incoming wastewater analysis is: pH = 8, N_T = 42 mg/L as N, C_T = 10 mM, $[NO_3^-] = 0$.

a. Calculate the alkalinity of the incoming wastewater. Choose the alkalinity reference species wisely.

b. What is the concentration of the charged species of the ammonia system in the incoming wastewater?

c. Write a balanced redox reaction for the process and calculate the nitrate concentration in the effluent.

d. What is the required dosage of a strong acid or base in order to keep a constant pH in the process?

Solution

a. The pH and total concentrations (C_T and N_T) are known. The alkalinity toward the most acidic species can be calculated:

$$C_T = 10^{-2}M, \quad N_T = \frac{42 \frac{mgN}{L}}{14000 \frac{mgN}{mol}} = 3 \cdot 10^{-3}M$$

$$Alk_{H_2CO_3^*, NH_4^+} = 2[CO_3^{-2}] + [HCO_3^-] + [NH_3] + [OH^-] - [H^+]$$

$$= \frac{(2 \cdot 10^{-6.3} \cdot 10^{-10.3} + 10^{-6.3} \cdot 10^{-pH}) \cdot C_T}{10^{-pH \cdot 2} + 10^{-pH} \cdot 10^{-6.6} + 10^{-6.6} \cdot 10^{-10.3}} + \frac{10^{-9.25} \cdot N_T}{10^{-9.25} + 10^{-pH}}$$

$$+ \frac{10^{-14}}{10^{-pH}} - 10^{-pH} = 0.01001 \left(\frac{eq}{L}\right)$$

b. Using the pH and N_T, and the correct equation from Table 2.3 (monoprotic system, [HA]):

$$[NH_4^+] = \frac{10^{-pH} \cdot N_T}{10^{-pH} + 10^{-9.25}} = 0.00284[M]$$

c. Nitrate concentration at the outlet of the plant can be calculated based on a 1:1 molar ratio with the removed ammonia:

$$[NO_3^-] = -\Delta[NH_4^+] = -\left(\frac{7\frac{mgN}{L}}{14000\frac{mgN}{mol}} - 3 \cdot 10^{-3}\right) = 2.5 \cdot 10^{-3}M$$

The balanced two half reactions:

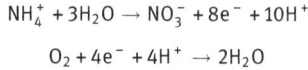

$$NH_4^+ + 3H_2O \rightarrow NO_3^- + 8e^- + 10H^+$$

$$O_2 + 4e^- + 4H^+ \rightarrow 2H_2O$$

In order to balance the number of electrons released (first half reaction, oxidation of ammonium) and the number of electrons accepted (second half reaction, reduction of oxygen) the second reaction should be multiplied by 2. After multiplication, summing up the two half reactions, one gets:

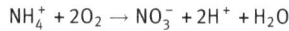

$$NH_4^+ + 2O_2 \rightarrow NO_3^- + 2H^+ + H_2O$$

d. According to the balanced reaction, for each mole of ammonium consumed 2 moles of protons are released. Consumption or releasing of ammonium does not change $Alk_{H_2CO_3^*, NH_4^+}$.

However, the release of protons reduces the alkalinity. In order to maintain a constant alkalinity in the solution, while also keeping the C_T constant one should dose a strong base equal to twice the ammonia removed (equivalent ratio). The ammonia removed is also equal to the nitrate released to the solution:

$$\text{base dose} = 2\frac{eq}{mol} \cdot [NO_3^-] = 2\frac{eq}{mol} \cdot 2.5 \cdot 10^{-3}M = 5 \cdot 10^{-3}\frac{eq}{L} \cdot 100,000\frac{L}{day} = 500\frac{eq\ Base}{day}$$

Note that in case one balances the redox reaction with ammonia instead of ammonium, the required dosage of the base would remain the same.

Question 21

1 liter of 10 mM NaOH solution ($N_T = C_T = 0$) was put in a 2 liter sealed beaker. The initial NH_3 partial pressure in the vessel's headspace was Pp (NH_3) = 0.05 bar. The two phases were given sufficient time to arrive at gas-aqueous phase equilibrium at 25 °C.

a. Calculate the partial pressure of $NH_{3(g)}$ at equilibrium.
b. Calculate the value of N_T in the water at equilibrium.
c. Calculate the alkalinity value in the aqueous phase (Specify the alkalinity term and write it explicitly)?
d. Would the partial pressure of NH_3 in the gaseous phase at equilibrium be lower or higher if instead of 10 mM NaOH, distilled water was put in the beaker (compare with the result of section (a) above)?

Solution

a. The initial total NH_3 in the system is equal to the final mass. In general, the total mass is the concentration in each phase times its volume (1 liter):

$$N_T = NH_{3(g)} \cdot 1L + (NH_{3(aq)} + NH_{4\ (aq)}^+) \cdot 1L$$

Since the aqueous phase is a 10 mM NaOH solution, that is, at pH = 12, which is ~3 pH units above the pK of NH_4^+, it can safely be assumed that in the aqueous phase, $N_T = NH_{3(aq)}$.

$$N_T = NH_{3(g)} \cdot 1L + (NH_{3(aq)}) \cdot 1L$$

At the initial state:

$$N_{T,initial} = \frac{0.05bar}{RT} \cdot 1L + 0 \cdot 1L = \frac{0.05bar}{0.082 \frac{bar \cdot L}{mol \cdot K} 298K} \cdot 1L = 2.05 \, mmol.$$

At equilibrium, this mass is divided between the two phases according to Henry's law

$$N_{T,equilibrium} = \frac{P_P}{RT} \cdot 1L + P_P(1/H) \cdot 1L = 2.05 \, mmol$$

$$\Rightarrow P_P = 3.48 \cdot 10^{-5} bar$$

b. As mentioned, at pH = 12, the concentration of ammonium and $N_T = NH_{3(aq)} = PP/H = 3.48 \cdot 10^{-5}/$ 0.017 = 2.04 mM can be neglected.
    This value means that practically all the ammonia mass is found in the aqueous phase (2.04 out of 2.05), this can also be concluded from the low value of H (NH_3).

c. The initial alkalinity of the water was only a result of the NaOH, since $N_T = C_T = 0$:
    Alkalinity $_{initial}$ = 10 mN.
    After adsorption of ammonia, the alkalinity increases at a 1 : 1 ratio with the adsorbed ammonia, since

$$alkalinity \, (NH_4{}^+) = [NH_3] + [OH^-] - [H^+]$$

Thus, Alkalinity $_{final}$ = 10 mN + 2.04 mM·1 eq/mol = 12.04 mN

d. At pH 7, which is 2 pH units below the pK of $NH_4{}^+$, more ammonia can be adsorbed in the water since a greater fraction of N_T will be converted to the ammonium species.

References

[1] Kemp, PH. Chemistry of natural waters—I: Fundamental relationships. Water Res. 1971; 5(6):297–311.

[2] Langelier, WF. Effect of Temperature on the pH of Natural Waters. J. Am. Water Works Assoc. 1946;38:179–85.

[3] Benefield, LD., Judkins, JF., Weand, BL. Process chemistry for water and wastewater treatment. Englewood Cliffs, New Jersy, USA, Prentice-Hall, 212 pp. 1982.

[4] Stumm,W., Morgan, JJ. Aquatic chemistry : chemical equilibria and rates in natural waters. NY, USA, Wiley, 1996.

[5] Bard, AJ, Chemical Equilibrium. NY, USA, Harper and Row, 134 pp. 1966.

[6] Gran, G., Johansson, A., Johansson, S. Automatic titration by stepwise addition of equal volumes of titrant. Part VII. Potentiometric precipitation titrations. Analyst. 1981;106, 231–242.

[7] Langelier, WF. The analytical control of anti-corrosion water treatment. J. Am. Water Works Assoc. 1936;28:1500–21.

[8] Larson, TE., Buswell, AM. Calcium carbonate saturation index and alkalinity interpretations. J. AWWA. 1942; 34, p. 1667–84.

[9] Dye, JF. The calculation of alkalinities and free CO_2 in H_2O by the use of homographs. J. Am. Water Works Assoc. 1944; 36, 895–900.

[10] Dye, JF. Calculation of effect of temperature on pH, free carbon dioxide, and the three forms of alkalinity. J. Am. Water Works Assoc. 1952;44, 356–372.

[11] Dye, JF. Correlation of the two principal methods of calculating the three kinds of alkalinity. J. Am. Water Works Assoc. 1958;50, 800–820.

[12] McCauley, RF. Calcite coating protects water pipes. Water & Sewage Works. 1960;107, p. 276–81.

[13] Prepared by members of the Accident Prevention Committee. Safety practice for water utilities. 3rd Edition Denver, Colorado, USA, Am. Water Works Assoc. Manual. 1977.

[14] Loewenthal, RE., Wiechers, HNS., Marais, GvR. Softening and stabilization of municipal waters, Water Research Commission. Pretoria, South Africa, Creda Press, 1986.

[15] Loewenthal, RE., Ekama, GA., Marais, GvR. STASOFT: a user-friendly interactive computer program for softening and stabilization of municipal waters. Water SA. 1988;14[3], 159–62.

[16] Guidelines for Drinking –Water Quality 2008 2nd addendum to 3rd Ed [Internet]. World Heal. Organ. 2008 [accessed 2018 Okt 31];Available from: http://www.who.int/water_sanitation_health/dwq/secondaddendum20081119.pdf?ua=1.

[17] Cotruvo, J., Bartram, J. Calcium and Magnesium in Drinking-water, Am. Water Works Assoc. Geneva, Switzerland, 2009.

[18] Gabbrielli, E. A tailored process for remineralization and potabilization of desalinated water. Desalination. 1981;39(C):99–110.

[19] Glade,H., Meyer, JH., Will, S. The release of CO_2 in MSF and ME distillers and its use for the recarbonation of the distillate: a comparison. Desalination. 2005;182(1–3).

[20] deSouza, PF., du Plessis, G., Mackintosh, GS. An evaluation of the suitability of the limestone based sidestream stabilization process for stabilization of waters of the Lesotho highlands scheme. Water SA – Special Edition, Wisa Proceedings. Stellenbosch, South Africa, 2002;10–5.

[21] Withers, A. Options for recarbonation, remineralisation and disinfection for desalination plants. Desalination. 2005;179(1–3.):11–24.

[22] Seacord,TF., Singley, JE., Juby, G., Voutchkov, N. Post Treatment Concepts for SW and Brackish Water Desalting, Am. Water Works Assoc. Membr. Technol. Conf. 2003; Atlanta, Georgia. 1–15.

https://doi.org/10.1515/9783110603958-011

[23] Marangou, VS., Savvides, K. First desalination plant in Cyprus – Product water aggresivity and corrosion control. Desalination. 2001;138(1–3).

[24] Fritzmann, C., Löwenberg, J., Wintgens, T., Melin, T. State-of-the-art of reverse osmosis desalination. Desalination. 2007;216(1–3).

[25] El Azhar, F., Tahaikt, M., Zouhri, N., et al. Remineralization of Reverse Osmosis (RO)-desalted water for a Moroccan desalination plant: Optimization and cost evaluation of the lime saturator post. Desalination. 2012;300:46–50.

[26] Birnhack, L., Voutchkov, N., Lahav, O. Fundamental chemistry and engineering aspects of post-treatment processes for desalinated water-A review. Desalination. 2011;273(1).

[27] Ben-Gal, A., Yermiyahu, U., Cohen, S. Fertilization and Blending Alternatives for Irrigation with Desalinated Water. J. Environ. Qual. 2009;38(2):529.

[28] Glueckstern, P., Priel, M., Kotzer, E. Blending brackish water with desalted seawater as an alterative to brackish water desalination. Desalination. 2005;178(1–3SPEC.):227–232.

[29] Dreizin, Y., Tenne, A., Hoffman, D. Integrating large scale seawater desalination plants within Israel's water supply system. Desalination. 2008;220(1–3).

[30] Duranceau, SJ. Desalination post treatment considerations. Florida Water Resour. J. 2009; 4–19.

[31] Ghermandi, A., Messalem, R. Solar-driven desalination with reverse osmosis: the state of the art. Desalin. Water Treat. 2009;7,285–296.

[32] ElDin, AMS. Three strategies for combating the corrosion of steel pipes carrying desalinated potable water. Desalination. 2009;238,166–173.

[33] Duranceau, SJ., Wilder, RJ., Douglas, SS. A survey of desalinated permeate post-treatment practices. Desalin. Water Treat. 2012; 37 (1–3): 185–99.

[34] Almulla, A., Eid, M., Cote, P., Coburn, J. Developments in high recovery brackish water desalination plants as part of the solution to water quantity problems. Desalination. 2003;153(1–3),237–243.

[35] Greenlee, LF., Lawler, DF., Freeman, BD., Marrot, B., Moulin, P. Reverse osmosis desalination: Water sources, technology, and today's challenges. Water Res. 2009;43(9):2317–48.

[36] NationalResearch Council. 2008. Desalination: A National Perspective. Washington, DC: The National Academies Press. Accessed Oktober 31, 2018, at http://www.nap.edu/catalog/12184.

[37] Glueckstern, P. Design and operation of medium- and small-size desalination plants in remote areas: New perspective for improved reliability, durability and lower costs. Desalination. 1999; 122 (2–3): 123–40.

[38] Catling, LA., Abubakar, I., Lake, IR., Swift, L., Hunter, PR. A systematic review of analytical observational studies investigating the association between cardiovascular disease and drinking water hardness. J. Water Health. 2008;6(4).

[39] Rowell, C., Kuiper, N., Shomar, B. Potential health impacts of consuming desalinated bottled water. J. Water Health. 2015;13(2).

[40] Nriagu, J., Darroudi, F., Shomar, B. Health effects of desalinated water: Role of electrolyte disturbance in cancer development. Environ. Res. 2016;150:191–204.

[41] Ovadia, YS., Gefel, D., Aharoni, D., Turkot, S., Fytlovich, S., Troen, AM. Can desalinated seawater contribute to iodine-deficiency disorders? An observation and hypothesis. Public Health Nutr. 2016;19(15).

[42] Shlezinger, M., Amitai, Y., Goldenberg, I., Shechter, M. Desalinated seawater supply and all-cause mortality in hospitalized acute myocardial infarction patients from the Acute Coronary Syndrome Israeli Survey 2002–2013. Int. J. Cardiol. 2016;220:544–550.

[43] Yermiyahu, U., Tal, A., Ben-Gal, A., Bar-Tal, A., Tarchitzky, J., Lahav, O. Rethinking Desalinated Water Quality and Agriculture. Science (80-). 2007;318(5852):920–1.

[44] Barron, O., Ali, R., Hodgson, G., et al. Feasibility assessment of desalination application in Australian traditional agriculture. Desalination. 2015;364:33–45.

[45] Martínez-Alvarez, V., Martin-Gorriz, B., Soto-García, M. Seawater desalination for crop irrigation – A review of current experiences and revealed key issues. Desalination. 2016;381:58–70.

[46] Hermony, A., Sutzkover-Gutman, I., Talmi, Y., Fine, O. Palmachim Seawater desalination plant—seven years of expansions with uninterrupted operation together with process improvements. Desalin. Water. Treat. 2015;55(9).

[47] Birnhack, L., Shlesinger, N., Lahav, O. A cost effective method for improving the quality of inland desalinated brackish water destined for agricultural irrigation. Desalination. 2010;262 (1–3).

[48] Migliorini, G., Meinardi, R. 40 MIGD potabilization plant at Ras Laffan: Design and operating experience. Desalination. 2005;182(1–3).

[49] Plummer, LN., Wigley, TM. The dissolution of calcite in CO2-saturated solutions at 25 °C and 1 atmosphere total pressure. Geochim. Cosmochim. Acta. 1976;40(2):191–202.

[50] Mackintosh, GS., de Villiers, HA. Treatment of soft, acidic, ferruginous groundwater using limestone bed filtration, Water Institute of South Africa Water Institute of South Africa Biannual Conference, 1998.

[51] Lehmann, O., Birnhack, L., Lahav, O. (2013) Design aspects of calcite-dissolution reactors applied for post treatment of desalinated water. Desalination, 314:1–9.

[52] Al-Rqobah, HE., HA-M, A.. A Recarbonation Process for Treatment of Distilled Water Produced by MSF Plants in Kuwait. Desalination. 1989;73(C).

[53] Birnhack, L., Fridman, N., Lahav, O. (2009) Potential applications of quarry dolomite for post treatment of desalinated water. Desalin. Water Treat. 2009;1(1–3):58–67.

[54] Lahav, O., Nativ, P., Birnhack, L. Dolomite dissolution is not an attractive alternative for meeting Ca^{2+}, Mg^{2+} and alkalinity criteria in desalination plants' post treatment step. Desalin. Water Treat. 2018;115:194–8.

[55] Birnhack, L., Lahav, O. A new post treatment process for attaining Ca^{2+}, Mg^{2+}, SO_4^{2-} and alkalinity criteria in desalinated water, Water Res. 2007;41(17), 3989–3997.

Index

https://doi.org/10.1515/9783110603958-012

www.ingramcontent.com/pod-product-compliance
Lightning Source LLC
Chambersburg PA
CBHW061357210326
41598CB00035B/6008